Ableton Live 8 and Suite 8

Ableton Live 8 and Suite 8

Create, Produce, Perform

Keith Robinson

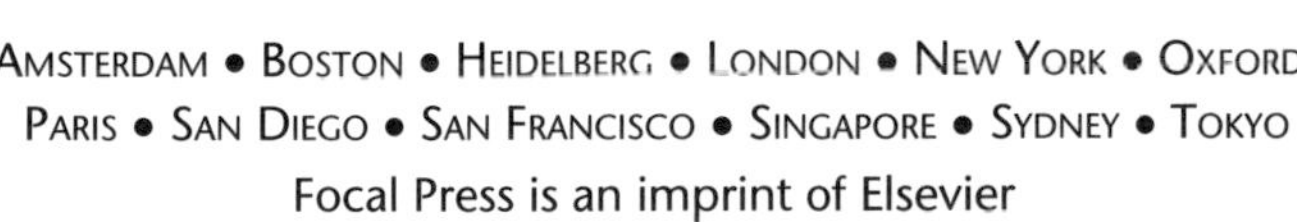

AMSTERDAM • BOSTON • HEIDELBERG • LONDON • NEW YORK • OXFORD
PARIS • SAN DIEGO • SAN FRANCISCO • SINGAPORE • SYDNEY • TOKYO
Focal Press is an imprint of Elsevier

Focal Press is an imprint of Elsevier
30 Corporate Drive, Suite 400, Burlington, MA 01803, USA
Linacre House, Jordan Hill, Oxford OX2 8DP, UK

Notices
Knowledge and best practice in this field are constantly changing. As new research and experience broaden our understanding, changes in research methods, professional practices, or medical treatment may become necessary.

Practitioners and researchers must always rely on their own experience and knowledge in evaluating and using any information, methods, compounds, or experiments described herein. In using such information or methods they should be mindful of their own safety and the safety of others, including parties for whom they have a professional responsibility.

To the fullest extent of the law, neither the Publisher nor the authors, contributors, or editors, assume any liability for any injury and/or damage to persons or property as a matter of products liability, negligence or otherwise, or from any use or operation of any methods, products, instructions, or ideas contained in the material herein.

Library of Congress Cataloging-in-Publication Data
Application submitted

British Library Cataloguing-in-Publication Data
A catalogue record for this book is available from the British Library.

ISBN: 978-0-240-81228-1

For information on all Focal Press publications
visit our website at *www.elsevierdirect.com*

Typeset by: diacriTech, India

Printed in the United States of America
12 13 5 4 3

Contents

About the Author *xi*

Acknowledgments *xiii*

Scene 1 Ableton Live 8: Create, Produce, Perform **3**

1.1 Introduction 3
1.2 Approaching Live in three intuitive ways 4
1.3 Immersing yourself in Live 4
1.4 How to use this book! 7
1.5 Taking advantage of Hot Tips 8

Scene 2 Overview: Live 8, Suite 8 **13**

2.1 Introduction 13
2.2 The concept 13
2.3 Important preference tweaks! 14
2.4 The Live Browser 19

Scene 3 The Quick Way to Start Making Music! **31**

3.1 Introduction 31
3.2 Starting a project 33
3.3 Audio in Live 34
3.4 MIDI in Live 41
3.5 Essential operations and tasks 48
3.6 Performance to arrangement 54
3.7 Finishing your work 55

Scene 4 Global Record: Capturing Arrangements on the Fly **65**

4.1 Introduction 65
4.2 The Global Record concept 65
4.3 Music on the fly 67
4.4 User interfacing: two parallel worlds 67
4.5 The linear approach 74
4.6 The nonlinear approach 80

Scene 5 Arrangement View Concepts **85**

5.1 Musical timeline 85
5.2 Layout 85
5.3 Navigating 86
5.4 The Arrangement View 92
5.5 Working with automation 101
5.6 Arranging concepts 105

Scene 6 Session View Concepts **109**

6.1 Real-time "launching base" 109
6.2 Layout 110
6.3 Clips 115
6.4 Tracks versus Scenes 120
6.5 Track Status Display 124
6.6 Working in Session View 124
6.7 Sessions into Arrangements 131
6.8 Musical concepts 133

Scene 7 Clips **137**

7.1 Musical building blocks 137
7.2 Clip View 138
7.3 Clip Box 142
7.4 Launch Box 143
7.5 Sample Box 146
7.6 Notes Box 154
7.7 MIDI Note Editor 156
7.8 Envelope Box 162
7.9 Envelope Editor 164
7.10 Musical concepts 170

Scene 8 Groove **175**

8.1 Introduction to groove 175
8.2 Grooves 176
8.3 Groove Pool 176
8.4 Commit groove 179

	8.5 Extract groove	181
	8.6 Musical concepts	182
Scene 9	**Recording**	**185**
	9.1 Recording MIDI clips	185
	9.2 MIDI overdub recording	190
	9.3 Freezing and converting MIDI clips into audio clips	196
	9.4 Recording audio clips	198
	9.5 Exporting and printing	207
	9.6 Musical concepts	211
Scene 10	**Working with Scenes**	**215**
	10.1 Musical foundation and structure	215
	10.2 Scene launch preferences	217
	10.3 Tempo and time	219
	10.4 Capture and Insert Scenes	220
	10.5 Musical concepts	222
Scene 11	**Grouping Tracks**	**229**
	11.1 Group Tracks	229
	11.2 Launching Group clips	232
	11.3 Mixing concepts	235
	11.4 Musical concepts	237
Scene 12	**Controlling Your Universe**	**241**
	12.1 Remote Control	241
	12.2 MIDI Mapping	243
	12.3 Key Mapping	247
	12.4 The Relative Session Mapping Strip	248
	12.5 Mapping Browser	249
	12.6 Musical control	251
Scene 13	**Warping Your Mind!**	**255**
	13.1 Elastic time	255
	13.2 Warp Modes	262
	13.3 Warping samples	266
	13.4 Musical concepts	276

Scene 14 Loops, Slicing, and More Looping **279**
14.1 Loops demystified 279
14.2 REX loops 280
14.3 Slice to new MIDI track 282
14.4 Working with loops 288
14.5 Looping in the Arrangement View 289
14.6 Loops with Unlinked Clip Envelopes 290
14.7 Looping concepts 292

Scene 15 Instruments and Effects **297**
15.1 Introduction to Live Devices 297
15.2 Working with Live Devices 298
15.3 Live 8 Instrument basics 303
15.4 MIDI effects 311
15.5 Audio effects 313
15.6 Device chains 325
15.7 Plug-in devices 332
15.8 External (MIDI) Instruments 336
15.9 Working with devices 338

Scene 16 Device Racks **345**
16.1 Introduction to Racks 345
16.2 Rack interface and layout 346
16.3 Drum Racks 350
16.4 Creating Device Racks 353
16.5 Racks in Session View 365
16.6 Working with Racks 367

Scene 17 Suite 8 **373**
17.1 Overview of Suite 8 373

Scene 18 Video with Live 8 **377**
18.1 The possibilities of video with Live 8 377
18.2 Working with video 377
18.3 Video clips 379

18.4 Synchronizing music with video 382
18.5 Aligning video clips on the track display 385

Index *389*

On the Website

Suite 8

ReWiring the Digital World

Max for Live 101: The New Max User

AKAI APC40: The Ableton Performance Controller

Share: Collaborative Live Sets

Essential Keyboard Shortcuts/Commands

About the Author

Keith Robinson is a composer, sound designer, and audio engineer living in New York City. He is also part of the adjunct faculty at *The Clive Davis Department of Recorded Music,* New York University, where he specializes in the *Fundamentals of Audio Workstations* (Pro Tools, Digital Audio, MIDI) and *Producing Music with Software & MIDI* (Live, Logic, Reason). Keith's music has been used in numerous commercials and television shows and he has composed additional music and sound design for multiple feature films. Currently Keith serves as Vice President and co-developer of the sample library company *Sample Logic LLC* (www.samplelogic.com) and music production company *Sample Logic Studios* (www.samplelogicstudios.com). His sample company's award winning products *A.I.R., The Elements EXP, Synergy, WaterHarp,* and *Morphestra* are heavily used throughout music for Film, TV, and gaming world.

Acknowledgments

I would like to thank Amy Baer for all her help, support, and encouragement, Jim Anderson, Joe Trupiano/Sample Logic LLC, Jeremy Serkin, Andres Levin, and Mom. Special thanks to Huston Singletary and Justin Hegburg for their contributions.

I would like to thank Catherine Steers both for making this book a possibility and for seeing it through with me till the end.

In this chapter

1.1 Introduction 3
1.2 Approaching Live in three intuitive ways 4
1.3 Immersing yourself in Live 4
1.4 How to use this book! 7
1.5 Taking advantage of Hot Tips 8

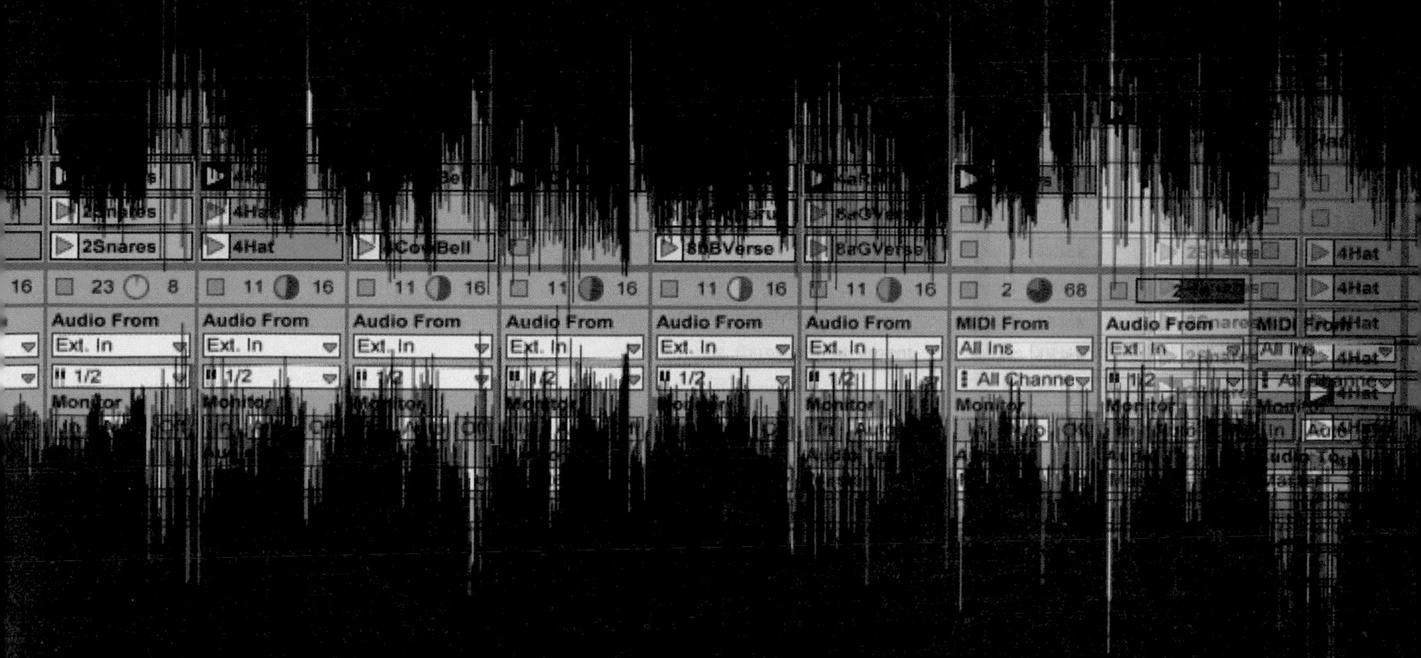

Ableton Live 8: Create, Produce, Perform

1.1 Introduction

Never has there been a more versatile, useful, or intuitive music production software that so closely emulates the human mind and the demand for delivering music through a computer. In less than a decade, Ableton has delivered a software package that meets the needs of the composer, producer, songwriter, DJ, and beyond. With an imaginative design and a forward thinking mission, Live 8 and Suite 8 drive music production into the twenty-first century. With such a progressive approach to its development, some of you may feel a bit disoriented or even intimidated at first sight of Live's unconventional design, especially those coming from a traditional digital audio workstation (DAW) background. If you are new to DAWs, DJ-style programs, or software music production in general, then you'll soon be right at home with the "parallel concept" of Live's Session and Arrangement Views. For the rest of us, we have to rethink our approach to composing, arranging, and producing music, but it will be a worthwhile adjustment. That is why this book has been written; to help reinvent the music production software user and to unleash the new user. For the current Ableton Live user – yes you – there is plenty here to unlock! After all, there is still a little "new user" inside all of us. So, it is the goal of this book to build and cultivate a strong understanding of Live 8's concepts and to provide material that will engage all sequencer and workstation users alike. With that goal in mind, at the end of each reading, you should feel that your current skills and knowledge base have been elevated to the next level.

Now it is time to learn how to *create, produce*, and *perform* with Live 8 and Suite 8. All you have to do is decide what your needs are because it's all here.

1.2 Approaching Live in three intuitive ways

Live naturally presents itself as a multipurpose tool for the composer, producer, and performer. For this reason, we have taken on the task to provide an accessible way for both new and intermediate Live users to obtain critical information that will act as a constant guide and reference for all things relating to Ableton Live 8 and Suite 8. We present to you a balance between creative techniques, tips and tricks, and the vital "nuts and bolts" of Live by concentrating on its three main facets: create, produce, perform.

With this approach, we educate, inspire, and propel you to new heights, synergizing creativity, production, and performance. In this manner, we hope you will use this book as a tool to uncover specific features and gain conceptual and practical knowledge that you can then integrate and apply in real-world situations. Rediscover yourself in the way that fits you best, whether through creation, production, or performance.

Now take a look to see where you fit within the following three content areas we've defined in our categorical approach to Live. Each area attempts to identify your unique needs and maximize your time spent with Live 8 and Suite 8. Use them as a guide to tailor-make your experience while reading the book.

Create: composers, songwriters, sound designers

Produce: producers, mixers, remixers

Perform: studio and onstage performers, real-time performances and improvisations, interactive music installations, choreographies

1.3 Immersing yourself in Live

The heart and soul of this book is about communicating the idea of making "music on the fly." This means an uninterrupted workflow that allows for multitasking the most complex ideas in real time. Therefore, this book is all about immersing yourself into the world of Live. We want you to think like Ableton and have the freedom to customize your learning experience. This is the ideology supporting every scene (chapter). The content of the text should be used as a launching pad for navigating to and from related concepts, skills, and unique features of Live 8 and Suite 8. That being said, you'll find that the book has been engineered in the same way that Ableton has conceived Live's own "nonlinear" musical environment. This means that not only can it be read front-to-back in a traditional fashion, it can also, and should, be read and used as a "nonlinear" resource for unleashing new and efficient skills for music creation, production, and performance. Simply put, read it in any order you desire.

To deliver the content in this way while staying true to Live, we have integrated the key elements of the Live software into the book itself. This includes the same creative features found in Live, transformed into a textbook format. We have explicitly adopted and literally implemented the terminology, labels, and features of the program, treating the book as if it were part of Live's graphical user interface (GUI) – in other words, as Ableton Live 8. The concept is to have the book look and feel like the software. Think of it as a new way of using a computer program and guidebook in tandem. Unlike a manual, this book will foster new and creative ways to use Live.

1.3.1 How the book has been constructed

This book does not use the "traditional" structural labeling system, rather it incorporates Live's nomenclature and terminologies to draw a parallel to Live itself. Our goal is that while reading, you will be learning the program without realizing it – thinking in "Live" language. The following legend identifies the terms, labels, and images that are used throughout the book to parallel Live – with some liberties of course.

- ***Scenes***: chapters
- ***File Browser 1***: table of contents
- ***File Browser 2***: index
- ***Clips***: topic titles
- ***Info Boxes***: sidebar definitions, concepts, topic highlights (Note that these resemble Live's Info View located at the bottom left corner of the main Live user interface.)

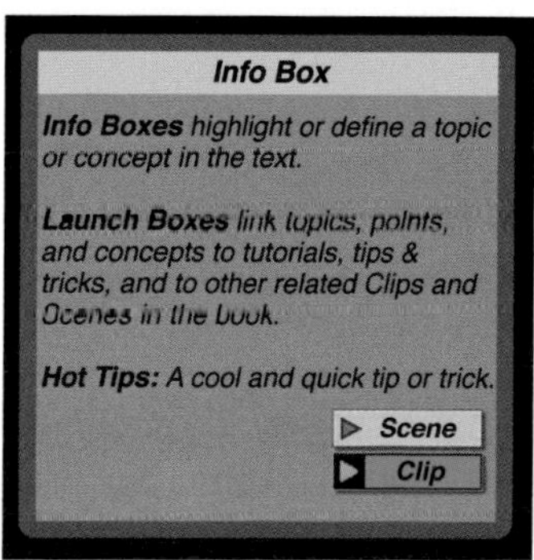

Figure 1.1 Info Box containing Launch Boxes.

- ***Hot Tips***: a cool and quick tip or trick. Hot Tips appear in two different formats (Note that neither of these physically exist in Live, but they resemble its Info View):
  - Hot Tip icon Hot Tip
  - Hot Tip box

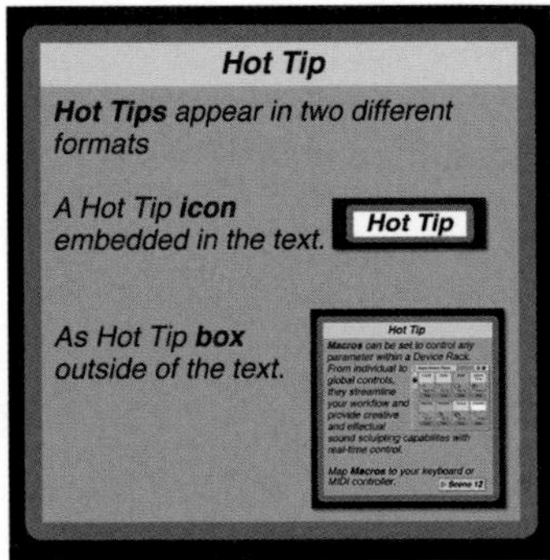

Figure 1.2 Hot Tip box.

- ***Launch Boxes***: "launch points" for linking Scene/Clip topics and concepts to other related Scene or Clip in the book. They can be found throughout the book embedded in the text and in Info Boxes.
- Scene Launch Boxes are yellow with a gray launch button ▷ Scene .
- Clip Launch Boxes are green, blue, or red ▶ Create ▶ Produce ▶ Perform .

We have also included screenshots from Live 8 and Suite 8. Be on the lookout for graphics and pictures that will help solidify concepts. "A picture is worth a thousand words."

1.3.2 General Scene layout

Each Scene (chapter) is constructed around the following general outline:

▶ Concepts and technology

- Features
- Functions
- Techniques
- Tutorials/examples

▶ Create concepts

- Composing/arranging

- Songwriting
- Sound design

▶ Produce concepts

- Producing
- Mixing
- Remixing

▶ Perform concepts

- Real-time "on the fly"
- Onstage performance
- Improvisation

▶ Information/Hot Tips

- Hot Tips
- Definitions

▶ Launch points

- Contextual Clip Launch Boxes
- Interconnected topics, techniques, features, tips, and tricks

1.4 How to use this book!

We have designed this book to be read in any order you choose, not necessarily from cover to cover. Customize your own path "on the fly" as you read. We recommend reading through the first three Scenes. This will take you through the basic essentials. At that point, you will be ready to take off on your own journey through the rest of the book. Feel free to approach your reading as an efficient interlinked adventure, tying together the concepts you need to enhance your workflow and unlock Live 8's or Suite 8's power. If you're new to Live, then you might choose to read straight through or attack the areas that address your immediate needs. Here is how we suggest using the book:

- Locate information and concepts that interest you, then let the book take you on a journey through its related topics.
- For those of you looking for basic principles, the "nuts and bolts," you can read straight through cover to cover. It's all here!

- If you haven't the time for a long read, dive right into the content you need, then launch from Scene to Clip, Clip to Clip as needed. Use the File Browsers (contents and index) or Launch Boxes within a Scene.
- If you have a little more time on your hands, you will find testimonials, concepts, definitions, and hot tips that are tailored to creating, producing, and performing with Live.
- Many of our exercises, tutorials, walkthroughs, and examples reference specific media. This content has been made available for you on our book Web site: http://booksite.focalpress.com/AbletonLive8. Download sets, audio, MIDI, clips, and more from the site when recommended. You will also find instructional tutorial videos and additional reading on Max for Live, AKAI APC40, Keyboard Shortcuts, Suite 8, ReWire, and much more.
- Visit http://www.ableton.com/downloads for additional Live Packs (presets and sample content) that will enhance your experience with Live 8 and Suite 8.

1.4.1 Summing it all up!

Just like the program, we have provided a customizable learning environment, allowing you to move freely throughout the text. Our strategic pointers (launch points) open up pathways to new features and concepts, tying together other related features and concepts. We like to refer to this as a lateral learning approach. In this way, the book becomes a tangible testament of Live, freeing you from solitary methods to better service your unique needs. Make reading it part of your daily routine and apply a new tool or concept to a composition, production, or a performance. The point is you can focus on one or all of Live's three facets – create, produce, perform – all while navigating freely throughout the book. Choose where to start, stop, and go with the freedom to launch to and from any section in a nonlinear fashion. This allows you to "choose your own adventure" and find answers and justifications that define how music and technology converge with Live. Every Scene is written with the singer/songwriter, studio composer, sound designer, DJ, producer, and real-time studio and stage performer in mind. Remember, launch points can be found inside Info Boxes, Hot Tips, and embedded within the text. They will guide you to related material that will help build your practical knowledge of Live. Note that some concepts are described in depth while others serve more as an overview. Of course, if you ever desire more information, refer to Live 8's documentation.

1.5 Taking advantage of Hot Tips

Strategically placed throughout the book, Hot Tips are invaluable cool insider tips that save you time while enhancing your workflow. They have been crafted and designed to provide important information that will streamline your experience with Live 8. Some of these are hidden tricks of the trade, whereas others are newfound work-arounds. Hot Tips are quick time-saving devices that can be put into practice right away! Some also act as launching points to interconnect skills, tips, tricks, and topics throughout the book where indicated.

Keep an eye out for Hot Tip boxes and icons Hot Tip.

In this chapter

2.1 Introduction 13

2.2 The concept 13

2.3 Important preference tweaks! 14

2.4 The Live Browser 19

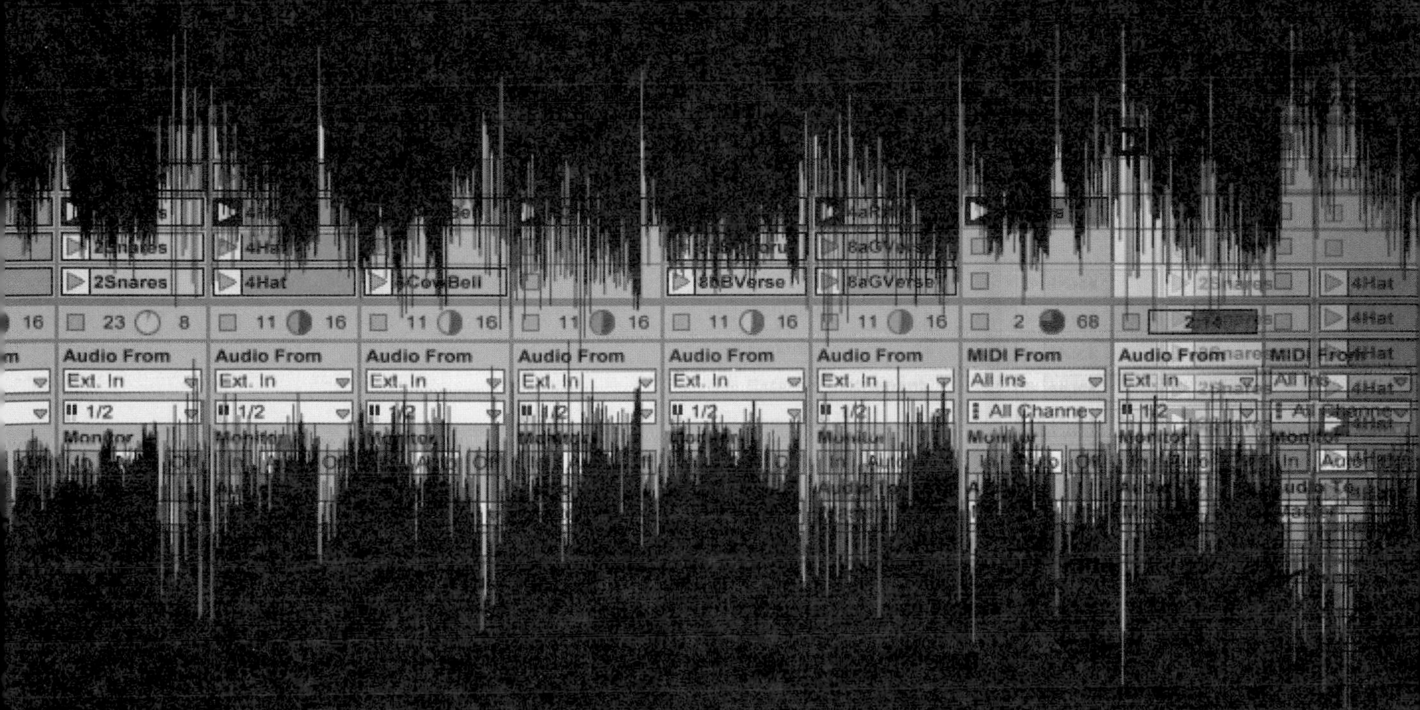

SCENE 2

Overview: Live 8, Suite 8

Live 8 asks you to re-imagine your music and invent new ways to present it to your audience. It needs to become a standard in commercial studios' software arsenals as much as that "other" program.

Brian LeBarton, composer, DJ & keyboardist/musical director for Beck

2.1 Introduction

In its eighth incarnation, Live has quickly become a mainstay in the music production and performance industry. The Ableton team brings a fresh approach to creating and producing music in the studio and on the stage. When you think of a digital audio workstation (DAW), you don't normally think of an instrument for performance and song creation. Nevertheless, Live can be used in precisely that way. It's the one thing that makes the program so special and unique. At the same time, Live's differences continually push the envelope and comfort level of the tried and true traditions causing resistance by some. "Dark and mysterious it is"... sounds intriguing! By now, all you want to do is launch Live and start working on your music, right? Who has time to read? In this Scene, we'll guide you through a few quick steps to get optimized and running in Live 8 and Suite 8, but first we should point out that there are currently three versions of Ableton Live available: Ableton Suite, Ableton Live, and Ableton Live LE. Suite is the complete package that bundles Live 8 with 10 Ableton instruments. Live is the standard version of Live 8 and comes with a basic set of sounds. LE is a slimmed down version of Live with only the essential tools and sound content.

Ok, now we can begin. Let's start out by addressing various software settings. Once your system is streamlined, we'll create a Live Set together and get you on your way.

2.2 The concept

Live 8 is a one-of-a-kind real-time musical instrument, sequencer, and production tool. It's all about nonstop creative flow. At first glance you will notice that Live is designed around two parallel views, or windows if you will: Session View and Arrangement View. Each view has its own unique musical purpose and functionality. They cannot be viewed simultaneously, but they do in fact work together in representing your audio. There will be much more on this "parallel" concept throughout the book, but for now we'll focus on the general concept. It is from one of these two views that you will create, produce, or perform your musical ideas. In general, you will use the Session View as a musical sketchpad and launching base, and the Arrangement View as a traditional linear music timeline for recording audio and sequencing your music, the latter of which is like any other DAW or sequencer software. This is in no way set in stone, but more often than not it's the most common method of using Live; whichever approach you chose, Live 8 can accommodate. More importantly, the real beauty is in Live's ability to realize and humanize your approach to creating "music on the fly" and enable real-time improvisations. Of course it also provides all of the traditional features of a DAW – powerful audio effects, instruments, intuitive editing tools, and features – but it's Live's nontraditional real-time flexibility and creative clutter-free workflow that sets it apart.

Figure 2.1 Arrangement View – right; Session View – left.

2.3 Important preference tweaks!

A little knowledge about how Live and your computer interact can go a long way. The last thing anyone wants is to be performing on stage when suddenly playback pops, glitches, drops out, or even worse, experiencing a black screen, blue screen, or the dreaded Pinwheel of Death. Although we cannot foresee those tragic system crashes, we can help you work through CPU, Audio, MIDI, and other related performance optimizations. In a perfect world, we'd all own the latest and greatest computer with the fastest CPU, tons of RAM, and all of the disk space in the world. Unfortunately, that is never the case. That being said, you should be sure that your system at least exceeds the minimum recommended system requirements of any production software you use! Here's something that software developers don't tell you: they rate their particular system specifications as low as they can before the software will not run. That means that as soon as you load a third-party plug-in or multitask, your system will stutter, pop, glitch, overload, error, you name it. In any case, keep an eye out for system requirements and if relying on Live for your bread and butter, you're better off exceeding Ableton's minimum requirements. Here are a few ways to smooth out your experience in Live 8 and Suite 8.

2.3.1 CPU load

When it comes to CPU, you will want to keep an eye on the CPU Load Meter at the top right of the Live's Main Screen.

This shows the amount of processing your computer is taking on at any given time. When this meter reaches 100% it is maxed out. Fortunately, Ableton has dedicated its first priority to audio, meaning that audio processes are the last to be sacrificed for processing power. In other words, noncritical processes and functions are the first to be dropped. This way your music can remain theoretically uninterrupted when it comes to a power struggle.

What will cause strain on your CPU? The higher your track count and devices, the more strain on your system. Effects of processing (third-party or native) also pulls on system resources, but the biggest offender will be third-party instruments and plug-ins. They can eat up a lot of system resources from your CPU, hard disk, and even RAM. Instruments in general can be CPU hogs. Fortunately, Live is very efficient in dealing with instruments, tracks, and devices. Only active instruments and effects contribute to the CPU load. When they are idle there is little to no pull on resources.

2.3.2 Disk load

The number of simultaneous audio channels used in a Set is directly related to disk load. Therefore, Live's performance is dependent on your computer's hard drive speed. Most optimized desktop computers have sufficient hard drives (7200 rpm) for operation with Live. On the other hand, laptops generally use slower drives (5400 rpm). Rest assured that Live 8 is laptop friendly and most systems work without a hiccup. Track count and virtual instruments will dictate disk performance of your Live Set. The *hard disk overload indicator* will light up if hard disk overloading occurs. In this situation, you may need to reduce the track count in your Set or use RAM Mode for certain clips. Launch to Clip ▶ 7.5.5 for more details on Clip RAM Mode. If a virtual instrument sample player is the culprit, you could switch its internal sample loading to a RAM-based mode as opposed to *direct-from-disk mode* (DFD) – if applicable. In short, in DFD mode samples are read directly from the hard drive. Using RAM will free up the hard disk by loading the samples into your computer's memory instead. The goal is to take the strain of reading and writing off the hard disk. Each plug-in handles this differently, so familiarize yourself with their unique options.

2.3.3 Track freeze

Freezing tracks (Edit Menu>Freeze Track) is a great way to free up system resources, both CPU, RAM, and Disk usage, when you are simply pulling too much power with your Set. Once a track is frozen, it plays back from its Freeze File stored in the project sample folder. This file is a render of the entire track's contents along with its effects. This allows you to use more instruments, devices, effects, and tracks, even if your computer is low on processing power. It is a simple, quick, and rather practical way to improve your system performance, and any track that can hold clips may be frozen – Audio or MIDI. Track Freeze is located under the Edit Menu or the clip's contextual menu.

Although freezing is a very intelligent and helpful feature, it has a few minor limitations. It limits some editing features such as note and audio changes or manipulations. On the other hand, Mixer functions are still available as well as launching clips, copy, paste, rearranging, volume, panning, sends, and so on. Ideally you won't have to use the Track Freeze command, but if you do, you can unfreeze tracks at any time. Track Freeze is a nondestructive process, but you always have the option to *flatten* a frozen track permanently, making any edits to frozen audio clips destructive and converting a frozen MIDI track into audio track/audio clips. The Flatten command is located just under the Freeze command in the Edit Menu. That being said, conserving CPU is not the only reason to use Track Freeze. There are a few very useful – actually mind boggling – ways to use Track Freeze with MIDI in which you can turn your MIDI clips into audio clips in a blink of an eye. What? Yes it's true: MIDI clips converted into audio and in real time. To learn more about Freezing and Converting MIDI clips into audio clips, launch to ▶ 9.3. To be honest, it doesn't stop with MIDI. Any frozen audio clip can be instantly rendered when dragged from a frozen track to an audio track.

When working with frozen tracks, keep in mind that editing track clips while they are frozen may result in a surprisingly different playback effect when you unfreeze them. This happens when dealing with tracks that use delays, reverbs, or any other time-based processing. Effects such as these are dependent and influenced by note or audio events that obviously are now part of the freeze file and cannot be influenced by a new sequence of clips or edits, etc.

2.3.4 Audio preferences

It's always a good idea to verify your audio device setup. To do this, go to the Preferences>Audio tab. Make sure that your audio input and output device is configured correctly and efficiently. Any unused input/output channels should be turned off. This will reduce the load on your CPU. As for sample rate, you will determine that yourself, but we must recommend using High-Quality SR & Pitch Conversion unless your system cannot handle the load. Change this to Normal if you must, but why settle for less? It only makes sense to choose the best quality. As for any audio latency issues, they should be addressed with the Buffer Size settings. For Plug-in Buffer Size go to the CPU tab.

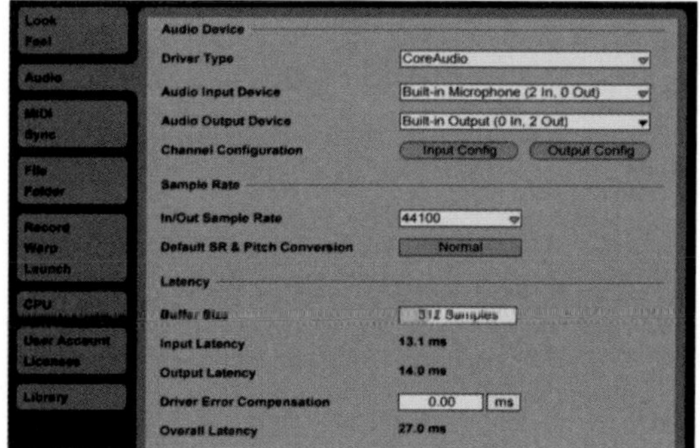

Figure 2.2 Audio preferences tab.

2.3.5 MIDI preferences

To set up your external MIDI devices, go to the Preferences>MIDI Sync tab. From there you can configure Control Surfaces, MIDI interfaces, or other MIDI devices that are connected to your computer. In the top area of this tab, you can choose up to six control surfaces and set its MIDI input/output ports via the available chooser menus. To the far right is the Dump button, short for *preset dump*. This feature will send preset information to a controller so that you can map its physical controllers (slider, faders, knobs, pads, etc., correctly) if and when needed according to the specific control surface. Note that Dump is only needed for certain hardware surface controllers/devices. In general, you won't need to use this feature for the devices natively supported by Live. Once again, this is dependent on the device you are using. Below this you can set the Takeover Mode for your device. This is a very important preference! It determines how Live will respond to continuous controller MIDI values sent from a physical knob or slider on your MIDI controller. There are three modes: None, Pickup, and Values Scaling. Each option determines how Live behaves and reacts to sudden jumps in values sent from your controller. Refer to Setting up Control in ▶ 12.1.1.

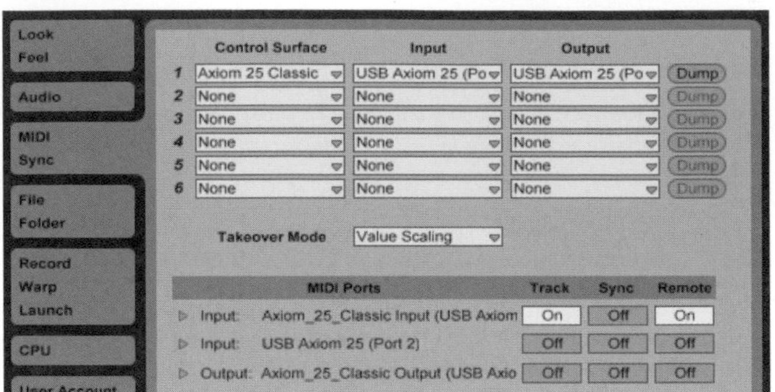

Figure 2.3 MIDI preferences tab.

The bottom section of the MIDI/Sync tab is the MIDI Ports List where you will find all of your available MIDI input and output ports. The left column lists the actual MIDI in/out Ports. To the right of these are three columns and their associated On/Off switches: Track, Sync, and Remote. To send MIDI to Live's tracks, you will need to activate Track (On) for the input port(s) you wish to use for sending MIDI. You can also send MIDI from Live tracks to external MIDI devices by activating the Track switch for the output port you wish to use for sending MIDI out. For synchronization with Live, activate the Sync switch for input ports that are to be used for syncing Live. To send sync messages from Live, activate the desired output ports that are going to be used. If you wish to remote control Live with a controller using customized mapping assignments, then Remote should be activated for the input ports that are to be used for remote controlling Live. For control surfaces with motorized faders or real-time displays, Remote should be activated for the associated output ports. Live can then send control update messages to the control surface so that its controllers, faders, or display will mirror Live's-related controls.

2.3.6 Record, warp, and launch preferences

Here you can set Record File Type (.wav or .aiff), Bit Depth (16, 24, or 32), and Record Count-In (1, 2, and 4 bar). These should always be set prior to recording audio. The Exclusive Arm and Solo options establish arming (record-enable) and solo restrictions so that only one track at a time can be armed or soloed. Click+hold the [⌘] Mac/[Ctrl] PC key while clicking track Arm and Solo buttons to temporarily bypass these exclusives.

Another important preference is Warp and Launch, which help to streamline your creative workflow. These will help automate how Live handles imported audio clips and the behaviors associated with launching clips. If you are just starting off with Live 8, the default settings are fine. For more information on warp settings and concepts, refer to Clip ▶ 7.5.9 and ▶Scene 13. For launching clips, refer to ▶ 6.3.2.

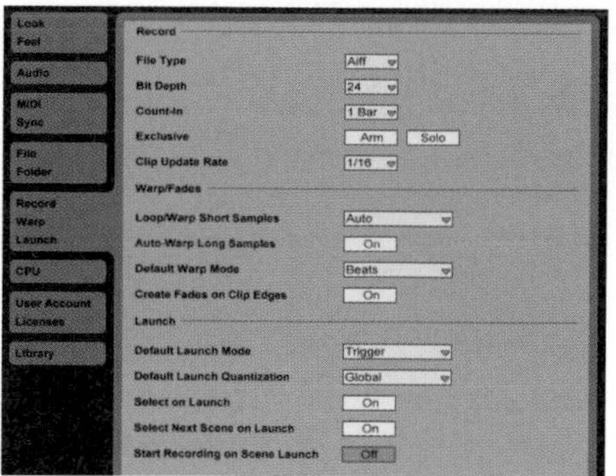

Figure 2.4 Record, warp, launch preferences tab.

2.4 The Live Browser

Before you dive into making music, you should understand how Live's Browser system works. Everything you will need to make music can be found within Live's Browsers. Yes of course, you can always drag and drop from your Mac Finder or Explorer Window, but utilizing Live's built-in Browser is a no brainer and an important asset to your workflow in Live 8. The truth of the matter is, you should never have to leave the Live 8 interface for any reason when you're working on your music other than to, well, to check your e-mail.

The Live Browser window is the interactive column located on the left side of the main Live screen. The Live Browser can be shown or hidden by clicking the unfold button (triangle) at the upper left directly under the TAP button ([⌘+opt/alt+B] Mac/[Ctrl+opt/alt+B] PC).

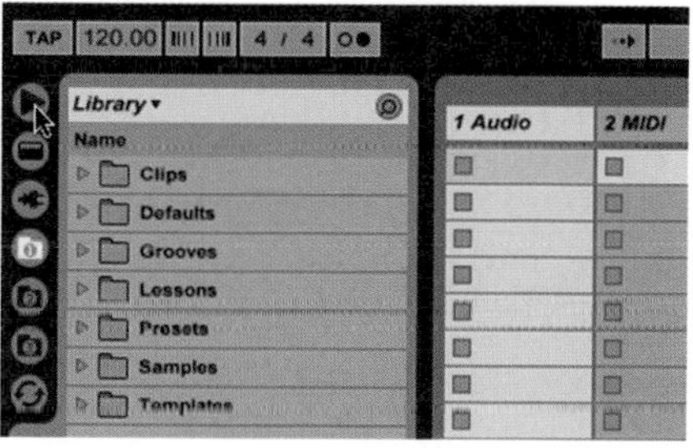

Figure 2.5 Show/hide browser.

The Live Browser is an intuitive way to quickly access all of your instruments, effects, plug-ins, samples, and much more without leaving Live's main screen or interrupting your workflow. Each icon on the Browser pane identifies the individual subbrowser categories.

With the Live Browser open and in view you can access:

Live devices: instruments, MIDI effects, audio effects.

Plug-in Device Browser: AU/VST instruments and effects.

File Browser 1, 2, 3: customizable file directories that mirror your hard drive(s) and all of its contents.

Hot-swap browser: hot-swap files.

In addition, you may notice two additional icons that will appear, the Mapping Browser's icon, (when in a Map Mode) and a Browser button that becomes available when sharing Live 8 Sets (currently in beta):

Upload/download browser: hot-swap files ▶ Website.

The Browser concept is simple and exciting! You'll notice in both the Session View and Arrangement View an area labeled Drop Files and Devices Here. You can insert a Live device into your Set by dragging and dropping it directly from the Device Browser to the Drop Area, to a track, or by double-clicking on a device Name inside the Device Browser. When dropping a device outside of a track in one of the given Drop Areas or double-clicking, Live will automatically load up and create the necessary tracks. To learn all about Live devices (instruments and effects) launch to ▶Scene 15. In addition to easy access, Ableton has provided the ability to change out instruments and effects on the fly with the Hot-Swap buttons. With Hot-Swap Mode activated, you can instantly change out an instrument, preset, or effect by double-clicking on a new one in the Browser. Gray = off. Orange = on. Follow any one of the following Clip and Scene launch links below to learn more about Hot-Swap Mode.

Hot-Swap buttons are incorporated for use with

- Devices/presets ▶ 15.2.2
- Grooves ▶ Scene 8
- Sample-based Live instruments/device chains ▶Scene 15
- Device racks ▶Scene 16
- Replacing/managing files referenced in a Set ▶ 3.7.3

2.4.1 Live Device Browser

The Live Device Browser grants you access to all of your Ableton instruments, audio effects, and MIDI effects. This includes device, specific factory, and custom user presets located under each device folder structure. Note that there is also a Defaults Folder available in the Live Library that is used to manage the various factory settings and behaviors of devices and related actions. This folder is accessible from the File Browser. Refer to Clip ▶ 15.2.4 for presets and customization.

2.4.1.1 Instruments

Both Live 8 and Suite 8 come with a core set of instruments and presets; the difference being that Suite 8 includes all of Ableton's software instrument collection as shown in our example here.

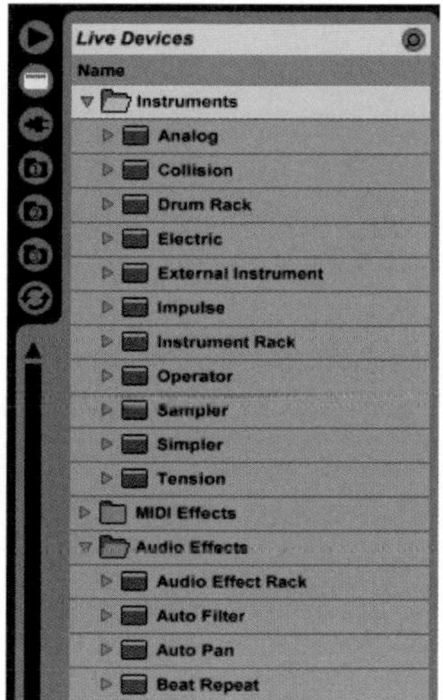

Figure 2.6 Live instrument devices.

On that note, we recommend that you seriously consider upgrading or buying Suite 8. It comes with all of Ableton's instruments (except for the Orchestral Instrument Collection), which are essential in creating, producing, and performing with Live. The instruments and content included in Suite 8 are a must for the dedicated Live user! For a detailed description of the instruments included in Suite, launch to ▶ Website and also check out the Ableton Website. For now we'll point out the two core instruments included with Live 8 located in the Instruments folder in the Live Device Browser: Impulse and Simpler. Impulse is primarily used for triggering and playing back sampled drums and percussion one-shot audio samples, while Simpler is a basic sample player instrument. These will get you up and running in no time either as standalone instruments or under the hood of an instrument or drum rack. This makes it very easy to begin making music without investing in third-party virtual instruments. Both instruments come with presets, so feel free to start exploring their capabilities as soon as you can. For explicit details on Impulse and Simpler, launch to ▶ 15.3 and for more on Device Racks launch to ▶ Scene 16.

2.4.1.2 Audio Effects

The Audio Effects folder contains all of Ableton's audio effects. As of now, there are over 30+ unique audio effects located in the Device Browser. Some of these are labeled and considered by many to be the "classic" effects used for mixing and tracking purposes while others can be used to wildly enhance and manipulate your audio and MIDI instruments. In addition, with the ability to effortlessly "chain" together effects into a custom Device Rack, the possibilities are endless.

Potentially, all of the essential effects you'll need are located in the Device Browser within the Audio Effects folder along with their included presets. No reason to run out and buy a huge and costly third-party effects bundle. After digging around a little deeper with each of the Audio Effect's presets, you'll find Ableton effects that are so unique and mind-blowing that it makes you wonder what the teams at Ableton were thinking when they devised them. This includes effects such as Erosion, Vocoder, Frequency Shifter, External Audio Effect, Looper, and the stand out classics like Beat Repeat and Filter Delay. For a description and explanation of how audio effects work and apply to creating music with Live 8, launch to ▶ 15.4. If you already have a handle on audio effects and are feeling ambitious, launch to ▶Scene 16. There you will find out how these monster effects can be linked together, utilizing another Ableton original: Device Racks.

Figure 2.7 Live audio effects.

2.4.1.3 *MIDI Effects*

This folder contains Ableton's included MIDI-based effects that can both alter and enhance any type of MIDI information you throw at it. These effects are inspiring devices that literally manipulate and grow your music into dynamically complex and rich arrangements. No need to dive into a crazy interface window or environment to program MIDI cabling (isn't that a relief). It's all here and ready for you to use. Each effect includes a set of discrete presets. Start with these to get a handle on the instant result of Live's MIDI effects. Reveal their presets by clicking on the arrow left of the names to navigate through the file tree, then load them up on a MIDI track and instrument and let the fun begin. MIDI effects have a unique and bold ability to alter your MIDI tracks and patterns. Just like audio effects, MIDI effects can also be turned into custom effects racks. Launch to ▶ 15.5 to learn more about MIDI effects.

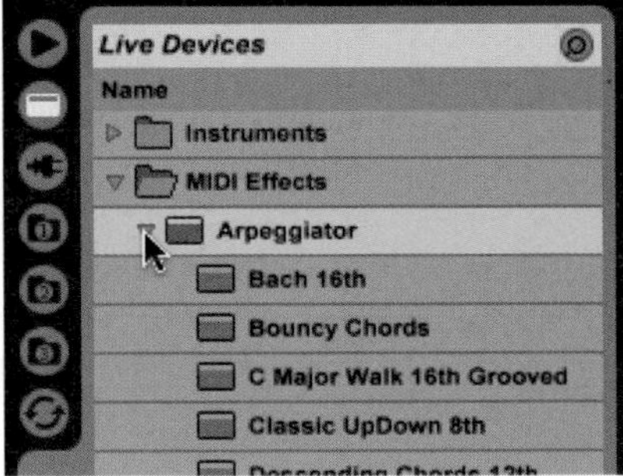

Figure 2.8 Live MIDI effects.

2.4.2 Plug-in Device Browser

As you can see, Live 8 and Suite 8 include many fantastic and well-integrated instruments and effects. Once you have exhausted all of them, you may consider branching out and investing in some additional third-party instruments and effects. For this, Ableton has you covered. Another great attribute of Live is the inclusion of third-party plug-in access built into the Live Browser. This gives your third-party plug-ins the same seamless integration as your Live 8 and Suite 8 instruments and effects.

Why do you need more? These are your tools, and you can never have enough! There are new plug-ins made available every day from numerous developers, created by some amazing programmers and musicians that specialize and dedicate every waking hour to their development. So when you are ready and in need, search the Web or your local music store for third-party plug-ins that are compatible with Live, and don't forget to continually check the Ableton Website for new instruments, effects, and other add-ons as well.

Before you can use third-party Audio Units (Mac only) or VST plug-ins with Live, you will need to turn them On in your Live Preferences menu (Preferences>File Folder>Plug-in Sources). Once activated, Live will scan and locate all of your AU and VST Plug-ins, thus making them accessible in the Plug-in Device Browser.

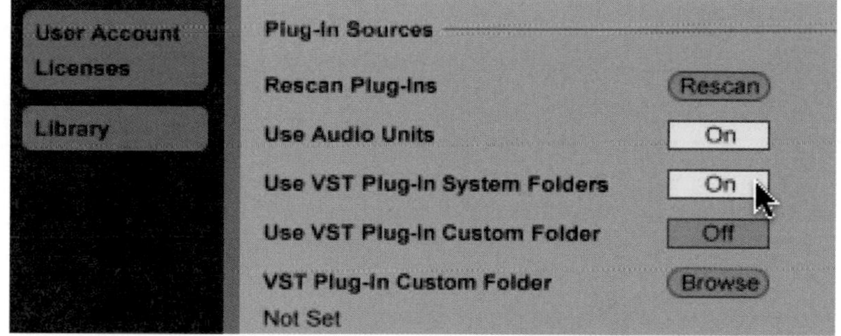

Figure 2.9 File folder preferences tab: activate AU and/or VST plug-ins for use in Live.

This includes both virtual instruments and audio effects. They have all been placed together into one single browser listed alphabetically. To learn how to use third-party plug-ins in Live 8 and Suite 8, launch to Clip ▶ 15.7.

2.4.3 File Browser

The File Browser is an intuitive and adjustable data hierarchy that gives you quick access to all of the information located on both your internal and external hard drives for instant access and use in Live 8. Why have one File Browser when you can have three? Ableton has conveniently included three independent File Browser buttons on the Live Browser pane for easy organization and accessibility. They are located directly below the Plug-In Browser button labeled "1, 2, 3."

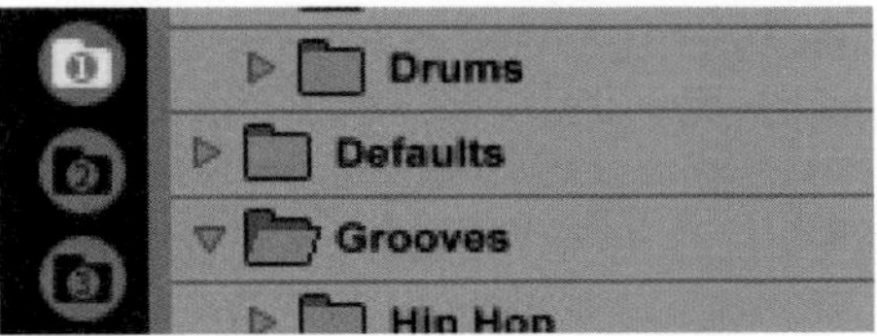

Figure 2.10 Three individual File Browsers for quick access to the Live Library, entire file folder structure of your computer, and external storage devices.

Each one offers an independent displayable hierarchy that always remembers its last selected location. This is an incredibly crafty way to view your common and most used information while working in Live 8. This way, your data is always at the touch of a button and always right in front of you, ready to be accessed. Because the File Browsers are part of Live's Main Screen, they are always accessible in both the Session and Arrangement Views so long as the Live Browser is open and in view.

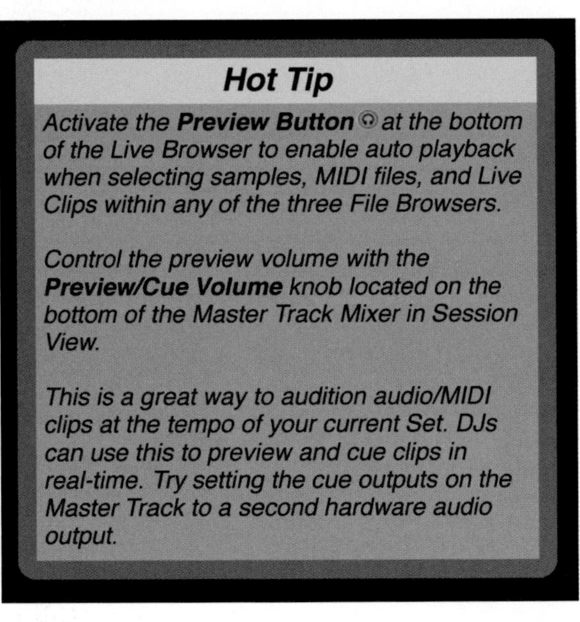

Hot Tip

Activate the ***Preview Button*** *at the bottom of the Live Browser to enable auto playback when selecting samples, MIDI files, and Live Clips within any of the three File Browsers.*

Control the preview volume with the ***Preview/Cue Volume*** *knob located on the bottom of the Master Track Mixer in Session View.*

This is a great way to audition audio/MIDI clips at the tempo of your current Set. DJs can use this to preview and cue clips in real-time. Try setting the cue outputs on the Master Track to a second hardware audio output.

Live's File Browsers are used to import a variety of media, most commonly samples (audio files), MIDI files, preexisting and clips into a Live Set. All of your audio/MIDI files may be previewed/auditioned directly from the File Browser prior to importing. This is very useful for auditioning loops, samples, and other clips without having to commit them to your Set or increase your track count. This is also great for auditioning complex tracks or entire songs for DJ Sets as preview material can be monitored through headphones or through a secondary set of audio device/interface outputs. Note that previewing is possible in real time and is in sync with Live's current tempo. Test groove against groove on the fly without interrupting playback. To activate Preview, click on the Preview button gray/blue headphones icon located at the bottom of the Live Browser. Gray = off. Blue = on. This is the Preview Tab.

Once you become familiar with browsing around your computer and content via Live's File Browsers, you'll stop using those extra Finder/Explore windows that

Figure 2.11 Preview tab: Listen to and audition audio/MIDI files and clips.

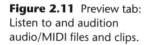

can clutter your desktop workspace. As you already know, each File Browser will recall its last location, even after shutdown and startup until it is manually changed. This is really efficient as you begin to rely on Live's Browsers to help you create and arrange your music without the distraction of locating files outside of the interface on your desktop or remote drives. Taking this a step further, customize your browsing experience by using Bookmarks. At the top of each File Browser window is a pull-down chooser menu allowing access to Live's Bookmarks. This menu lists both default and custom bookmarked folders for quick navigation. By clicking on the Bookmark chooser window, you have choice of selecting from the following bookmarks: Library • Current Project • Desktop • All Volumes • Home. On a PC, the names are slightly different, as you can see.

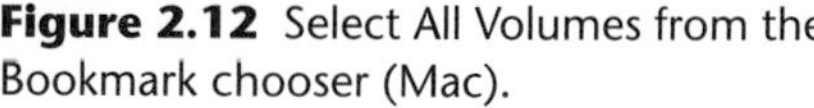

Figure 2.12 Select All Volumes from the Bookmark chooser (Mac).

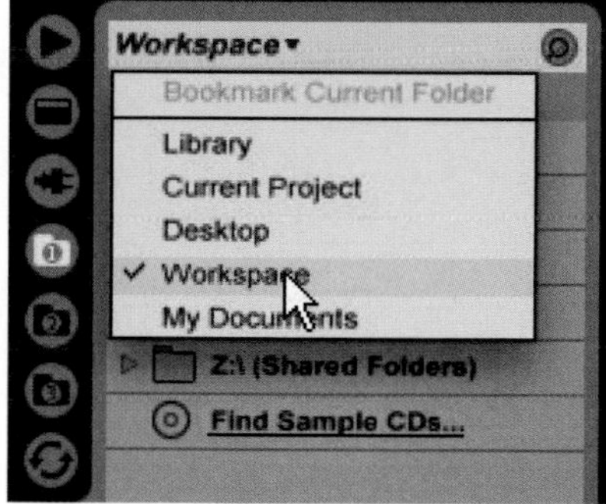

Figure 2.13 Select Workspace from the Bookmark chooser (PC).

The All Volumes (Mac)/Workspace (PC) folder is very useful because it displays all drives attached to your computer system. Use Bookmarks to customize each File Browser button or to quickly navigate through your system directories by selecting the arrow button next to the Drive's folder to open it, revealing its contents.

The Library is where all of Live 8's and Suite 8's included and expanded content (Live Packs, Clips, Grooves, Lessons) is located, which includes an array of key elements such as samples (sounds), MIDI clips, Sets, and much more to be used with all of your Live projects. This Library is installed along with your Live 8 or Suite 8 installation. From the Library bookmark you will find folders dedicated to

Clips, Defaults, Grooves, Lessons, Presets, Samples, and Templates. In addition, the Library can store all of your own customized device presets, clips, and Sets. Launch to 15.2.4 for more on this efficient feature.

Figure 2.14 Library bookmark set to File Browser 1.

Figure 2.15 The Live Library.

The remaining Bookmark categories, Desktop, Home (Mac)/My Documents (PC), and Current Project are self-explanatory in that they all display either your current desktop hierarchy, root drive hierarchy, or your currently displayed Live Set along with its currently used clips and devices.

Now that we've explained the File Browser system, let's create our own Bookmark in the following example:

1. Click on the File Browser 2 button. We'll leave 1 for the Live Library content.
2. Click on the Bookmark Chooser menu.
3. Click on All Volumes/Workspace.

Figure 2.16 Click on the All Volumes (Mac) or Workspace (PC) bookmark.

If you have an external drive(s), this is the best way to view them. Two, three, or even a dozen external FireWire/USB drives is not uncommon in a Live user's studio arsenal, which can get quite confusing. Using the All Volumes/Workspace view will really sort things out for you.

Figure 2.17 All volumes.

4. Open the folder of your choice by clicking the triangle on the left to expand the hierarchy, preferably one with audio or MIDI content.

Now you should be looking at all of the folders and files of that particular drive. If these are sound files, you will be able to click and audition each file in succession. If you prefer, you can double-click on a favorite volume folder to see only the contents of that particular hard drive. As you can see, there is no adjacent window to open or separate audio window to display the file or work area. This makes it very easy to audition and listen or simply access your data files in real time.

5. (Optional): Double-click your favorite folder or volume then select Bookmark Current Folder from the chooser window.

Figure 2.18 Use the triangle to expand/unfold and view folder contents.

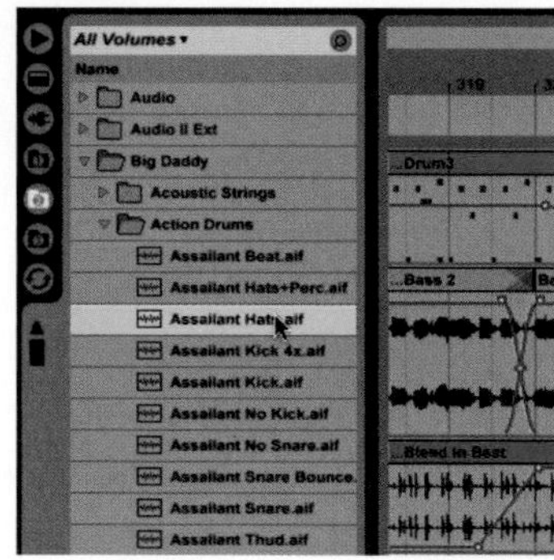

Figure 2.19 Audition or access your files in real-time.

Custom bookmarks will appear at the bottom of the Bookmark Chooser window below a dividing line.

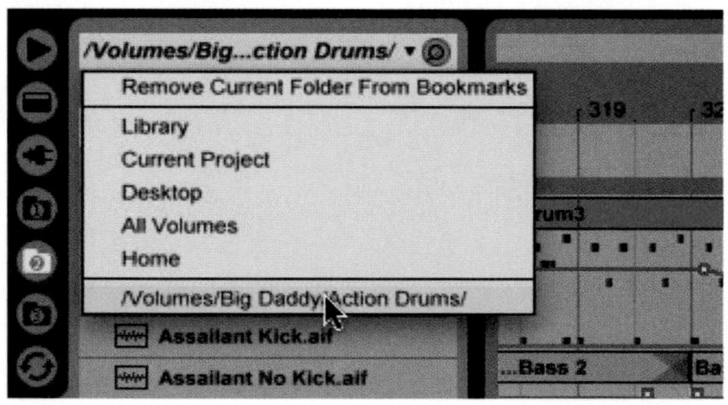

Figure 2.20 Custom Bookmarks appear at the bottom of the Bookmark's menu.

In this chapter

3.1 Introduction 31
3.2 Starting a project 33
3.3 Audio in Live 34
3.4 MIDI in Live 41
3.5 Essential operations and tasks 48
3.6 Performance to arrangement 54
3.7 Finishing your work 55

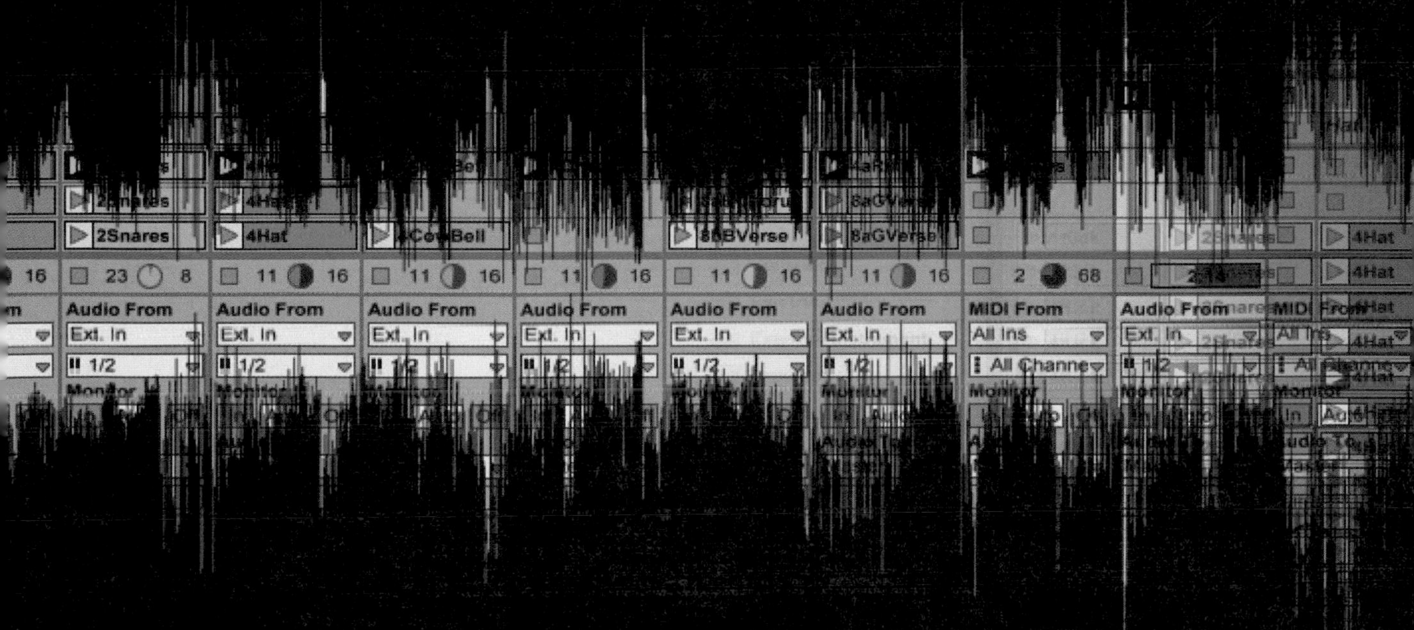

SCENE 3
The Quick Way to Start Making Music!

I've reliably used Ableton Live on tour for three years now. Live 8's new workflow enhancements ensure it'll be the only program I open before the curtain falls.

Steve Ferlazzo, Producer & Programmer. Keyboardist for Avril Lavigne

3.1 Introduction

So, you've installed and authorized Live 8 or Suite 8 and now you're wondering how to get started. Whether or not you have experience with a sequencer or a digital audio workstation (DAW), sometimes you just don't have the time or patience to dig in and read the entire manual from front to back. For that reason, this Scene is a quick-start guide to get you into Live 8 in the least amount of time possible. It underlines the general concepts without too much theoretical and technical jargon. When you are done with this scene (chapter), you will be able to execute the basic fundamentals of Live 8 with a general conceptual understanding of how it works. You may not always know why something does what it does, but that's ok. Our goal is to build a foundation that you can rely on later while saving you time now. As you read this book, keep in mind that this is intended to get you up and running quickly. If you feel like you still need more help getting started then go ahead and launch to the various locations referred to within this scene. There you will find dedicated and in-depth information pertaining to each topic. You can also take a look at the interactive Live Lessons that come preinstalled with Live 8 and Suite 8. These are incredibly useful and are a must for all Live users! Lessons can be found in Help View located on the right-hand side of your Live Set. If it's hidden from view, select Help View from the View Menu. Help View is also useful for quickly accessing various folders within Live's Library such as Instruments, Drum Kits, Audio Effects, Clip/Loops, Construction Kits, and Templates.

Now, before you get started, rest assured that Ableton always has you, the Live user, in mind. You can find helpful information displayed inside the Info View in the lower left corner of Live's main screen (your Live Set). Click on the arrow button in the extreme lower left corner of your screen to open it. Nearly every feature, button, and object in Live 8 is methodically tagged with a brief description of its function.

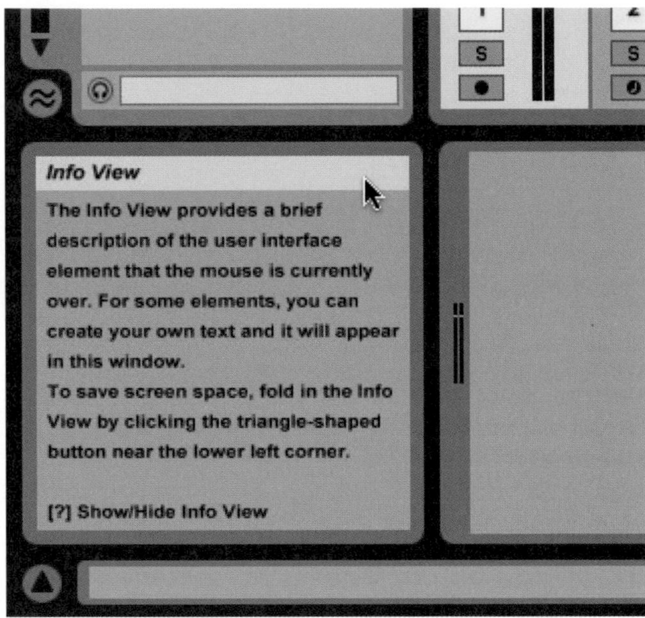

Figure 3.1 Info View.

As you mouse over and around any area of Live's interface, the descriptions will automatically pop-up in the Info View. This is a true time-saver for the curious or those of us who forget certain functions and parameters now and again.

Hot Tip *If you're that really organized producer or engineer, add your own info text to items in Live, such as tracks, scenes, clips, and devices. Place your mouse over any one of these items and right click or ctrl+click to bring a contextual menu, and then select Edit Info Text. Whatever you type will then be viewable in the Info View box when you hover your mouse over the item you named. Think about using these as comments or notes. You could also use them to leave a friend a little surprise.*

3.2 Starting a project

All right, it's time to get you on your way to creating and producing music, so open up Live 8 and let's begin. When you open Live 8 the default Set will appear in Session View consisting of one audio track and one MIDI track – the same default setup occurs when you create a new Set. This is just to get you started. You can always add or delete tracks as needed. If you want to design a different default Live template, you can do so by saving any Set you create as default from the Preferences>File Folder tab under Save Current Set as Default. If you decide that you want your default setup to always start at 105.0 BPM, for example, you can save a new default setup with this new parameter. You could also add several more audio or MIDI tracks as well to your blank Set then select Save Current Set as Default. It's that easy.

Set is the name given to your work file in Live. Other programs use the terms Session, Song, Project, and so on. Looking at your new Set, be sure to familiarize yourself with how to show/hide the Live Browser on the left-hand side of the main Live screen by clicking the arrow tab in the upper left just under the Tap Tempo switch. This is where you will find all of the media needed to work on your music, for now at least. Later you will use the Live Browser for a number of tasks, including loading Live Sets and more. That being said, your new best friend will be the Live Library that is installed with Live 8 and Suite 8. We recommend setting File Browser 1 to the Live Library while getting started.

Figure 3.2 File Browser 1 assigned to the Live Library.

Go to File Browser 1 then choose Library from the Bookmark chooser. This contains all of Live's content, such as clips, grooves, lessons, presets, samples, and templates. On the opposite side of the main Live screen is *Help View*. As we mentioned earlier, you can show and hide this window from view in the View Menu, or by dragging the right edge of the main Live screen to the left. From Help View you can quickly access specific areas of Live's Library, all of Live's interactive tutorials, view the Live 8 manual, and get extra help setting up your audio and MIDI hardware.

Back to your Set! The last thing you need to be aware of is Live's two views for creating, producing, and performing music. Currently, you are looking at the Session View. If you hit the Tab key on your computer keyboard, the view will switch to the Arrangement View. These two views are independent working environments that reside within Live's main screen. For now, we will keep

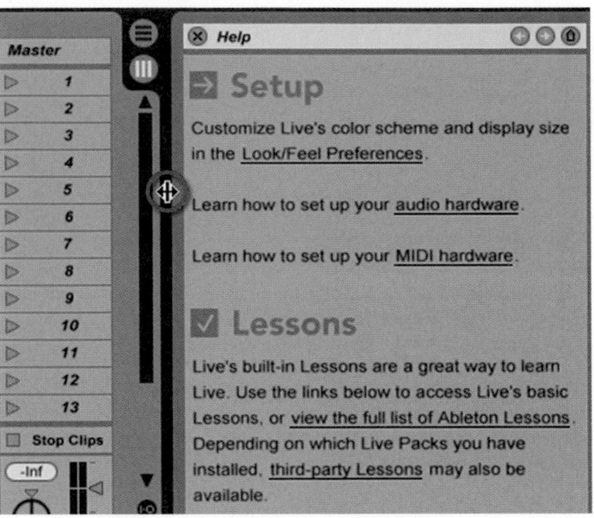

Figure 3.3 Click+drag the right edge of the main Live screen to access Help View.

things basic. Use the Session View to construct your musical ideas by importing and recording audio and MIDI into tracks. When your ideas are ready, you can record them to the Arrangement as a sequenced song (arrangement). This is the fundamental concept of Live, but the Arrangement View can also be used for traditional importing and recording audio and MIDI too. For that reason, we'll get you up to speed on how to work in both. Later we'll go over recording your Session clips into the Arrangement View.

Now that we've covered the basics, hold onto your questions, and let's start working in Live 8! If this moves too fast, you can skip to the next Scene where content is broken down in more detail.

3.3 Audio in Live

The first thing you will need to do is create a New Live Set from the File Menu ([⌘+N] Mac/[Ctrl+N] PC). Now for your empty Set you'll need an audio source to generate sound. You can import preexisting audio files or record your own acoustic sound from an external input. Let's start out simple by importing samples to an audio track. Before importing audio you have to consider which view to work in, Session View or the Arrangement View. Your Set should be currently open in Session View. As we mentioned earlier, one of the most common approaches is to import or record audio into Session View Clip Slots first, then later record a performance of these clips into the Arrangement View, thus creating an arrangement. Although this is the intended design of Live, by no means do you have to work that way. For this reason, we'll stick with importing and recording directly into each view for now and save the performance aspects for later ▶ 3.6 ▷ Scene 4 ▶ 6.7. Keep in mind while working through each of the following demonstrations that there are a number of ways to execute tasks in Live 8.

3.3.1 Importing audio

To import an audio file to the Session View, choose one of the following: (If you need audio samples to get started with, you can download some example samples from our Website.)

(a) Drag or double-click an audio file directly from the File Browser into an audio track Clip Slot or to the Drop Area in the middle of the Session View. Double-clicking on an audio file will import it directly into a new audio track.

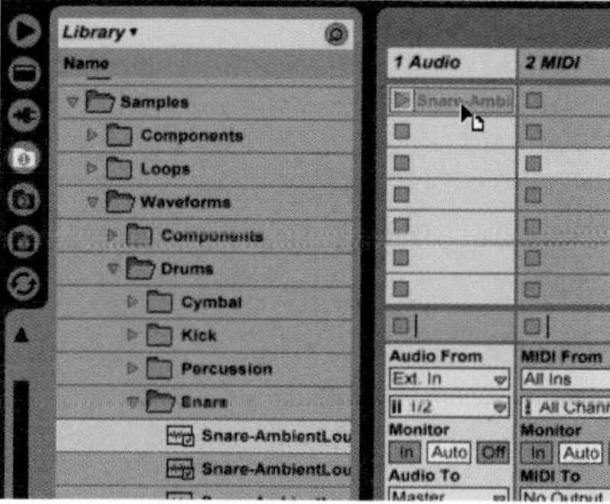

Figure 3.4 Session View: (Import) Drag and drop audio file (audio clip) directly from the File Browser into an audio track Clip Slot.

To import audio into the Arrangement View, follow the same procedures as listed earlier. The main difference is that you cannot double-click an audio file in the File Browser to automatically import it. You must click on the file and drag it into the designated track display area along the timeline in the Arrangement View.

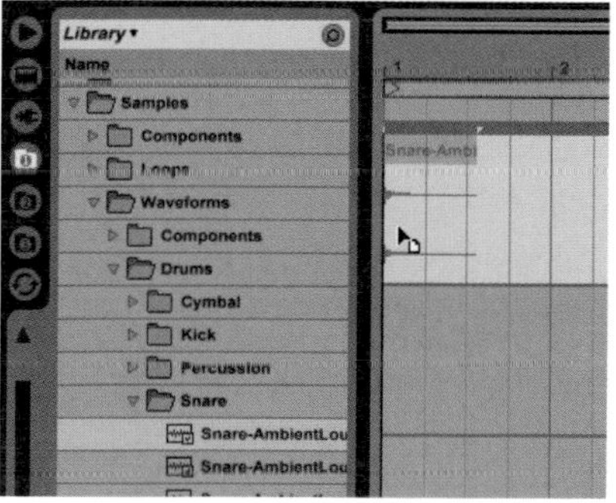

Figure 3.5 Arrangement View: (Import) Drag and drop audio file (audio clip) directly from the File Browser into an audio track's track display.

Take notice that your Live Browser can be opened and accessed in both the Arrangement and Session Views. It's part of Live's Main Screen. If it not visible, simply click on the show/hide arrow button on the upper left.

(b) Drag an audio file directly from a Finder/Explorer Window into an audio track Clip Slot or to the Drop Area in the middle of the Session View.

Figure 3.6 Session View: Drag and drop files directly from a Finder/Explorer Window into a Clip Slot.

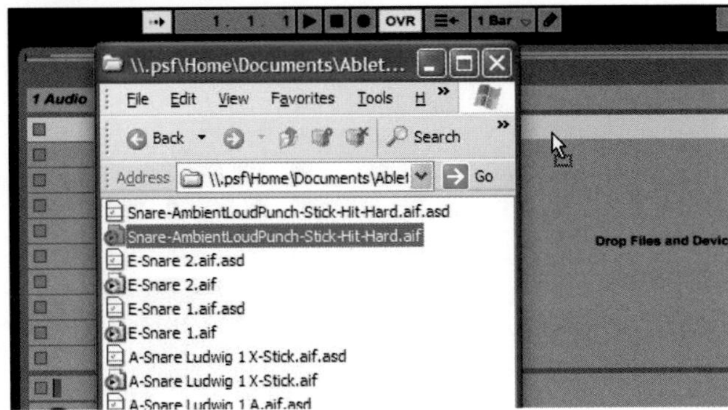

Figure 3.7 Arrangement View: Drag and drop files directly from a Finder/Explorer Window into the track display or Drop Area.

For the Arrangment View, drag directly from the Finder/Explorer Window to the track display or Drop Area and place your audio file at the left most edge of the track display (the beginning of the Set's timeline).

Once you have imported an audio file into the Session View Clip Slot, you can click on its Clip Launch button to activate playback. Our example is a short snare hit, so it's a one-shot sample that won't be looped. After you launched the clip and heard the sample, click the Stop button on the Control Bar Transport to stop the arrangement.

To listen back to a clip that has been imported into the Arrangement View, first double-click the Stop button on the Control Bar Transport to set playback at the

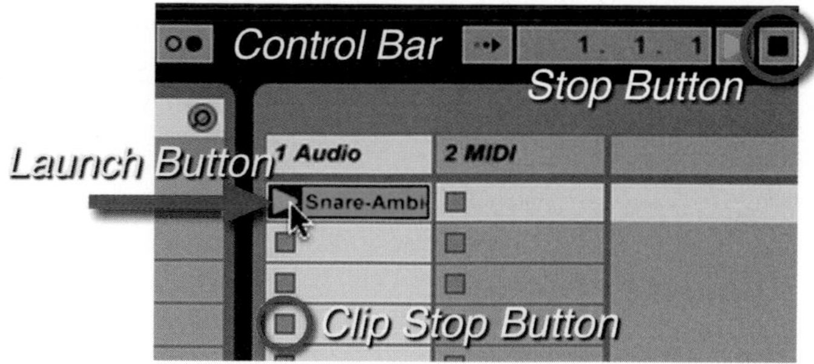

Figure 3.8 General Session View playback controls.

beginning of the Arrangement to ensure the proper playback position. Then if there is a red-lit button (Back to Arrangement button) on the Control Bar, click that too so it is no longer illuminated red, and then click the Play button on the Control Bar. The Back top Arrangement button indicates that the current playback state differs from the stored/captured arrangement. Turn it off to revert playback to the exact stored/captured arrangement.

Figure 3.9 Make sure the Back to Arrangement button is not illuminated red. Click it to turn it off to revert playback to the exact stored arrangement.

After your clip plays back, click the Stop button to stop playback.

Figure 3.10 Arrangement View playback controls.

3.3.2 Recording audio

You can record an audio input source directly into Live's Session View or Arrangement View. It's completely up to you. Each has its own benefits. No matter which you choose, you must always record-enable (arm) and audio track from its Arm Session/Arrangement Recording button directly on the track. Now, before recording an external audio source, verify that your audio and record preferences are configured to your specifications 2.3.4. Once you have set them up, follow the next exercise steps in succession:

1. Using an audio track in the Session View, set Audio From to Ext In (if not already set), and then set the Input Channel chooser to your desired audio input channel. We'll use input "1" because we are recording a mono source.

Figure 3.11 Set audio input channel.

Later when you start working in the Arrangement View, you may need to set the I/O section to be shown in view (In/Out is checked from the View Menu) so that you can access it on the Arrangement View track itself. Otherwise it's often hidden from view. Note that the I/O section and other related display attributes are unique to each view (Session/Arrangement) and are therefore shown/hidden independently for each view.

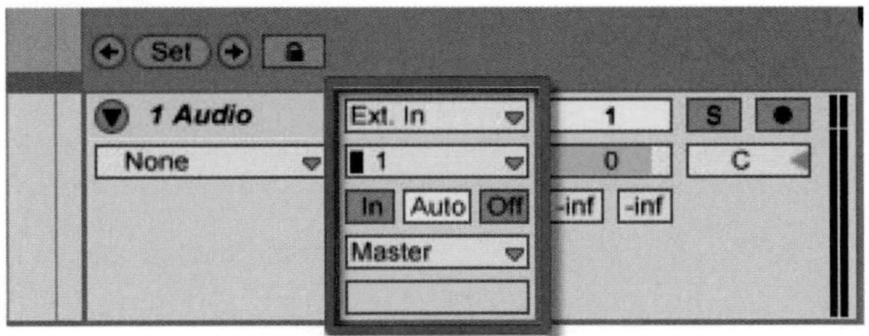

Figure 3.12 Arrangement View In/Out section mirrors the Session View In/Out.

In addition, be aware that there are no vertical track volume faders in the Arrangement View; instead there are horizontal track volume sliders. They belong to the Session View only. It is not possible to view the Session View faders and the Arrangement View simultaneously.

2. Monitor should be set to Auto and Audio To set to Master (Master track) then arm your audio track for recording by clicking on the track's Record button.

Figure 3.13 Arm Session recording (record-enable).

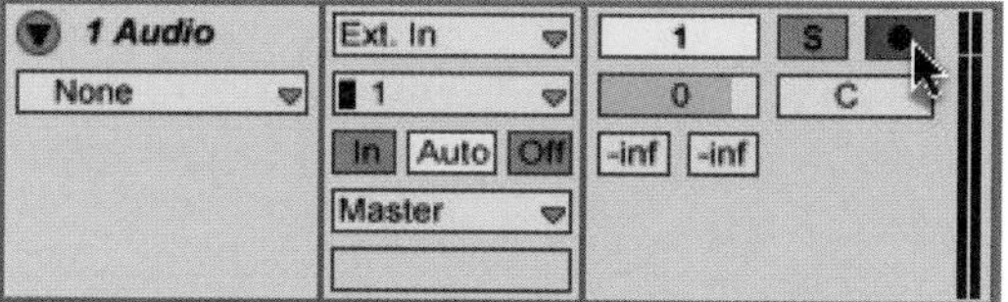

Figure 3.14 Arm Arrangement recording (record-enable).

3. Now that your track is armed (record-enabled) you can begin recording. If you wish to use Live's Recording Count-In, simply right click or ctrl+click on the Metronome in the Control Bar and choose a setting. This can also be set as a default setting in the Preferences>Record tab. The count-in will be both audible and viewable in the Control Bar when activated. Note that this is not a click track, only a "preroll" to recording.
   (a) *Session View*: click the Record button in the first Clip Slot on your audio track. This button will appear as a gray circle until activated (red triangle).

Figure 3.15 Record button.

Figure 3.16 Record button activated and recording.

Once the Clip Record button is clicked, count-in will begin (optional) and then recording will commence.

(b) *Arrangement View*: click the Global Record button on the Control Bar to turn it ON (arm) then click the Control Bar Play button or spacebar.

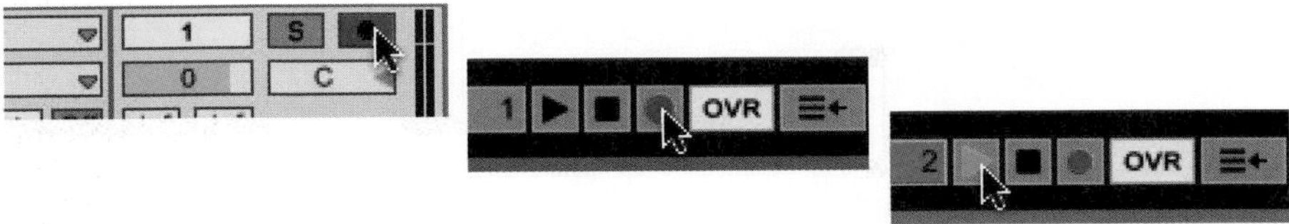

Figure 3.17 Arrangement View recording – three-step process.

Once the Play button is clicked, count-in will begin (optional), and then recording will commence. To stop recording, simply press the spacebar (stops everything) or the Global Record button (disengage recording) at any time.

If you need a click track during recording, activate the Metronome. Adjust the click's volume by selecting the blue highlighted Preview/Cue Volume knob located at the bottom of the Session View's Master track.

Figure 3.18 Metronome On/Off switch.

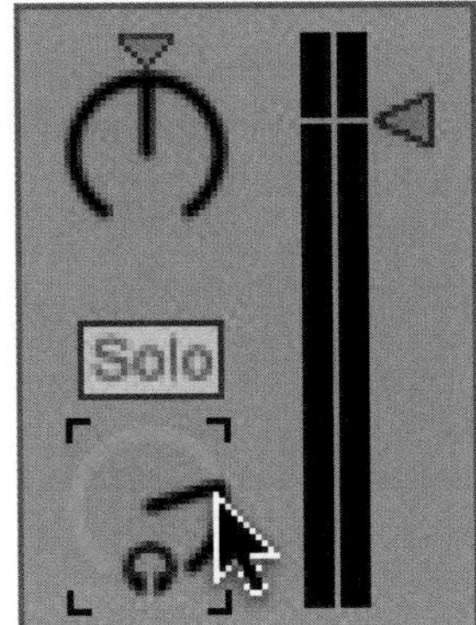

Figure 3.19 Adjust Metronome click volume.

You can also "loop record" in the Arrangement View by selecting a specific range (timespan) in the track display and then recording multiple takes in Loop Mode 9.2.2. To do this, click+drag horizontally within the track display following along the Beat Time Ruler to create a yellow highlighted selection then select Loop Selection from the Edit Menu or right click to bring up a contextual menu ([⌘+L] Mac/[Ctrl+L] PC). After you have created a Loop Selection, click on the Loop switch to activate Loop Mode.

Live will now record in Loop Mode. Your record passes will be stored inside the Clip and will show up in the Clip View's Sample Editor as one continuous audio

Figure 3.20 Activate the Loop switch to loop a timespan selection in the Arrangement View.

sample. To hear and playback your various "take," move the Start Marker to the point in the sample where your desired take(s) occurred during loop recording. For an in-depth look at recording concepts processes refer to ▶ Scene 9.

3.4 MIDI in Live

The concept of MIDI in Live is no different than any other DAW's sequencer. MIDI is still just note information without sound, which requires that you have some type of MIDI instrument to generate sound and MIDI data to tell that instrument what and how to create the sound. Therefore, you need to import a MIDI file or input your own MIDI note data. To input MIDI note data, you will either use the mouse, a MIDI hardware controller with a trigger pad or keyboard, or the clever computer MIDI keyboard feature in Live. Whichever you use, a MIDI instrument must translate the data into sound. This can be a Live 8 or Suite 8 software instrument, virtual instrument plug-in, or an external MIDI hardware instrument. For now, let's stick to Live 8 or Suite 8's built-in instrument devices and choose the one that best fits our needs. If you prefer to import a MIDI file or MIDI Clip, then you can create an instrument after the file or clip has been imported. The order does not matter. Obviously, if you input your own data, then you will normally choose an instrument first, and then input notes to a track when you're ready.

MIDI files are imported via the File Browser. They can be dragged directly to a MIDI track, the Drop Area, or double-clicked in the File Browser, the latter of which will auto create a MIDI track for that file. Use this same process to create an instrument when ready. Depending on the type of MIDI files you're working with (single or multitrack), Live may autocreate additional tracks upon import to accommodate. For multitrack MIDI files, keep in mind that you will have to route any additional MIDI tracks' MIDI outputs channels to a software instrument to be translated into sound. By design, if you use Library Clip Presets, Live will automatically create the associated instrument upon import of the clip. This is very useful, especially if you are creating your own clip presets, which will behave in the same way. To learn more about these unique MIDI Clip features, launch to ▶ 15.2.3.

For the following walkthrough we'll use an Impulse instrument preset to playback our MIDI. Our Impulse preset will load up as an Instrument Rack. Depending on which version of Live 8 you are using, your presets may vary. If you have no

Impulse presets available, download the free Live Packs: Impulse, Live 7 Legacy, and Basics from Ableton's Website: http://www.ableton.com/downloads. You will also want to have your downloaded CPP content from our site readily available for use.

Installing Live Packs is easier than tying your shoe. Simply drag & drop them anywhere in the main Live screen. Live will take over from there.

To load an Impulse Instrument Rack preset into your Live Set first click on your MIDI track's title bar, then navigate to Live Devices>Instruments>Impulse in the Live Browser and double-click on a preset's name or drag the preset from the Device Browser directly to a MIDI track or Drop Area. Choose a preset such as Backbeat Room or All Purpose that is well-suited for standard acoustic MIDI drum kit loops.

Figure 3.21 Choose an Impulse preset from the Live Device Browser.

You will find similar presets in the Acoustic folder if you downloaded the Live Packs or already have that folder in your library.

3.4.1 Importing MIDI

There are two ways to consider MIDI in Live: through MIDI clips and MIDI files. Clips are exclusive to Live, whereas MIDI files are universal across most MIDI compatible software and other related platforms. Among MIDI files, there are two types of files that you may encounter: multitrack (format 1) or single track (format 0). We will go over both because this dictates how many tracks will be needed and how MIDI signals are routed. Keep in mind that MIDI clips and MIDI files are different, but as far as Live MIDI clips are concerned, they are formatted as single-track data.

To import a single-track MIDI file or MIDI clip to the Session View or Arrangement View, try the following: (feel free to use the CPP MIDI files available on our Website.)

1. Click on File Browser 1, and navigate through the Library hierarchy, Library>Clips>Drums and drag any MIDI clip from the File Browser into the first Session View Clip Slot of your Impulse track or to its track display in the Arrangement View.

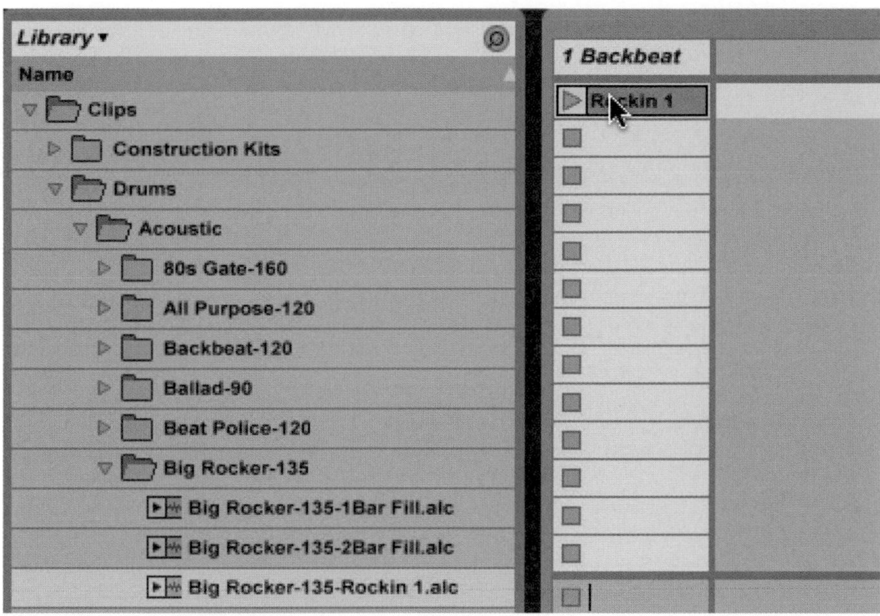

Figure 3.22 Drag any MIDI clip from the File Browser into the first Session View Clip Slot of your Impulse track.

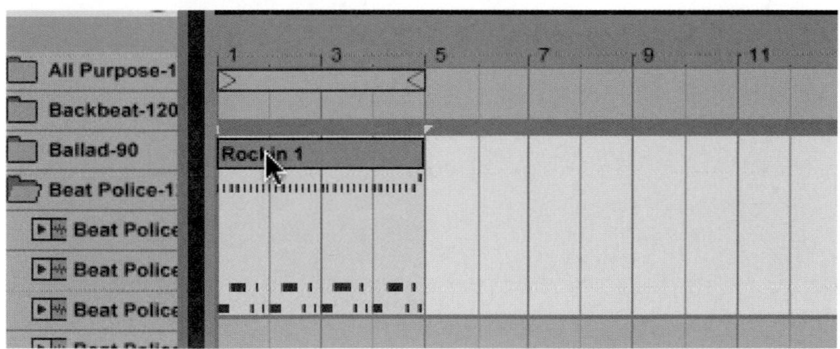

Figure 3.23 Drag any MIDI clip from the File Browser into the Impulse track display.

2. Navigate through the same hierarchy, but this time choose a different MIDI clip and double-click the clip to import it into the Session View, or drag it to the Drop Area in the Arrangement View. Either way, Live will automatically create a new MIDI track and the instrument associated to the clip!

To import a multitrack MIDI file to the Session or Arrangement View, execute the following steps in succession:

1. Create a new Live Set.

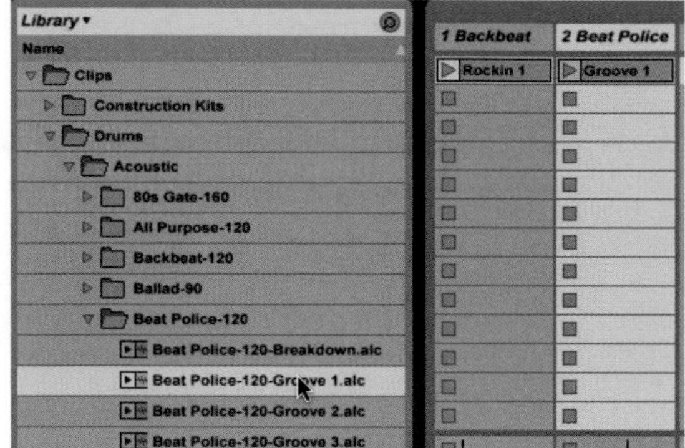

Figure 3.24 Double-click a clip to import it into the Session View. This will automatically create a new MIDI track and the instrument associated to the clip.

2. Select the default MIDI track then load up a Drum Rack preset onto the track (Live Devices>Instruments>Drum Rack). We'll use Kit-BrightRoomLite-Stick.

Figure 3.25 Select the default MIDI track in Session View and then double-click on a Drum Rack preset to load it up.

3. Locate the MIDI file MT-KitLoop-1 that you downloaded from our Website through Live's File Browser then double-click or drag it into the Session View Drop Area. (You could also drag it to the Drop Area in the Arrangement View). This will create the necessary MIDI tracks to accommodate the multitrack MIDI file, four in this case. Notice that MIDI files have different icons than MIDI clips.
   Note: when importing to the Arrangement you may be prompted to import tempo and time signature data.

4. Now, you will then need to route each MIDI track to the track that contains your Drum Rack. To do this, make sure that the In/Out section is shown in

Figure 3.26 Double-click or drag MIDI file MT-KitLoop-1 into the Session View Drop Area.

view (View Menu) then select all of the new MIDI tracks by holding shift and clicking each one's title bar.

5. While still selected, use one of the multiselected tracks to route the MIDI To chooser (Output Type) and Output Channel to your Drum Rack's name. The Output Channel chooser can also be routed to Track In to achieve the same result. It is possible to route each MIDI track to an individual drum sample in your Rack, but we'll keep it very simple and route to all channels for now.

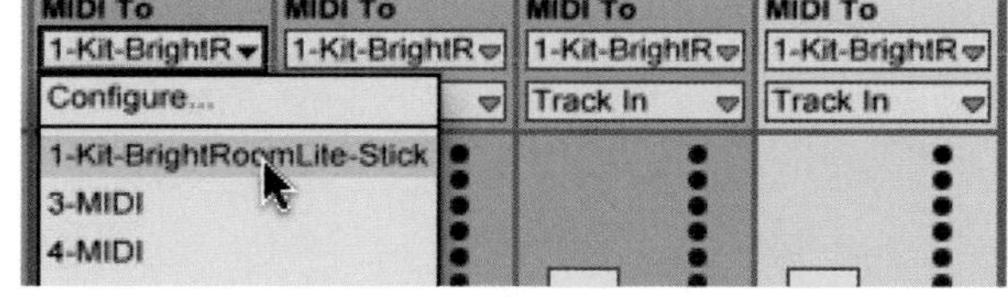

Figure 3.27 Route the MIDI To chooser (Output Type) to your Drum Rack.

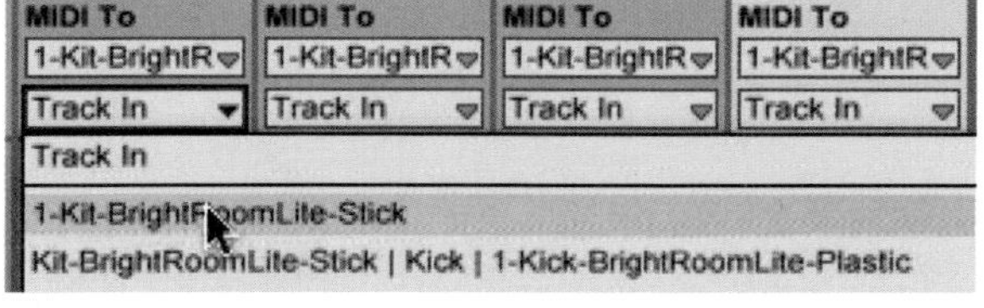

Figure 3.28 Route the MIDI To Output Channel to your Drum Rack.

Your Set should now look something like our example.

Take a moment and launch the entire row of clips from the Scene Launch button on the far right of the screen as well as trigger some of your Drum Rack's samples with your keyboard or controller. Remember, record-enable your Drum Rack's track to trigger it with a controller.

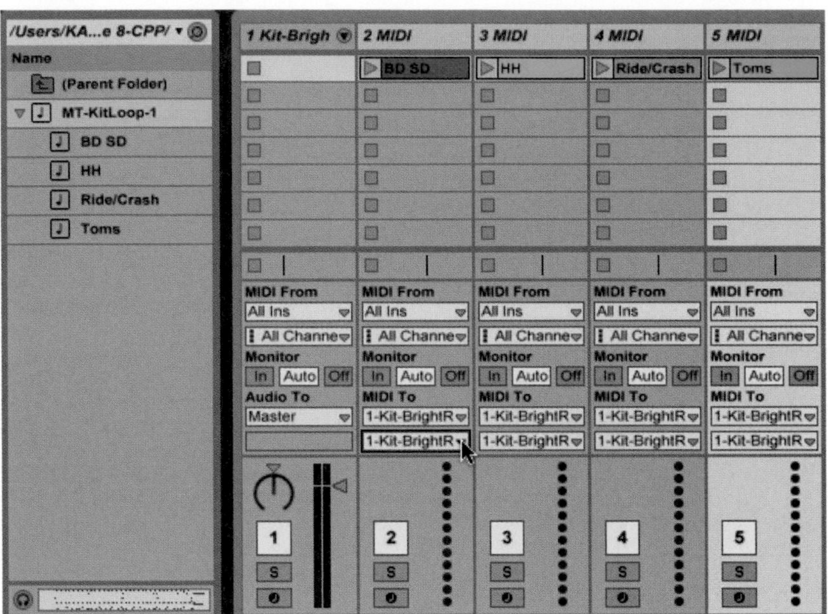

Figure 3.29 Example of multitrack MIDI routing.

3.4.2 Recording MIDI

To record MIDI directly into the Session or Arrangement View, follow these steps in succession:

1. Starting with a new Set, load one of the same Impulse drum kits from earlier into a MIDI track and set MIDI From to All Ins (if not already). You can also set this to your MIDI input device name instead.
2. Directly below MIDI From, set the MIDI Input Channel chooser to All Channels. Make sure the In/Out section is in view so that you can access the I/O section on the track (View Menu).

Figure 3.30 Session View.

3. Monitor should be set to Auto, Audio To set to Master, and then arm your MIDI track for recording by clicking on the track's Record button.

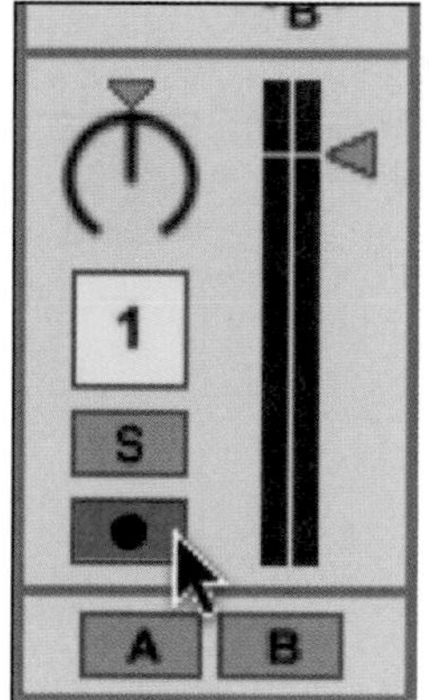

Figure 3.31 Session View.

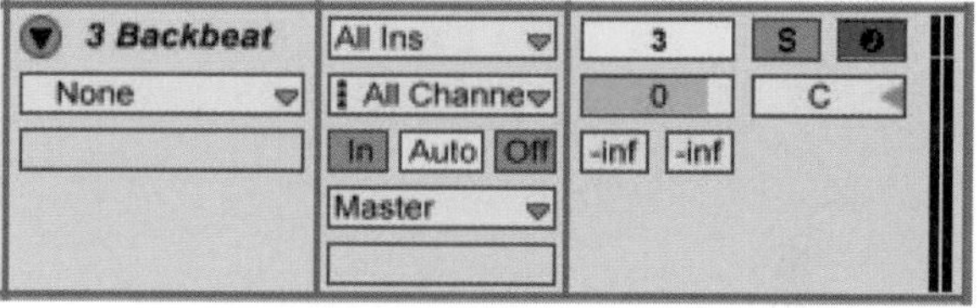

Figure 3.32 Arrangement View.

4. Now that your track is record-enabled you can begin recording.
   (a) *Session View*: Click (activate) the first Clip Record button in your Impulse track. This will initiate count-in (if enabled) and then recording will begin.

Figure 3.33 Activate Clip Record button.

   (b) *Arrangement View*: Click the Global Record button on the Control Bar to turn it ON (arm/record-enabled) then click the Play button or spacebar. This will initiate the count-in (if enabled) and then recording will begin.

Figure 3.34 Arm Global Record and then click Play to begin recording MIDI directly into the Arrangement View.

Activating the Overdub switch OVR (highlighted in yellow when activated) will allow you to continually record MIDI notes to a MIDI clip, building it up with multiple notes or voices – a drum kit for example – without overwriting the existing notes recorded in previous takes or loop cycles, hence "overdub." In this way, Live is able to continually record your MIDI input, looping over and over until you tell it to stop recording (Clip Stop button), stop playback entirely (Control Bar Stop button or spacebar), or disengage recording (deactivating the Track Record button or Global Record depending on which view you are working in). The Overdub feature/technique is great for inputting selective notes one at a time, pass after pass, or whenever you feel inspired to add more notes to your current recording. For more on looping, launch to ▶ 3.5.1 and for recording and loop record, launch to ▶ Scene 9.

3.5 Essential operations and tasks

Now that you have had a chance to go over importing and recording audio and MIDI, let's take a moment to look at how to use a few of the more important and essential everyday operations and tasks for working in Live. Throughout the following group of exercises we'll use the same Set and media, so keep your work open as we go.

3.5.1 Looping your loops

There is no doubt that at some point or another you will need to loop your MIDI or audio. The truth is, that looping is the underlying principal of Live's Session view, and so you'll need to understand how to loop audio and MIDI clips. As far as the Arrangement View is concerned, you may or may not have to loop audio and MIDI clips. This all depends on how you create and produce your music. In the following section, we will run through the quickest ways to execute looping operations with audio. As you read through this section start to conceptualize with how Live handles looping. You won't get it all right away (that's not what this section is for), but when you get to ▶ Scene 14, you will be well on your way. To that end, we leave you with this preface: clips loop in multiples ways. Remember, clips are containers, therefore, they loop as Session clips governed by their Clip View Loop Brace. They also loop as Arrangement clips governed in two ways, by their physical position on the track display and timeline, and their Clip View loop brace. Wow! That's heavy.

3.5.1.1 Session looping

Looping in the Session View is driven by the content of your clips. For example, it's quite common to see a vocal track in the session that never loops. Think of it as a one-shot recording intended to play through an entire song section without repeating. On the other hand, your rhythmic tracks are more likely to loop in two, four, or eight-bar phrases. Working with premade loops makes looping simple. When you drop one into a Clip Slot it will loop automatically.

This goes for both MIDI and audio loops. If you recall from the audio Preferences Tab, short samples are automatically looped upon import unless you change this setting. Go ahead and try this out for yourself. Locate S3-AudioLoop-1 and S3-MIDILoop-1 from your CPP content download using the File Browser. Drag Audio or MIDI Loop 1 into the Session View, then launch it and you'll see what we mean. While the clip is looping, take a close look at the Clip View towards the bottom of the main Live screen and you'll see the clip's contents (waveform or MIDI notes) and playback looping in the Sample Editor/MIDI Note Editor. This view displays a clip's waveform or MIDI notes aligned to a grid – bars and beats. Also notice the Loop Brace that bookends the clip's contents, establishing the loop cycle points of the loop.

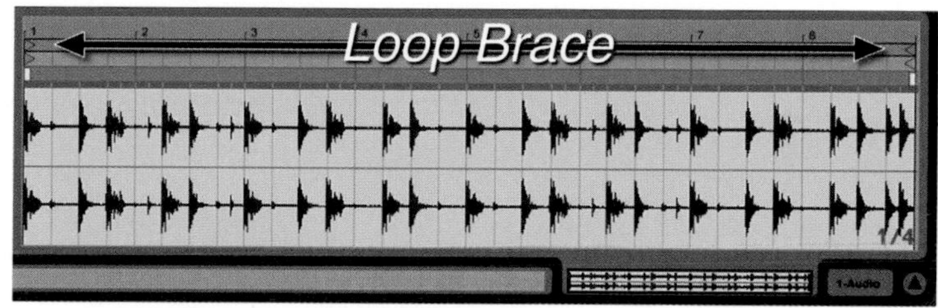

Figure 3.35 Loop Brace in the Sample Display/Editor.

Move the Loop Brace Loop Start or End (bracket) to customize and loop a specific section. You can also move the entire Loop Brace by selecting it (clicking and highlighting it to a solid black color) or move the Loop Start/End individually. Click+hold and drag to reposition the Loop Brace as a whole or just the Start/End points.

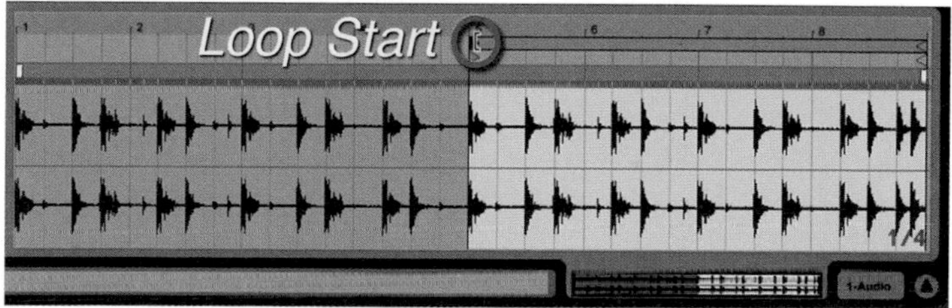

Figure 3.36 Loop Brace/Loop Start.

Taking this a step further, you can set a unique clip Start Point that differs from your Loop Start by moving the Start Marker on its own.

Keep in mind that it is possible for this to be done in real-time allowing you to hear what is happening with your Start and End points within the specific clip – a key element when used alongside playback of other tracks in your Set.

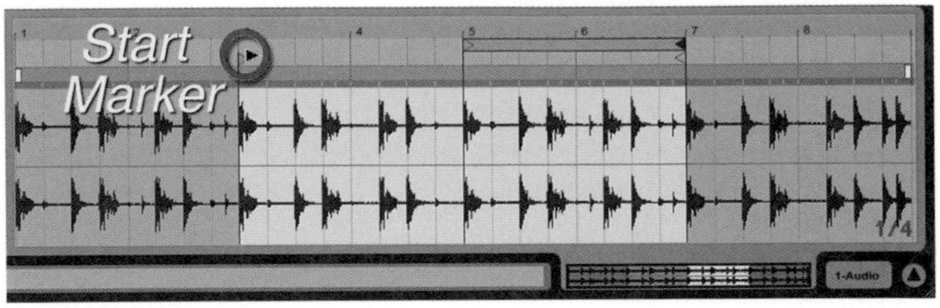

Figure 3.37 Start Marker determines where playback begins.

3.5.1.2 Arrangement looping

Looping in the Arrangement is very similar to any traditional DAW or sequencer. A clip can be set to loop in playback by setting the Arrangement's Loop Brace to a specific time range selection then activating the Control Bar Loop switch (activates Arrangement loop). You can also select a clip(s) and press [⌘+L] Mac/[Ctrl+L] PC to set the Loop Brace and activate the Loop switch all at once.

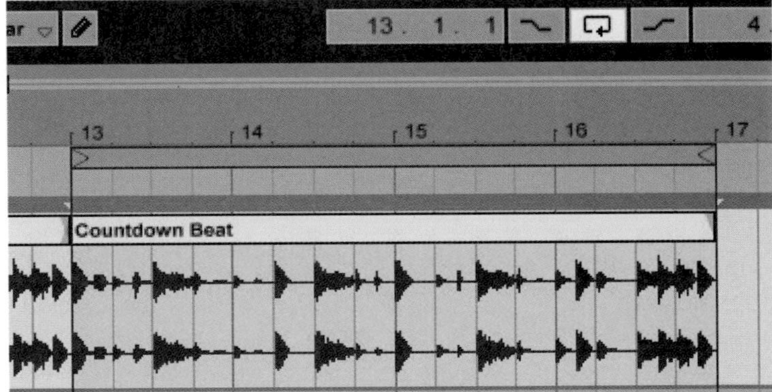

Figure 3.38 Arrangement Loop: Select a clip/timespan with the Arrangement Loop Brace then activate the Loop switch.

A clip can also be looped internally – as we just looked at previously – then looped infinitely along the main timeline by dragging the clip's edges to extend its length.

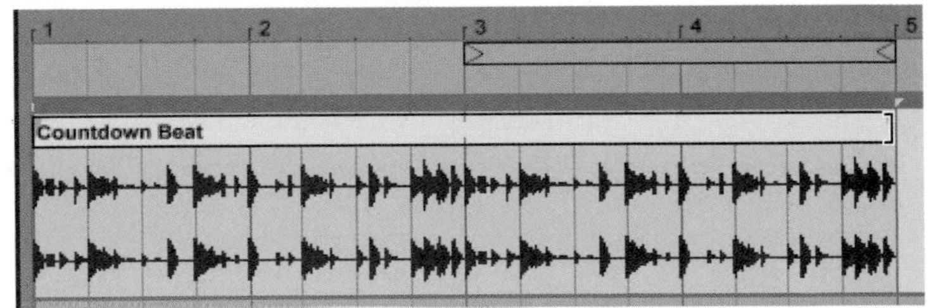

Figure 3.39 With a clip's Loop switch activated it can be looped infinitely by dragging its edge to extend its length.

If the clip's *Loop switch* is deactivated in its Sample Box (Clip View), the clip cannot be extended beyond its set region length. The alternative then is to simply duplicate the clip in the Arrangement View ([⌘+D] Mac/[Ctrl+D] PC) as many times as necessary along the timeline.

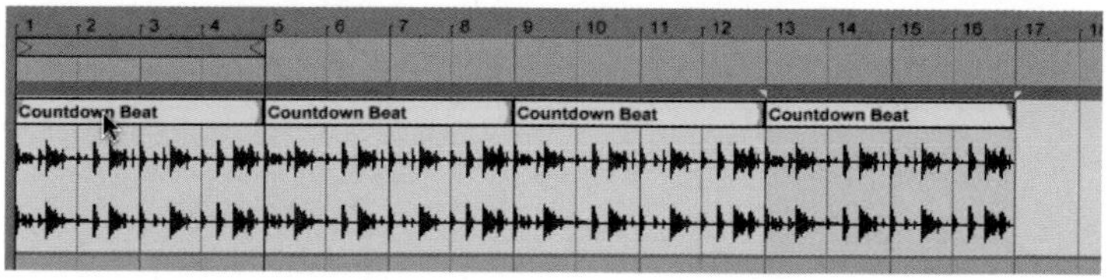

Figure 3.40 Manually duplicate a clip in the track display to simulate looping.

A big difference between looping in Live's Arrangement View and other DAWs is that when a clip's Loop Brace (accessed in the Clip View Sample Display/Note Editor) is changed, the actual audible selection changes with it in the Arrangement. This is because a clip's Loop Brace determines what regions/selection is looped and the Start Marker dictates where in the waveform/MIDI file playback begins. When a clip loop is deactivated the Start/End markers dictate the length of the playable waveform/MIDI file in the Arrangement View. For more information on this, read about Takes in Clip 9.2.3 which talks all about how Live handles Start Markers, and the Loop Brace, etc., in correlation to looping and Takes.

3.5.2 Adding effects

With all of the effects built into Live 8 and Suite 8, you'll definitely want to learn how to use them quickly and efficiently, keeping them ready at your fingertips. Audio effects can be applied to any audio track, which includes Return and Master tracks. They can also be easily connected to the output chain of any instrument. MIDI effects can be applied only to MIDI tracks, but must be placed before instruments in the device chain. Simply drag the effect you want from the Device Browser to the appropriate track or double-click it in the Browser. To put this into practice, let's add an audio effect to one of the audio loops located in the CPP_AudioLoops folder from our Website. Feel free to use your own audio if you like. Let's start with Live's *Auto Filter*.

3.5.2.1 Audio effects

1. Using a new audio track, drag any one of the audio loops into the first empty Clip Slot in the Session View.
2. Now launch the clip and make sure that it's looping (Loop switch should be highlighted yellow in the clip's Sample Box). Double-click on your audio clip to show the Clip View.

3. While the clip is running, drag the default Auto Filter from the Audio Effects folder in the Device Browser to your audio track or double-click it to insert the effect on the selected track, and then click on the Hot-Swap button in the upper right corner of Auto Filter ● and select Cut-O-Move L from the Auto Filter presets folder in the Browser. Pretty cool, huh?!

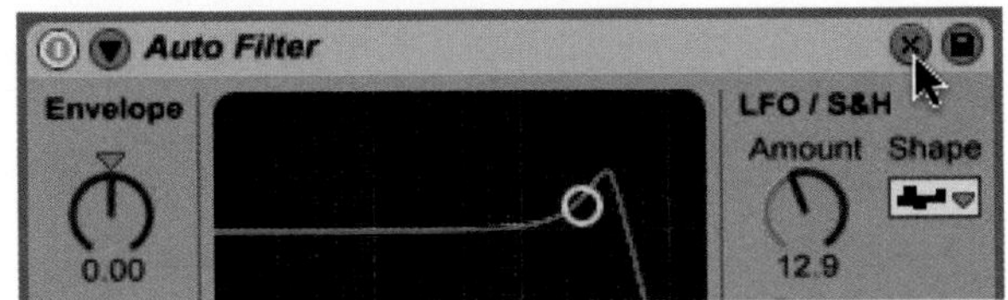

Figure 3.41 Hot-Swapping device presets.

4. Let the loop continue to play and click the Hot-Swap Preset button "On" again (if not already "On") so you can audition other presets within the Auto Filter preset folder.
5. Take a listen to a few different effects and presets by double-clicking each one. This will temporarily load them in place of the previous effect. During this process you will be so inspired that you'll want to stop and play around for a while. Before you get too distracted, let's add two more effects to create a chain.
6. Delete and reload Auto Filter. You can easily delete a device by selecting and highlighting from its title bar at the top and choosing Delete in the Edit Menu or using the Delete/Backspace key on your computer keyboard.
7. Once you have reloaded it, navigate to Flanger in the Device Browser and drag the default Flanger effect to the right of Auto Filter in your Track View (you could also double-click).
8. Select EQ8 and add it to the right of Flanger. Leave Flanger as is, and move the EQ8's settings around with your mouse.

Figure 3.42 Customize frequency band settings with your mouse within the graphic display.

9. Now listen to your loop with and without the effects. To do this, deactivate and reactivate each effect on the fly by *clicking* on the *Device Activator* button ● in the upper left corner of each of the effects. This should help you to get familiar with the interface while observing the audible differences. You can also switch the order and placement of the audio effects in Track View by

clicking+holding, then dragging the highlighted audio effect left or right to a new position alongside your existing two audio effects. A vertical yellow line will appear between the effects when you reach the correct Drop Area.

Figure 3.43 Three audio effects inserted on a single audio track.

3.5.2.2 MIDI effects

For adding MIDI effects, let's use a Live 8 software instrument and an optional MIDI clip SIMP-ARP.mid. If you have a MIDI keyboard controller or want to use the computer keyboard, then you may choose not to use the MIDI clip we have provided and play notes on your own. Being that your instruments and presets will vary depending on your version of Live 8, we recommend downloading the free Loopmasters' Solid Sounds Live Pack from Ableton's Website http://www.ableton.com/downloads. You'll find some nice instruments in there along with the other recommended downloads we have mentioned.

1. On a new MIDI track, load up Instrument Rack preset "Guitar-Wide Acoustic" (Instrument Rack>Guitars and Plucked>Acoustic Guitars). This uses Live's Tension Synth! Feel free to use any other plucked type staccato instrument for this exercise such as a Simpler or Loopmasters preset.

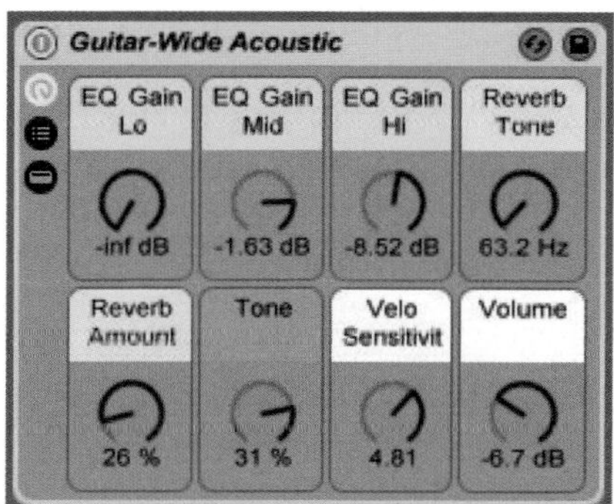

Figure 3.44 Instrument Rack Macros in Track View.

2. Add Arpeggiator preset ClassicUpDown 8th (MIDI Effects>Arpeggiator> ClassicUpDown 8th). Change the Rate to 1/16 and Gate to 200% from the Arpeggiator interface.

3. Now drag in SIMP-ARP.mid into a Clip Slot, launch it, and take a listen. If you rather play your own MIDI, then do so by playing long sustained chords. This way you will hear a nice arpeggiated effect on your MIDI instrument. The Arpeggiator is synced to the tempo of your Set, so feel free to adjust it to taste.
4. Lastly, we'll add an audio effect to the output of our Instrument. Add the default Ping Pong Delay audio effect by dragging it to the Guitar MIDI track or to the end of the instrument/output chain in the track display just after the Guitar-Wide Acoustic preset (Audio Effects>Ping Pong Delay).

Figure 3.45 Track View showing a MIDI effect, Instrument Rack Macros, and audio effect forming a customized instrument.

Now hold chords out again and enjoy the complex rhythmic effect.

Figure 3.46 Computer MIDI Keyboard On/Off switch.

Before we move on, we'll let you in on a little secret. If you don't have a MIDI keyboard controller then you can use the computer keyboard to supplement. This feature can be activated from the upper right corner of the main Live screen next to *Key* and *MIDI*. It will light up yellow when activated. Hold down multiple notes to hear the arpeggiated effects. The general layout is the row of "A, S, D, F, G, H, J, K, L" keys resulting as the white keys on a keyboard and the "W, E, T, Y, U, O" keys designating your black keys. This is a very useful tool for inputting MIDI notes on the road with your laptop, and it works for all MIDI instruments!

3.6 Performance to arrangement

The final step of the creation and production process is to record a session performance into the Arrangement View. This is one of the features that makes Live special. It is so inventive and powerful that you will definitely want to spend some time reading about it in ▶ Scene 4 Global Record. For now, we are going to just scratch the surface of this concept. For this you can use either the Guitar-Wide Acoustic instrument or the audio loop you used in the previous exercises.

To record your Session performance to the Arrangement View, you should follow these important steps to learn early on how to avoid confusion or launching of clips abruptly.

1. Click the Stop Clips button in the Master track of the Session View to clear any clips that are in stand-by. When in stand-by the launch button will be illuminated green while playback is stopped.
2. Double-click the Stop button in the Control Bar transport to return your Arrangement to at the beginning of the timeline (1.1.1 start time).
3. Click the Global Record button on the Control Bar transport to ready the Arrangement for recording.
4. Now, if you're ready, click the Play button on the Control Bar. After the count-in, the arrangement will begin recording and logging all of your actions made in Session View into the Arrangement View. Launch your audio or MIDI clip so that it plays while the arrangement is recording. Of course you could play Simpler in real-time if you like. Let your clips loop for about 15 seconds or so, then click a Clip Stop button on the same track to stop the clip.
5. Click the spacebar or Stop button on the transport to stop the Arrangement recording/playback.
6. Now, select the Tab key to flip over to the Arrangement View to see your performance and then press play or the spacebar to listen back to your music.

3.7 Finishing your work

Once you get to the final phase of your work or at least to the end of your workday, there are a few final considerations and operations you must deal with before calling it quits. Saving and organizing your work are essential operations to completing a successful experience with Live. This includes exporting an arrangement or audio tracks and managing your files.

3.7.1 Exporting your audio

When you are finished working on your song you will most likely want to share it or listen to it as a final two-track mixdown for a CD, portable music player, iTunes, Windows Media Player, or a Website. To do this, you will have to export your audio out of Live to disk. Other DAWs may refer to this process as "bouncing to disk." In Live, it is called "exporting" or "rendering." The rendering/exporting process offers many options, such as rendering an entire song, track(s), clip(s), or video. Although this may sound like it's an easy process, it can get confusing very quickly. It all begins with what output source you want to render from and what content you want to render (Session clips or Arrangement clips). Although you will see that each view presents a different protocol for executing the rendering process, ultimately the view you are rendering from has no bearing on the actual rendered output. The rule is that whatever is supposed to play will be rendered regardless of the view that contains the content.

3.7.1.1 Arrangement

To export an entire arrangement or clip(s) from the Arrangement View, select a timespan across the track display, and then select the Export Audio/Video command from the File Menu ([⌘+shift+R] Mac/[Ctrl+shift+R] PC). This brings up an Export Audio/Video window that allows you to choose the desired output source, method for exporting your audio, and export settings, i.e., bit depth, sample rate, and so on. Choose an output source to render: master track, return track, individual track, or all tracks. The rendering of all tracks individually is also known as creating stems. Whichever you choose, just be sure that what you are listening to is generated from the Arrangement View and does not include clips from the Session View, unless that's what you want. Without getting too deep into the conceptual relationship between the Session View and Arrangement View, understand that the views are separate from one another, but they are capable of sharing playback when clips in both views are playing simultaneously, as long as they don't share the same track. Session clips always have priority. That being said, what you hear during playback is what will be exported, unless you choose a single track or all tracks as the rendered track source in the export window. In such a case, the render outputs will correlate directly to what is to play from its track output. So when it comes to rendering only from Arrangement tracks, make sure that the Back to Arrangement button is not red. This will exclude the tracks/clips in the Session View, exporting only the stored arrangement.

3.7.1.2 Session

Rendering a Session clip or scene can be a bit trickier than in the Arrangement View. For this, you have to launch your clip(s) or scene, then stop playback with the Control Bar Stop button or spacebar (essentially pausing clips that are playing) so their clip launch buttons remain green. This is in lieu of clicking their Clip Stop buttons. Think of them as if they are in standby. The clips you have been listening to (now in standby) will be what is rendered, meaning that what you would hear during playback is what will be rendered, including Arrangement clips if they too would be heard during playback (dependent on the position/location of the Arrangement clips in relation to the chosen render length). This means that if an Arrangement clip is located outside of the Session selected length, which will commence from bar 1 of the timeline, it will not be exported with the Session. Of course all of this is contingent on what output you choose to render from, i.e., Master, track, all tracks, etc. Make sure to set the Length [Bars-Beats-Sixteenths] to render the selection of audio you wish to save to disk.

3.7.2 Set versus project

Creating, producing, and performing with Live begins with a Set. As you already know, this is the environment you have been working in all along. Sets are stored in Project Folders, which optionally store all of the media related to your Set(s). Sets can either refer to media located in various places, such as the Library, or from its Project folder. This includes samples (imported, recorded, and processed), Sets, instruments, clips, presets, and other pertinent information. When you save a Set for the first time, it will automatically be saved in its own new Project Folder named to match the Set name. Live Sets are identified by the *.als* file extension. As you work save as often as you like. This overwrites the current Set each time you save. You can also save your Set as a new name inside the same Project folder, or outside in a new location. The purpose of Project folders is to store the information for a particular body of work, i.e., a project with multiple drafts of the same Set. If you are saving it outside the folder, Live will again create a new Project folder matching the name of the new Set. If you save a Set with a new name outside the current Project folder structure that contains imported and or recorded audio, the samples will continue to reference the original project's samples folder. This is fine for now, but when you take your project to a different computer or delete the other project, you will be missing your samples. Not a good situation to be in. To avoid this, use Live's Collect and Save feature accessed from the File Manager or the File Menu before taking your project on the road or archiving it. It is imperative that you manage your Projects and Sets!

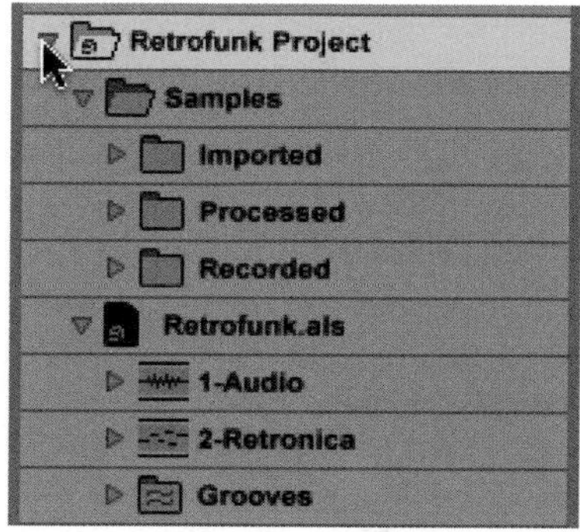

Figure 3.47 Project Folders store Set(s) and their related media.

Here's a quick example. Let's say that you have been working on a Set and you have saved and named it as usual. If you look inside its Project folder you will notice that if you have recorded audio it is there in a samples folder, but if you imported audio it is not. This is because Live is referencing the imported audio from its original file folder location. If you manipulate, or "process," the audio it will then show up in the Project folder. This is why you use the Collect and Save option before you take your project on the road. So, here is the general procedure: when you first create a new Live Set, choose Save Live Set from the File Menu. When you want to save as a different name within the Project folder then choose Save Live Set As..., and if you want to save it to a new location or as a separate back-up, choose Save A Copy… or Save Live Set as… . If you want any one of the Sets that you just saved to be self-contained, then open that Set up and choose Collect and Save. It's up to you how you want to manage. When you do this, Live will prompt you to select the files you wish to copy into your Project folder, which includes copying Library content if you so choose. You can also access the more detailed save options from the File Manager.

One key difference with Live in regards to the saving process is in the Live Browser itself. You can simply drag a clip directly into one of the File Browsers to save your clip as a Live Clip. You can also drag a track(s) or an entire selection of clips to a File Browser as a Live Set (.als). Choose how you wish to copy your samples and assorted files for this feature in the Collect Files on Export chooser in the Preferences folder. This is a very productive and strategic way to keep your Sets organized and ready to load at a moment's notice while you are running Live. Once you develop a system of dedicated folders for your projects, this method becomes a surefire way to save your Sets and all-in-one Clip Presets. For more on customizing and saving presets with the Live Browser, launch to ▶ 15.2.4.

As easy as it may seem, to close your Live Set, select Close Live Set from the file menu or use [⌘+W] Mac/[Ctrl+W] PC. This will close your Set and reopen the default Set. If you choose the Close Window button in the upper corner of the main Live screen, the Live program will quit as opposed to just the Set.

3.7.3 Managing files

The File Manager helps to manage and administrate all of the files contained in your Set, Project, and in the Library. This is necessary because the content in your Sets can be referenced from various locations rather than from the actual Project folder as mentioned earlier. This keeps Live running efficiently, but at the risk of having missing files. The File Manager provides a way to view all of your media, locate any missing files, and copy the external files – those being referenced from elsewhere – to your current Set, among other things. Managing your files is very important. On one hand, you want your projects to be self-contained and complete when you transfer them to another studio or computer, but you don't want to copy your entire Live Library into a single Project while working at your computer. That would be redundant and would use up a lot of hard drive space. To ensure that situations such as these don't arise, use the File Manager to manage your Sets and Projects, and Collect and Save. If a sample is missing, or unable to be located when you open a Set/Project, Live will warn you. Use the File Manager to search and rescue your files in conjunction with the Live Browser.

To use the File Manager, go to the View Menu and select File Manager. It will appear on the right-hand side of the main Live screen where the Help View would have been located. From there, choose what to manage. There are three

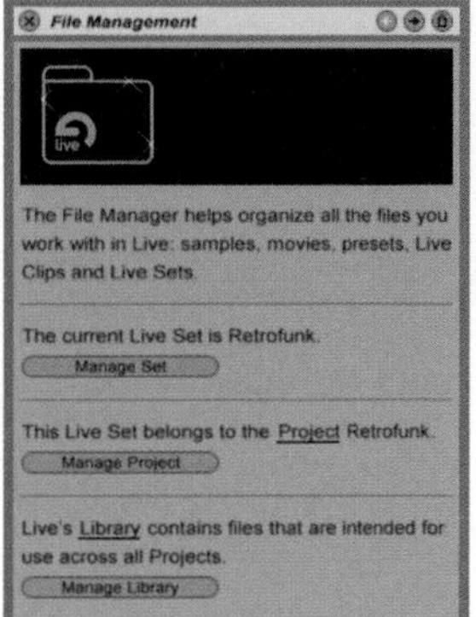

Figure 3.48 The File Manager, when in view, is located to the right of the main Live screen.

options: Manage Set is specific to the current Set; Manage Project catalogs all of the Sets and information included within the Project Folder; and Manage Library indexes and lists important information related to the Library.

3.7.3.1 Searching for missing files

At some point in your Live career you will run into a scenario where one of your Live Sets loses track of some important files that may have been moved or renamed, etc. Things like this do happen. How do you know when there are missing files? Live will notify you. You'll see the Status Bar at the bottom left of the main Live screen highlighted orange, displaying "Media files are missing. Please click here to learn more." Great, now what's the solution? Live has a built-in search feature that is accessed through the File Manager.

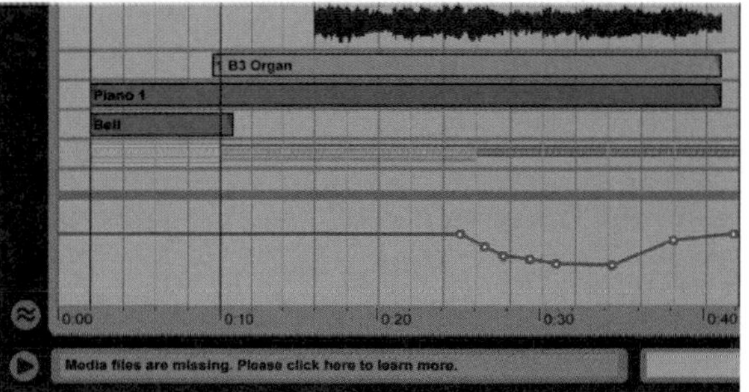

Figure 3.49 Status Bar: media files missing!

So there is no need to panic. Clicking on the orange prompt will immediately open up the File Management Menu to the right of the Main Screen. Here you have a few options to help you find your missing file(s). If you have a general idea of where they were last located, you can tell Live where to search by choosing a folder directory and activating the Search Folder button. If you don't have a clue, then let Live do all the work. Either way, select Go.

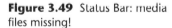

Look closely at the top of the File Management Menu and you'll see an exact number of files missing from your current Set. This count references the listing of files located further down in the menu, which Live will look for.

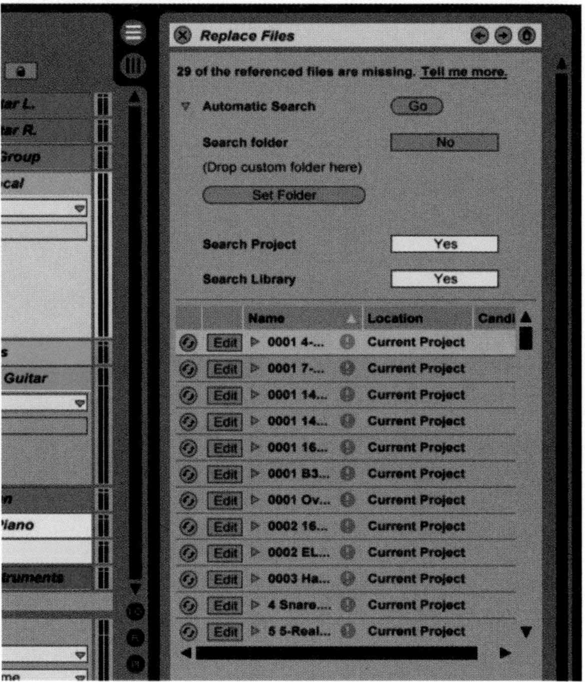

Figure 3.50 The File Manager Menu: Missing files.

You also have the choice to manually find and repair your files by clicking the Find File button. This will open up the Browser where you can start a search similar to Hot-Swapping. More information on this can be found by selecting the "Tell me more" prompt at the top of the menu next to the Missing Media File count. More often than not, Live will successfully locate your missing files. Whichever process you choose, it can take awhile depending on the size of your hard drive and various locations of your files in question. Be patient, nobody likes a scenario of missing or corrupt files, but Live's File Management will help to ease the pain. Once the candidates are located, select OK, then Save from the bottom of the menu. For additional information on managing and locating files, follow Live's help link or reference the manual.

3.7.3.2 Creating a Live pack

Don't forget that the File Manager is not just for correcting missing file associations, it's also a very powerful tool for collecting and saving your media into one location, especially with its ability to save your entire Set as a Live Pack.

This option is found within the Manage Project area within the File Manager. Click on Manage Project and Select the triangle unfold button named Packing where you can select Create Live Pack. In doing so, name it and choose a location to save it. This is used for copying and compressing the entire contents of your project into a single .alp file easy for archiving and sharing. Live will use a specific lossless compression process to minimize your Set to a smaller size. Not only does this make it easier to transfer but is also a safe way to archive your precious data.

For Live 7 users, you cannot save and overwrite a Live 8 Set with your Live 7 Set. You need to perform a Save As instead.

In this chapter

4.1 Introduction 65

4.2 The Global Record concept 65

4.3 Music on the fly 67

4.4 User interfacing: two parallel worlds 67

4.5 The linear approach 74

4.6 The nonlinear approach 80

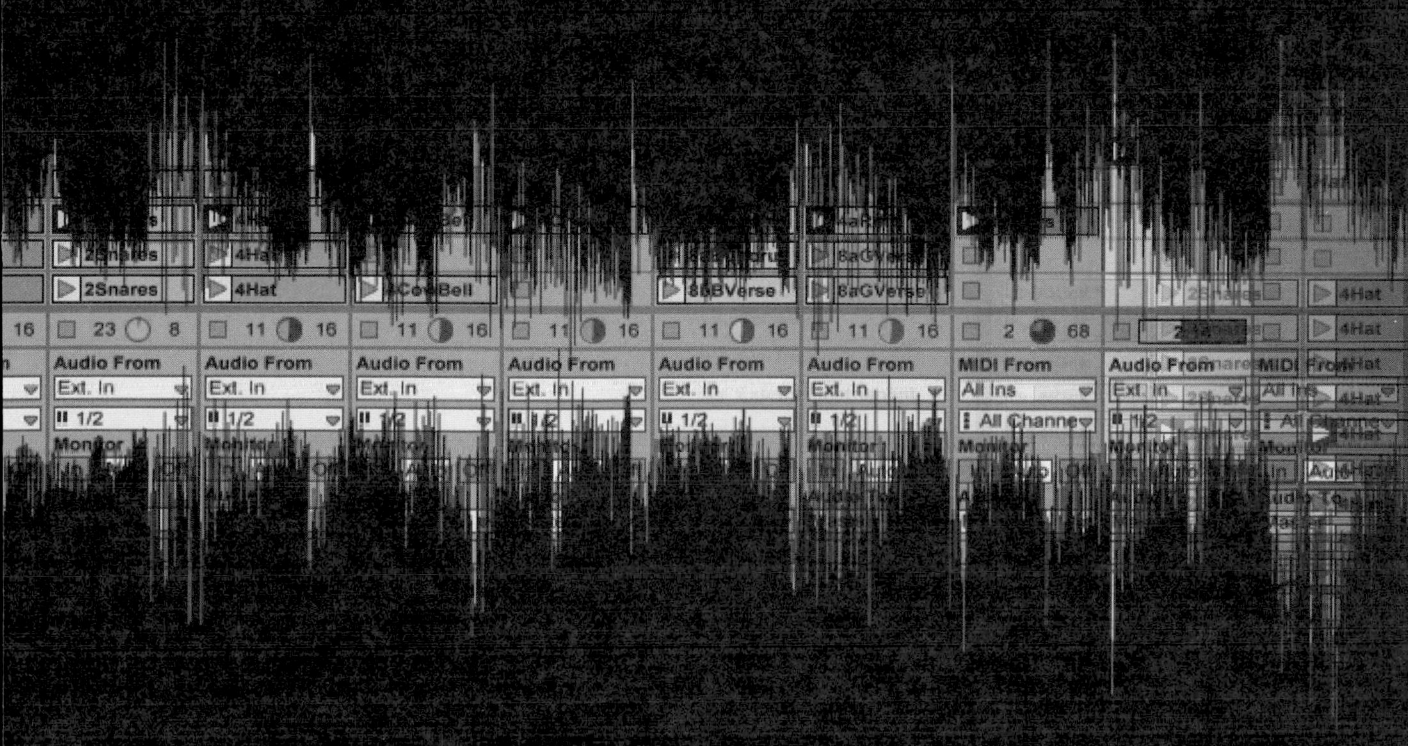

Global Record: Capturing Arrangements on the Fly

The thing I love most about Live 8 is the ability to keep creating on the fly when the creative juices start. With other sequencers, there's a stop–start mentality to it but with Live 8, you just keep going in real time until the idea that was in your head is finally coming back at you. Long live the jam.

Dave Spiers, Founder of G-Force, Producer and Musician

4.1 Introduction

Live offers so many ways to control your music that it can be overwhelming at times. As a musician, composer, or performer, Ableton Live 8 and Suite 8 provide you with the tools to literally launch your music in a forward direction. Forward and constant motion is a good thing! This is no different than a band of musicians jamming on stage, striving for that "perfect flow." In this scene, we will show you how to elevate your music production to the next level, making it more exciting and even better than it was before. Our focus here is on the concept of Live's Global Record and how to incorporate its unique features to *create*, *produce*, and *perform* music.

Remember to keep an eye out for launch points to scenes ▶ Scene 1 and clips ▶ 1.3.1 as you read. These will help tie in related concepts for expanding your knowledge and focus your area of concentration.

4.2 The Global Record concept

As a musician, arranger, or composer, it's likely you have used a digital audio workstation (DAW) or similar sequencing program that led you to Ableton Live. Whichever it may have been, one thing is for sure, they all share the same basic

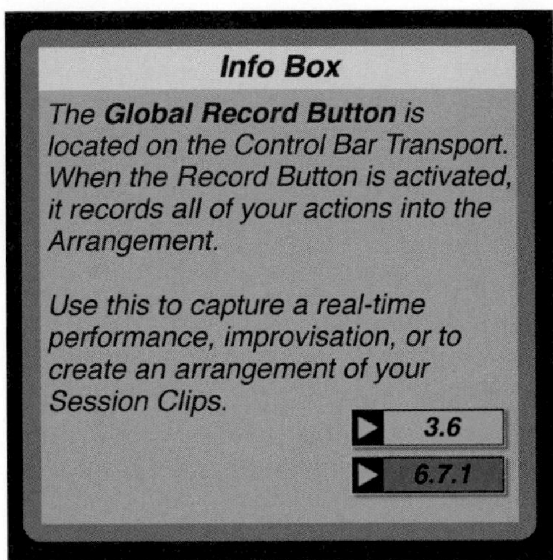

design concept of *"linear recording."* This means the program follows a horizontally aligned timeline that guides your music (audio/MIDI) in a strict sequence over time. However, Live goes beyond a linear timeline introducing you to *"nonlinear recording."* Nonlinear means out of sequence or free of, with no confining order. Think of it as a coloring book without lines or a clock without numbers. Everything is based on relative timing. So what does this have to do with the Global Record concept? Well, the Global Record concept is all about giving the composer, producer, and performer a bridge from relative time to linear time, in other words, capturing musical improvisation and limitless freedom into a recorded sequenced performance or composition.

Info Box
The **Global Record Button** is located on the Control Bar Transport. When the Record Button is activated, it records all of your actions into the Arrangement.
Use this to capture a real-time performance, improvisation, or to create an arrangement of your Session Clips.
3.6
6.7.1

Figure 4.1 Capturing a Session View performance into a sequenced arrangement in the Arrangement View.

This concept can be mind-blowing, but it is a powerful, inventive, and logical way of realizing music. So, how does this all work? Well, let's start by taking a look at one of the most unique features in Live: the *Global Record button*.

The Global Record button is located at the top of the Main Live Screen between the *Stop* and *Overdub* (OVR) buttons on the *Control Bar/Transport*. At first glance, this round button looks like a standard transport record button, but it's the concept behind it that unlocks a whole new way of approaching and arranging your music tracks and productions. It gives you the power and ability to create linear musical arrangements out of nonlinear musical sketches and improvisatory musical ideas as a performance in real time. Not only does it give you the power to create musical arrangements free of a linear timeline but also allows you to realize nonlinear improvisatory musical sketches as a performance in real time. Don't worry; in no way does Live abandon linearity, it just adds the ability to create music on the fly free and unconfined.

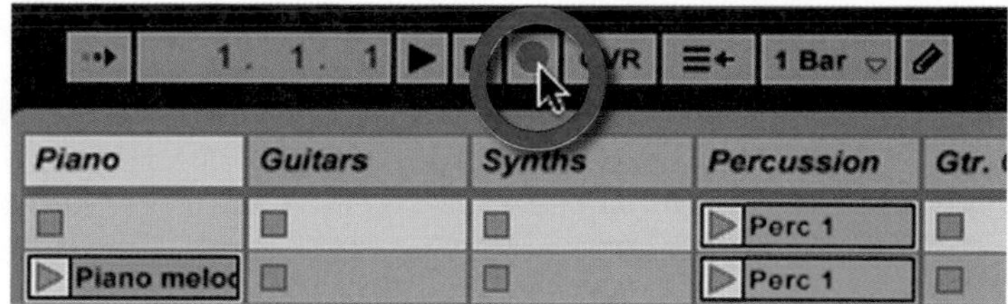

Figure 4.2 The Global Record button.

4.3 Music on the fly

As a musician, let's pretend that you are holding a large handful of musical phrases, index cards labeled A–Z if you will. You are happy with the phrases contained on the index cards, and putting them into an arrangement is a "no brainer," as far as you are concerned. Frankly, you have been arranging parts for years, right? What if you could just toss these phrases onto a large open table either one at a time or handful at a time, having them play instantly in the order they land on the table? Hmmm . . . sound interesting? Now, let's stand back and listen to the playback of the index cards in their new arrangement just as they fell and see what works. Ok . . . not bad, in fact it was quite inspiring the way card C transitioned into card A. Actually, it would be really cool to run that twice and then end with card D altogether repeating four times (for 16 bars lets say). Ok, toss these parts again on the table in our new order. Whether you realize it or not, we have just created a new arrangement in real-time without cutting, pasting, or moving any type of region or chunk along a timeline in any way. We simply threw the cards on the table in an order we felt could sound good as a whole. With the musical phrases right in front of us, we recorded an arrangement all in real time while launching any card we wanted at any time-in perfect sync too! This analogy is no different than working in Live's Session View. Just substitute the index cards of musical phrases for MIDI and audio clips.

4.4 User interfacing: two parallel worlds

It's very important to understand that Live consists of two main views that make up its overall interface. This is critical! These views, Session (vertical) and Arrangement (horizontal), share the same tracks, instruments, plug-ins, effects, and tempo, etc., but function independently in regards to musical performance and arrangement. Although they do work in tandem in regards to signal flow via the track structure, the Arrangement and the Session Views are completely different in functionality; that is, until the Global Record button bridges the two together. Global Record grants you the ability to launch audio and MIDI parts right off one page view and onto a new page view as a linear arrangement in real time.

This is how Live's Global Record concept is used for making music on the fly. It is the ability to launch your audio and MIDI parts in real time from the Session View directly into the Arrangement View, all while experimenting with effects, loops, and automation. With the unique interaction between the Session and Arrangement Views bridged by the Global Record button, the sky really is the limit. Once you have captured a musical arrangement from the Session View, you can take out your workstation tools and edit your tracks with a fine-tooth comb, and scalpel down to the barest detail, all within the Arrangement View. To get a better handle on this, let's examine both views. Our goal is to ensure that you grasp this parallel concept from the very beginning, since it is an entirely new way to record and sequence musical arrangements.

Hot Tip *To increase the overall display size of your Main Live Screen, go to Live's Preferences Look/Feel tab and scale the Zoom Display to increase or decrease your workflow environment. This applies to all views in Live 8.*

4.4.1 Arrangement View

Working along a linear timeline is great for sequencing, editing, copying, and pasting MIDI and audio. It's been done this way for years. In fact, the most common way to experience music is in a linear progression. This is because music is experienced through a set start, stop, and sequence of events that follow a strict pattern and timeline; the only exception being forms of improvisation or the old "jam session" where things may take a few turns before coming to an end.

The Arrangement View can function in two useful ways: (1) as a traditional DAW sequencer, providing a view where you can record and edit audio and MIDI in tracks as well as add prerecorded media, virtual instruments, and effects all along a linear timeline; (2) as a canvas for recording musical performances. It is totally up to you and your project's needs. Of course, this is only the beginning. The Arrangement View in Live actually gives you the ability to create an entire musical arrangement in real time as a predesigned or improvisatory performance rather than just piecing it together one region (clip) or track at a time. Sure, you could click and drag tracks or parts in across the timeline and paste them together, but one of the best things about using Live is that you don't necessarily have to. Live provides the freedom to choose between two unique ways of working: a "launch pad" style of creating a production and a more traditional linear way of recording and editing. There is no right or wrong way to work and as we mentioned earlier, the methods work great together as well as wonderfully on their own. You have to see it to believe it.

The general layout of a completed arrangement should look quite familiar in our Arrangement View example.

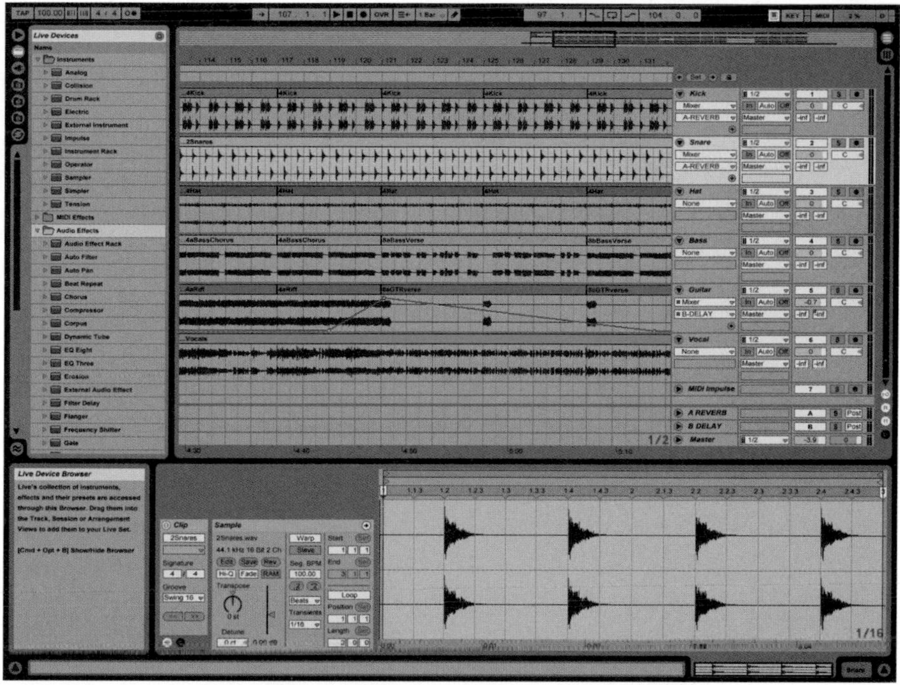

Figure 4.3 The Arrangement View displaying an arrangement.

You can see that the Arrangement View displays all the usual information found in any traditional DAW such as tracks, MIDI notes, audio waveforms/regions, automation, etc.

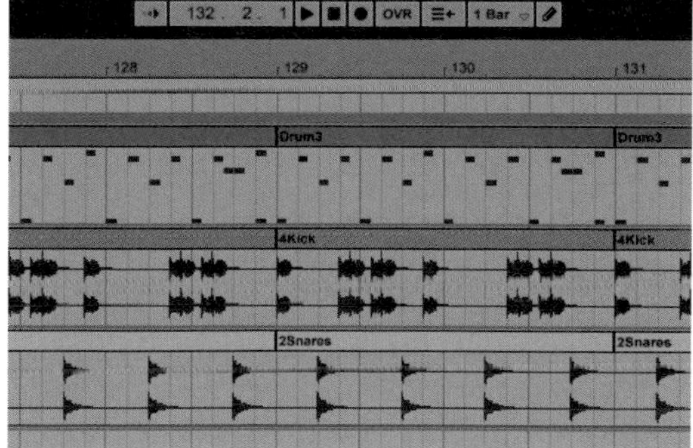

Figure 4.4 Arrangement View displays traditional linear/horizontal tracks that can store MIDI notes, audio waveforms, automation, and much more.

This is the direct result of launching clips in the Session View as a performance and capturing them into the Arrangement View by engaging the Global Record button. From here, you can alter the arrangement however you see fit. Now that you know what it looks like as a finished product, let's look at Global Record in action. For an in-depth look at the Arrangement View, launch to ▷ Scene 5. An arrangement is recorded into the Arrangement View in real time as a performance is generated from the Session View. This could either be a completely predesigned arrangement that is simply launched as a whole or an improvisation of MIDI and audio clips performed on the fly. Of course, MIDI and audio can be recorded directly into the Arrangement without ever existing as a clip in the Session View.

4.4.2 Session View

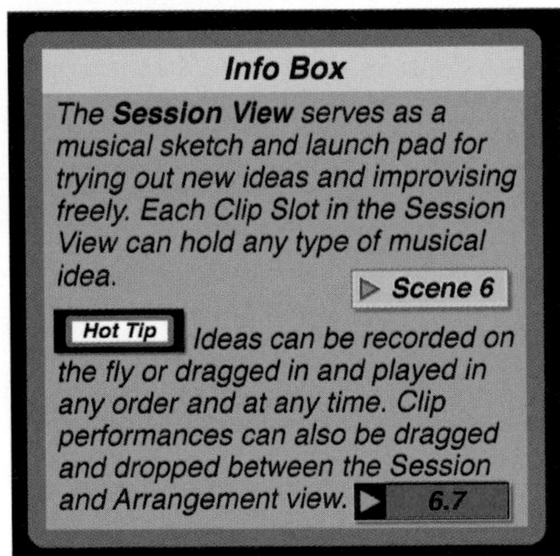

Info Box
The **Session View** serves as a musical sketch and launch pad for trying out new ideas and improvising freely. Each Clip Slot in the Session View can hold any type of musical idea.
▷ Scene 6
Hot Tip
Ideas can be recorded on the fly or dragged in and played in any order and at any time. Clip performances can also be dragged and dropped between the Session and Arrangement view.
6.7

Live's Session View acts as a musical sketch and launch pad, allowing you to try out new ideas and improvise freely, which we already hinted at earlier. Each Clip Slot in the Session View grid can hold a recording, predefined audio loop, or MIDI file and can be recorded on the fly or dragged in from the Browser and played in any order and at any time you wish. Pure flexibility!

Let's dive deeper into the Session View and look at how clips and scenes make our music flow. This will help to realize the different ways we can arrange our music on the fly with the Global Record concept. For an in-depth look at the Session View and how it works, launch to ▷ Scene 5.

As shown, Live's main screen is designed to accommodate both Arrangement and Session Views along with sidebars for the Browsers, Devices, and additional detailed overviews. Unlike the Arrangement View, the Session View has a totally original look and feel. This is different than your traditional DAW except for possibly the fader, record, mute, and solo buttons at the bottom of each track.

4.4.3 Clips

Live displays both audio and MIDI as a self-contained piece of information or container called a *clip* – audio is an audio clip and MIDI a MIDI clip. Each one of these clips holds a uniquely vast and invaluable amount of information.

For every new version of Live, Ableton has faced the challenge of keeping clip boxes clutter free while retaining a clip's full information. So much information, that it could easily take more than this entire book to cover it all. Nonetheless,

Figure 4.5 Session View, Browser, and Track View (Detail View).

a clip has a built-in activity button called a Clip Launch button. It is a triangular button located on the left of every clip. When it is pressed, the clip is activated (launched) and starts to playback.

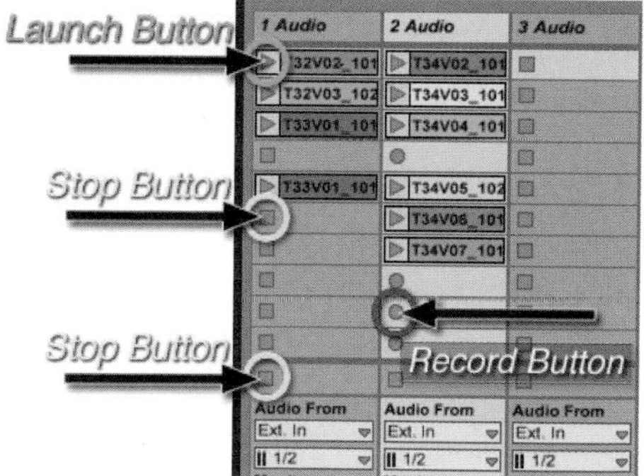

Figure 4.6 Session View Clip Launch and Stop functions.

Over the years, several terms have come about to describe the launch function – "Firing off a clip," "Playing a clip," or "Launching a clip." You get the idea. Every track can contain an infinite number of clips and has a unique Clip Stop button. These buttons are located in empty clip slots directly below the active clip – except when the track is record enabled – and in every track column just above the track Mixer section. Having the Stop button on each track is an integral part of the Live's real-time power! They will stop any running clip that is located in the same track as the Stop button itself. Clips can also be stopped with the Stop All

Clips button (Stop Clips) on the Master Track. A common mistake of "newbies" to Live is to use the Space Bar or Control Bar Stop button to stop playback of all the clips and deactivate their play status. Although both stop playback, they actually only pause the clips. You will notice that the Clip Launch buttons are still green (active), meaning that the clips are in standby (paused) and ready to continue upon relaunch. There are uses for this, specifically exporting/rendering your audio. Launch to 3.7.1 for more.

Figure 4.7 Stop All Clips button.

Figure 4.8 Control Bar Stop button. This is the same as the computer keyboard spacebar.

While launching clips, you will come across only one restriction: only one clip can be active at any given time within a track. So what does that mean? It means you'll use scenes. What are scenes? Let's read on.

4.4.4 Scenes

Figure 4.9 Scenes organize and launch entire horizontal rows of track clips all at once.

The Session is built on a grid system made up of rows and columns. Each horizontal row is a scene and vertical columns are tracks. Since tracks can only play one clip at a time, clips are often spread out across multiple tracks as rows of clips, also known as scenes. We already mentioned that clips are the basic musical building blocks of Live. These musical ideas – melodies, beats, and bass lines, etc. – can be constructed into larger scene structures. In this scenario, scenes form the musical structure of your songs and compositions in the Session View. Like clips, scenes are labeled, organized, and launched but rather from the Master Track column. Putting scenes in the context of Global Record, they are generally used to launch entire musical sections of a performance such as a verse, chorus, bridge, or outro, etc. In this way, you are able to activate an entire group of clips simultaneously as opposed to each individual clip one by one. Of course, a scene can be used in many different ways as determined by you the user. There will be much more on how to use scenes creatively in the sections to follow. In addition, you can launch to ▶ Scene 6 to see how scenes function within the Session View and to ▶ Scene 10 for an in-depth look at their advanced functions and musical concepts.

4.4.5 Session View as a Mixer

Even though the Session View and Arrangement View operate independently, their Mixer Sections are linked. This includes signal routing, I/O, Sends, and all the Mixer's functionality. The Arrangement and Session Views share all of the same controls, parameters, and information. The difference is that the Arrangement represents it in a horizontal view and obviously the Sends and Pan controls have been moved to the track display via the Control Chooser instead of knobs as in the Session View. In other words, the Session View Mixer Section with its traditional vertical faders is ideal for mixing your music regardless of which view it's generated from! The point is that you can work on your mix from whichever view you prefer. Most users will gravitate to the vertical track mixing of the Session View (which is great) and rely on the Arrangement View for fine-tuning and automation of a mix. For more on Track Automation, launch to clip ▶ 5.5.

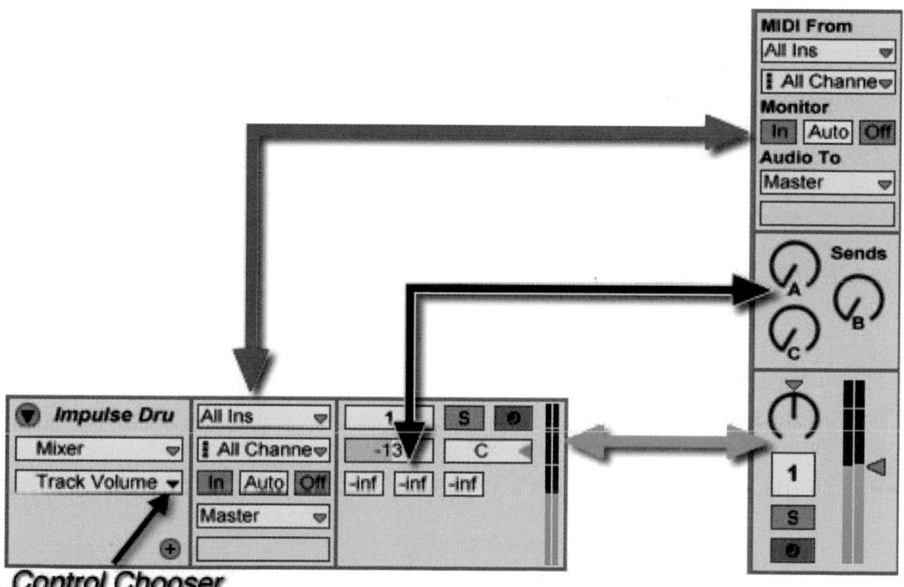

Figure 4.10 Tracks and Audio/MIDI signal flow is shared between the Session View and Arrangement View. This includes I/O, Sends, Returns, and Mixer Section.

4.5 The linear approach

By now, you have heard the term *linear* mentioned many times throughout this book. There is a good reason for that. Until the advent of the digital computer, we were unable to randomly move or process data outside of real time. Thus, the concept of linearity is engrained in our brains, (a good reason to not throw the concept out the window). In fact, we need timelines and the ability to string musical ideas and thoughts together so that they can reproduce in a consistent sequence. Sequencing is the heart of music programming and for that reason, Live still embraces the linear arrangement. There are two ways to work in a linear fashion in Live: (1) build and sequence your arrangements solely in the Arrangement View and, (2) build and perform a precise arrangement in the Session View.

4.5.1 Create

As a composer, you can use the Arrangement View to write out your musical ideas just like you always have. A great program inspires the composer and Live 8 and Suite 8 is definitely built to kindle/power your imagination, inspiring unlimited creative possibilities. The ability to record digital audio and MIDI events is the key building block for the digital composer. Ideas can be composed modularly in the Session View or Arrangement View as simple ideas. This includes adding prerecorded audio or MIDI clips in either view. From a composer's perspective, think of the two views as working together. The Session is your musical sketch, and the Arrangement is your finished work. Of course, nothing in a DAW is permanent. You can edit and make changes as much as you like. Another way to think of the concept is that the Arrangement is a multitrack recorder that is used to record your session sketches as they are played in real time, when you are ready to print them as a final sequence, song, arrangement, etc. That being said, there is a direct link between the Session and the Arrangement Views, which share a significant relationship with the composer. Let's take a closer look at specific ways to exploit this relationship. First and foremost, keep in mind that there is no one right way to work in Live. You must determine what works best for you!

4.5.1.1 Capturing a Session performance as an arrangement

Our first approach is to begin by composing in the Session View and then recording the Session performance as an arrangement into the Arrangement View. This is the way in which the Global Record process was intended. You will begin sketching and refining your ideas in the Session View, then recording your ideas as a logical sequence, arranged as a performance in the Arrangement View. For techniques on composing in the Session View, launch to ▶ 6.6. Let's look at the Session and Arrange through the eyes of Global Record.

The process begins with importing or recording audio or MIDI into track clips. Working track by track, you build musical phrases, riffs, licks, or whole musical sections. Of course, this really depends on how you compose your music. As you develop your ideas, you have the option to stack them up for each track (clip by clip) or assemble them into scenes (song sections). You then can stack those scenes into a sequence. Again, this will depend on how you prefer to compose. Once you have constructed your ideas or entire sequence, it's time to capture (record) them as an arrangement. Here, you have two choices. First, you must press the Global Record button followed by the Play button. Then you may either begin launching your individual clips, as you deem appropriate, or launch your constructed scenes as you see fit. Taking this a step further, you can also assign and automate Follow Actions that will execute designated clip launchings automatically 7.7.4. Either way, once you stop recording, flip over to the Arrangement View to see your new arrangement. If need be, feel free to edit and tweak any elements of your arrangement directly in the Arrangement View.

Figure 4.11 Active real-time performance in the Session View.

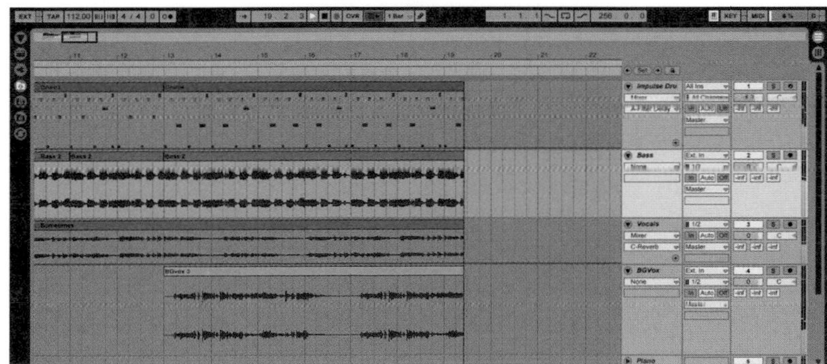

Figure 4.12 Capturing a Session performance into the Arrangement View on the fly.

4.5.1.2 Working in both views

The second approach to Global Record is more of a hybrid process that involves working back and forth between both views. When composing directly into Session Clip Slots, you may feel that loops and rhythmic patterns are naturally a perfect fit for recording into Clip Slots, but that you prefer to write out keyboard parts or record vocal tracks along a linear timeline. In this case, one way to work is to layout your rhythmic patterns first in the Session View. Then, record your keyboard, guitar, or vocal layers in the Arrangement View using the Session View to playback the rhythm section or backing tracks as a guide while recording the lead elements directly into the Arrangement View. Once you've finished recording your lead and harmonic elements, you could then capture the backing tracks into the Arrangement View or move either the lead or harmonic elements – keyboard, guitar, or vocal part – or rhythmic elements manually into either view using the "drag and drop from view-to-view" capability. Simply select the material you desire to move and click, hold + drag it over the specific View Selector or press the Tab Key and then place the material wherever you want. When dragging and dropping into the Arrangement View, you will need to duplicate and edit the clip's lengths along the timeline to fill out song sections to complete the arrangement.

In this way, you are constantly building your arrangement back and forth between views, using the strengths of each to piece together an arrangement. Through a combination of Global Recording, Arrangement Recording, and editing, you can create and produce your musical ideas into full song arrangements. To learn about Arrangement View Concepts, launch to ▶ Scene 5.

4.5.2 Produce

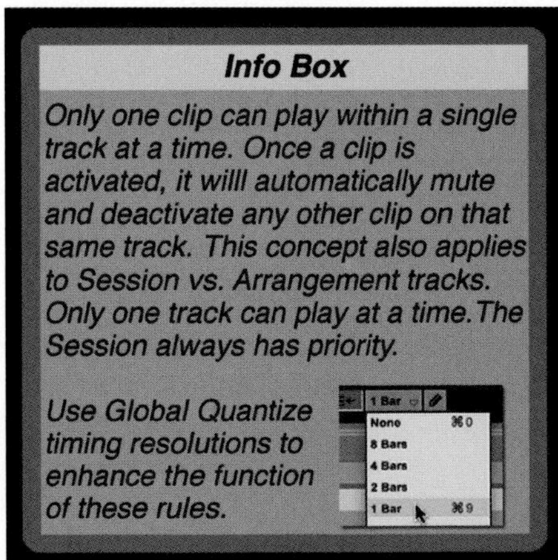

Info Box
Only one clip can play within a single track at a time. Once a clip is activated, it willl automatically mute and deactivate any other clip on that same track. This concept also applies to Session vs. Arrangement tracks. Only one track can play at a time. The Session always has priority.
Use Global Quantize timing resolutions to enhance the function of these rules.
1 Bar
None ⌘0
8 Bars
4 Bars
2 Bars
1 Bar ⌘9

As a producer, you need the ability to edit and rearrange tracks and takes to manipulate or remix a song. You will find that Live's linear Arrangement View allows you to do this as well as work and produce your music in the same way you always have. To that end, with any other DAW, the song edit you create is the end of the line. In Live, you are given new life for your productions: the opportunity to evolve your work beyond the traditional edit/remix by trying out (auditioning) different clips while your linear arrangement plays back.

Simply launch a clip or clips in the Session View while the arrangement is playing and the session clip/track will takeover playback priority, muting out their equivalent arrangement track(s).

If that sounds interesting, through the Global Record feature it's possible to record your track clips into the Arrangement and overdub them, making Live a powerful producer's tool. This is similar to the concept described in the section above. You can produce your edit or mix in the Arrangement View first – importing new material, editing audio stems, etc. – and then bring it back into the Session View as a whole or sectionalized via the drag-and-drop technique. This is a fantastic and powerful feature that adds flexibility to any sequenced arrangement, especially when you are working/remixing from preexisting audio stems. Imagine taking the entire studio multitrack as it was sequenced out as an arrangement, cutting out pieces, grooves, riffs, among others parts, then putting them into the Session View where you can launch them into a new order or over the original sequence. Pretty mind boggling stuff! When copying clips from view to view, you will have to get used to how the Session View interprets the physical layout of the arrangement as it relates to launching and recreating the identical arrangement in the Session View.

Take a look at the following walk-through to understand how to drag and drop between views. This can be executed from either view (Session to Arrange or Arrange to Session). For this exercise, we will go from the Arrangement to the Session.

1. Select all clips (or whichever ones you want) in the Arrangement View by holding down the *shift* key and clicking on each clip. To select all at once, press [⌘+A] Mac/[Ctrl+A] PC.
2. Click + hold on the selected clips and then drag across the Session or Arrangement View Selector in the upper right. You could also press the TAB key while clicking + holding on a clip or clips. This flips your current view over to the Arrangement View (the opposing View).
3. While still holding/hovering the clips over the Session View, line them up with their respective tracks then drop them (let go of the mouse click) into the Session View tracks.

In our example, you will see our Arrangement View and how it transferred into the Session View when we dragged and dropped it into the Session View.

You will see in our example, that the Session View now contains a vertical representation of our Arrangement clips. Notice that the Session clips are relative to where the Arrangement clips were located along the Beat Time Ruler (timeline). This is a very literal interpretation of the Arrangement clips. Each clip is laid out based on the individual clip lengths as originally shown in the Arrangement track

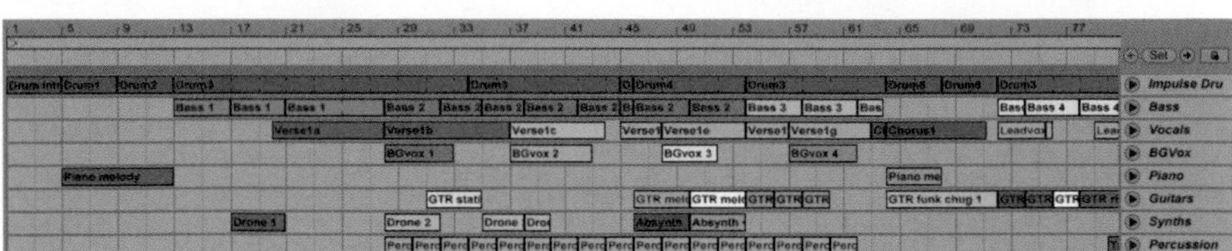

Figure 4.13 Arrangement in Arrangement View.

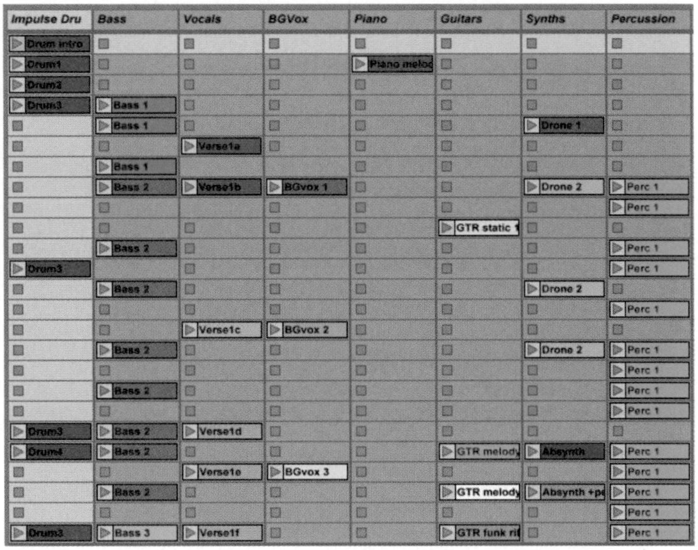

Figure 4.14 Drag and drop an arrangement into the Session View from the Arrangement View.

display. This means that no matter how long a clip looped for, it will only take up one Clip Slot in the Session View, leaving the Clip Slots below it empty. The actual clip itself will contain the exact amount of bars and beats that it played for in the Arrangement View. Since the clips are positioned in a relative relationship, we'll have to handle launching individual clips and scenes yourself if we want to recreate the exact arrangement we had. That's a little obsessive. The point really is not to necessarily recreate the arrangement *per se* but to use this feature to capture clips from the Session View in the Arrangement View and then bring them back into the Session View just the way you performed them. Maybe it was an impromptu performance or a deliberate sequence of clips. Reverse this process and you can build grooves and modular musical ideas, and then bring them into the Session View. No matter your reasoning, move clips in this manner in either direction from the Session to Arrangement View or vice versa. It's a simple but powerful process. Just reverse the outlined steps from above.

The great thing about all of this is it can all be done in real time without stopping playback; a true Ableton Live trademark.

4.5.2.1 Remixing

When it comes down to remixing, the possibilities with Live 8 are endless. If your song has already been assembled in the Session, then it's really easy to try your hand at multiple performances or passes of the song. Each time you might try different bass lines with different beat loops. Instead of launching entire scenes, maybe it would sound cool for the groove style clips of one scene row to playback with the clips of vocals of another scene row. No matter how you formulate your remix, you have unlimited flexibility. Feel free to work in either view or both to produce your remixes. Drag and drop your Session to the Arrange or vice versa. In the Arrangement, you can manipulate, edit, and loop in a linear fashion, rebuilding the song from the ground up. You can also use the hybrid approach by overdubbing ideas from the Session to the arrangement.

4.5.2.2 Beat Making

Making beats is an art form requiring patience, skill, and imagination – all character traits of a great producer. Coming from a linear approach, you can produce beats just like you always have using the Arrangement View. The concept is the same as any other DAW. Create and sequence along the timeline or ruler as you go. The most common method is to program beats with a drum machine approach by looping playback and adding a layer during each pass or take in Overdub Mode. Program beats with a MIDI controller or input (insert) them in Draw Mode ▶ 7.7.4. Creating beats in the Session View is just about the same process as far as programming MIDI goes. The real difference is in how you go about linking ideas, loops, and phrases together. As far as producing a beat, the process is still a linear concept, but you are working with vertical tracks and scenes as opposed to horizontal tracks along a timeline. In the Session View, there is no timeline. Instead, you will create clips that loop and scenes (rows of clips) to launch simultaneous clips (your loops in this case). When you wish to make forward progress or move to a new groove (next song section or new loop), launch the next clip(s) or scene. As you can tell, herein lies the departure from linearity. The point to understand is that producing a beat loop in the Session View is a linear process, but linking loops and song segments is not.

4.5.3 Perform

Enough about making music in your project studio, it's time to get out and perform! With your Live Set in hand, Session clips laid out into scenes, and a little inspiration, you can perform and even improvise your music from the stage – never missing a beat. Live is literally a playable instrument, but let's save the serious improvisations for the next section. For now, we'll focus on using Live as an onstage sequencer and musical trigger/synth device. Of course, you can use Live to play any number of virtual instruments in a performance situation, but Live is a lot more than that. The true linear performance aspect is in the ability to open a Set and playback a predesigned song from the Arrangement while performing with it as a band or solo performer. Another idea would be to playback your arrangement while performing additional parts and clips in real time from the Session View. Finally, you might launch your entire song from the Session View generating a predesigned arrangement in real time while making a few game-time decisions about the performance. You could even globally record your Session performance onstage while performing! Since the last idea borderlines on nonlinearity, we will move on, but before we do, understand that in the real world, you will always walk the line between the linear and nonlinear aspects of Live. That's the point of it all!

4.6 The nonlinear approach

When have you ever had the opportunity to take your linear composition and remix it on the fly, swapping bass lines, and beats without hesitation or stopping playback? How about composing and improvising on the fly? Now you do! We have already hinted at ways to perform with Live in a nonlinear fashion and are sure you are already dreaming up a million ways to do so. On the other hand, you might be thinking, how can I take advantage of this as a singer songwriter or producer? Before we get ahead of ourselves, let's break it down into our three categories: create, produce, and perform.

4.6.1 Create

When looking at the Global Record concept from a nonlinear perspective, things can start to get confusing, especially, since Live gives you the ability to create without ever having to stop playback. For that reason, we'll go into more detail.

One of the easiest ways to work in Live is to use the Session View to experiment and sketch out modular ideas. There is no timeline; therefore, you can just keep running the clips as long as you want. For example, you could launch a loop and let it keep running. While it's playing, you can experiment with other musical elements such as a bass line, more rhythm, or a melody. This could be an audio loop or a musical line you play in yourself. Either way, you can test them out together in any way you like. Drag in other loops or ideas from the browser and just start building music as you go. If you like what you hear, you can toggle the Global Record button and capture your idea as an arrangement or at least a fixed performance. Once you have your performance idea recorded, you can then bring the arrangement clips into the Session to create a scene while clips are still running and looping. In the same vein, if you already have an arrangement but want to change it or create new ideas to add to it, just use Global Record. Play your arrangement and when ready, toggle global record, launch your clip(s), and

Figure 4.15 Back to Arrangement button.

lay down your new takes. When you're done, click Global Record again and the arrangement will take over playback from the Session. Similarly, you can experiment with alternate clips and at any point return back to the arrangement by clicking the Back to Arrangement button.

4.6.2 Produce

As a producer, Live's nonlinear principles give you the ability to literally mold and shape any song on the fly. It's all about the way Live lets you drop in tracks, clips, loops, and effects without stopping playback. You can even Hot-Swap instruments and effects in real time . Ableton's intention is to let your musical ideas and talents flow without interruption.

4.6.2.1 Remixing

From a producer's point of view, you will find Live 8 "remixing friendly," especially when the linear shackles and chains are off. It is amazing how quickly you can begin to remix a song on the fly by simply starting playback and launching clips wherever you might want them to go. Even more intriguing, think about a song you want to remix. In an ideal situation, you'll have the multitrack session with each element or layer of the mix isolated or as stems. Normally, this would be on a timeline locked to a linear grid. Now, think of the Session View as the multitrack session, free of a timeline. There you will be able to listen to the song sections in any order you wish. Launch the hook whenever you like. Swap out the accompaniment or groove with a new one or one from another song. It's that simple and that powerful.

4.6.3 Perform

Ableton Live 8 has forever changed the way I play my "live" music. The real-time capabilities and control are unreal. It has opened new doors for me and how creative I can get on the fly. It has taken the concept of Jamming "live onstage" into whole new dimensions!

Richard Devine, Electronic Musician

When it comes to Live's nonlinear capabilities, it's all about performing and improvising. Ableton has made it very easy to use Live as an instrument. You cannot say that about any other DAW software out there. Sure you can playback sequences or use virtual instruments with other DAWs, but with Live, your music is free flowing and always in your control. For example, while playing a song from the Arrangement View, you can use the mouse pointer to move around and relaunch playback in real time without

Figure 4.16 Global Quantization Menu.

stopping the current playback position. On the basis of *Global Quantization* Menu settings, you can vamp and loop sections as long as you like.

There are simply no rules governing what you can do. On top of that, with Global Quantization, you never miss a beat! I'm sure you're thinking, "all this sounds great, but how does this apply to Global Record?"

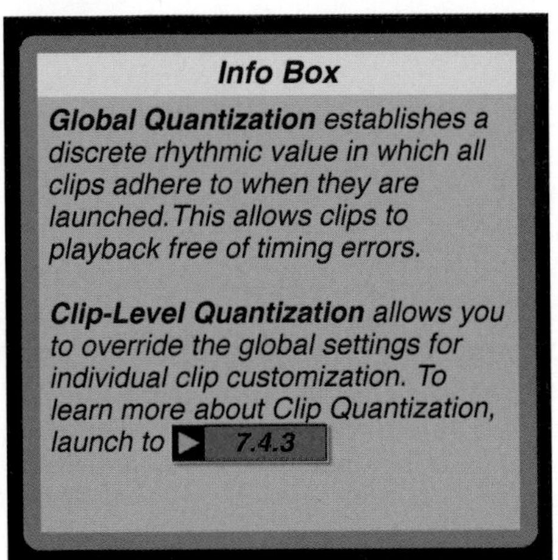
Info Box

Global Quantization *establishes a discrete rhythmic value in which all clips adhere to when they are launched. This allows clips to playback free of timing errors.*

Clip-Level Quantization *allows you to override the global settings for individual clip customization. To learn more about Clip Quantization, launch to* ▶ *7.4.3*

Performing and improvising onstage in real-time sounds almost impossible with a DAW, doesn't it? Well, it's not with Live. Whether working as a DJ or performing with a band, you can launch audio and MIDI clips while mixing and effecting your music in real time. Live has all the tools to tweak, automate, and sweep through real-time effects on the fly. It all starts out with preexisting Session clips. This can be a complete song, basic foundation of a song, or a bunch of loops (hopefully organized). Simply start playback and launch and manipulate audio, MIDI, and effects however and whenever you like. Interact and improvise with your music while it's playing. You can dial up instruments, effects, and drop clips of all sorts. Ok, this is great, but it really doesn't incorporate Global Record. The truth is, you will not perform with Global Record *per se* but rather employ it to predesign elements of your Set for use and interaction onstage. What you can do is... while all of this impromptu stuff is going on, you can be recording your performance through the Global Record feature. That means when all is said and done, you have your entire performance logged into the arrangement for use later! Thinking out of the box, you can also come up with creative ways to incorporate Global Record into your performance, such as recording an impromptu section then pasting it ahead or pulling it into the Session as a scene, etc. When all is said and done, you will have a sequenced performance.

In this chapter

5.1 Musical timeline 85

5.2 Layout 85

5.3 Navigating 86

5.4 The Arrangement View 92

5.5 Working with automation 101

5.6 Arranging concepts 105

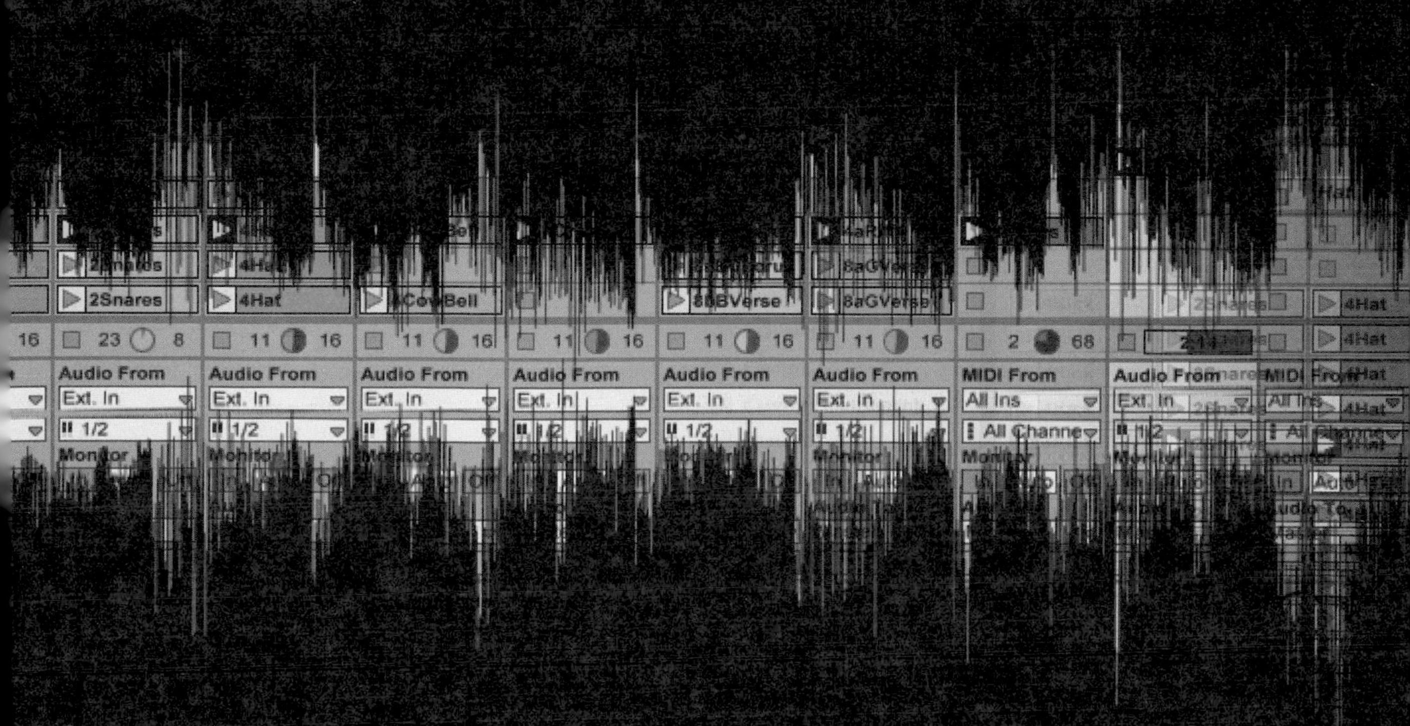

SCENE 5

Arrangement View Concepts

I have fun mixing, DJing, and producing in one powerful, clutter-free environment: Live 8. Best of all, I can still fit my entire musical life in a backpack!

Tyler Gusich, Artist, DJ, and Pro-Audio Consultant with B&H Photo

5.1 Musical timeline

The Arrangement View functions as a "musical timeline," or as some like to refer to it, "linear timeline" where audio and MIDI clips are sequenced into fixed arrangements. This is where Live records and stores all your actions, movements, and commands. You can record clips, Clip Launches, clip manipulation, mixer automation, and device automation. As a Live user, you will also edit and alter clips and clip sequences. The main purpose of the Arrangement View is to capture the musical performance of Session clips and provide a linear representation of songs to be edited, mixed, and exported as a final production.

5.2 Layout

To view an arrangement, toggle over to the Arrangement View by pressing the Tab key on your computer keyboard or toggling the View Selector on the upper right of the Main Screen. The Arrangement View layout is very straightforward. Just like any other digital audio workstation (DAW), it consists of horizontal tracks that host track clips and manage audio signals. This includes virtual instrument plug-ins, external audio input sources, insert returns – delays and reverbs, and a Master track. Track names are located to the far right of track display.

The triangle to the left of the track name is the Unfold Track button. Click this to view track contents (MIDI events, waveforms, breakpoint envelopes) and to make selections within a track. To the right of the track name, you will find the In/Out Section and the Mixer Section, when they are in view. The In/Out

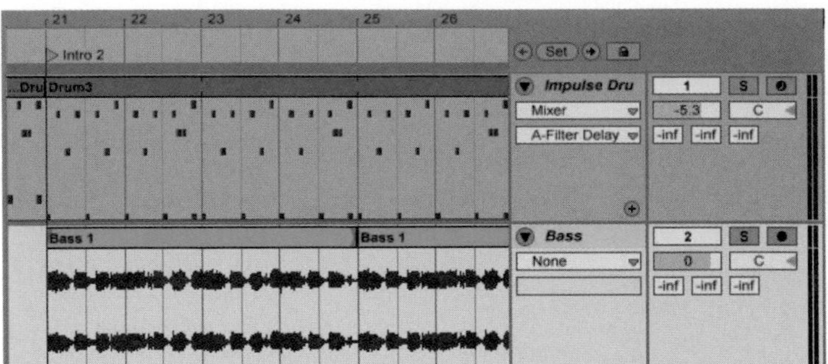

Figure 5.1 Arrangement View layout with MIDI and audio track.

Section contains track input and output choosers for routing audio and MIDI in Live. The Mixer Section contains three buttons and additional input fields and sliders. The numbered yellow button is called the Track Activator. It allows you to bypass/mute a track from being heard. Next to that is the Solo/Cue button. This solos a track, isolating it during playback. To learn about the *Cue* feature, launch to ▶ 6.2.2. The last button on the right is Arm Arrangement Recording, or as we call it, Track Record button (record-enable). It can have many names.

The Mixer Section is where volume, pan, sends, automation lanes, and envelopes are displayed and adjusted. If you do not wish to have these parameters in view, then you can hide them. Check or uncheck Mixer from the View Menu or from the show/hide button (M) to the lower right of the Main Live Screen. Also notice the other view options available from this menu.

5.3 Navigating

Arrangements are fixed along two linear timelines in the Arrangement View: the Beat Time Ruler measured in beats and bars, based on the designated time signature, and the Time Ruler measured in hours:minutes:seconds. The Time Ruler can also be set to display other standard Film/TV timecode formats (SMPTE), often important for those working with video in Live (Options Menu) ▶ Scene 18. In theory, the concept of navigation in Live is very similar to the edit/arrange windows layouts commonly found in most other DAWs, but is quite unique when actually put into practice. To that end, the Arrangement View consists of many parts, so let's start by looking at how to navigate along the timeline and take control of our music and playback.

5.3.1 Scroll and Zoom

Getting around in Live will take a little getting used to. Zooming and scrolling through the Arrangement View and displays is very different than most of the common zooming features found in word processing, Internet, and other audio

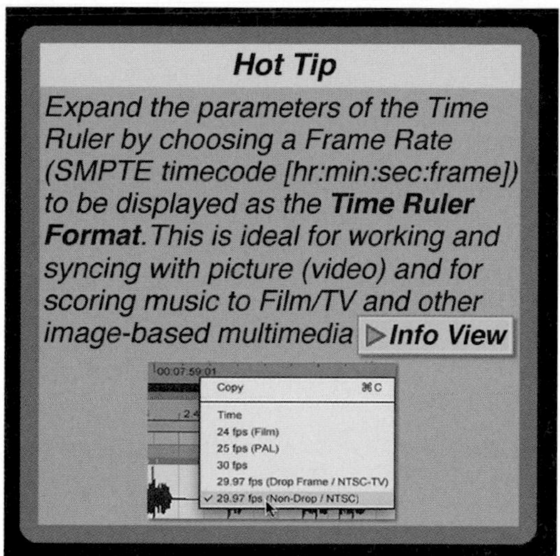

production applications. As mentioned above, the Arrangement View consists of two rulers for measuring time and location: Beat Time Ruler and Time Ruler. Navigation primarily takes place via these two rulers or from the Overview. The Overview is a horizontal display located directly below the Control Bar. It is subdivided into the Clip Overview/Zooming Hot Spot, a box-shaped outline indicating the area currently in view that is helpful for navigating the Arrangement.

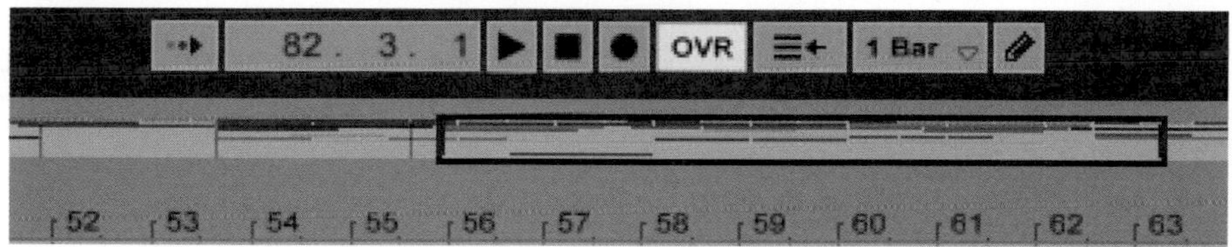

Figure 5.2 Overview of an arrangement. Also used for navigation and zooming.

The Beat Time Ruler is located just below the Overview in the Arrangement View. The Time Ruler runs parallel to the Beat Time Ruler and is located at the bottom of the track display. Placing your mouse over any one of these areas will cause the mouse pointer to switch to a navigation tool.

Figure 5.3 Arrangement View: Beat Time Ruler and Time Ruler.

A magnifying glass will appear when your mouse pointer is placed over the Overview or Beat Time Ruler. With the magnifying glass, click + drag horizontally (right/left) to scroll and vertically (up and down) to zoom. A single click anywhere along the Overview will refocus the display on that specific area. To zoom in closer on your arrangement, click + drag the magnifier downward towards the bottom of the display in the general area of the clips you want to focus on. Repeat this gesture to zoom in to the finest resolution possible. Keep an eye on the bars and beats displayed on the Beat Time Ruler to get a sense of where you are along the timeline. As you zoom in, more bar numbers will be displayed indicating your exact location. Double-clicking anywhere along the Beat Time Ruler will zoom back all the way out to fit the entire arrangement as long as there is no selection made in the track display. If there is a selection made, double-clicking will zoom in to show the selection. When zoomed out, clips can become mashed together making it difficult to identify the exact location or identities of the clips. Zooming around the Arrangement View will take practice!

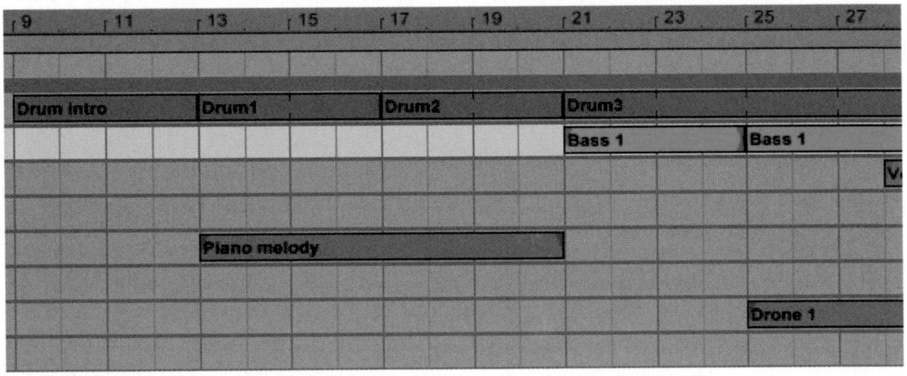

Figure 5.4 Moderate zoom level of Arrangement View track clips.

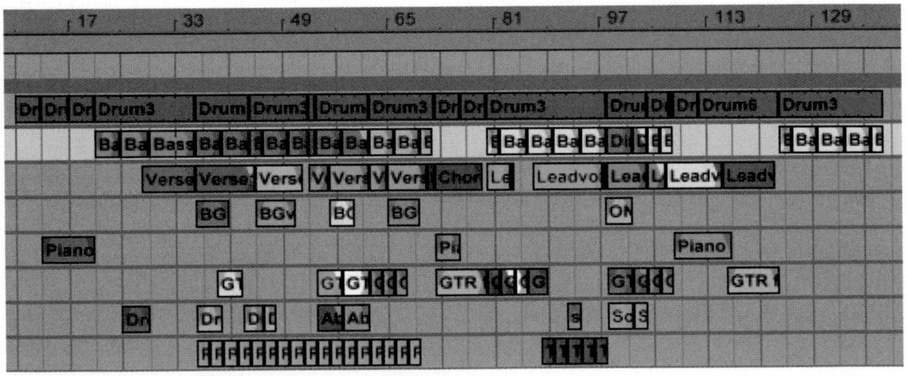

Figure 5.5 Tight zoom level showing Arrangement clips mashed together.

Scrolling through the Arrangement View is very quick and easy. As mentioned, you can use the Overview, Beat Time Ruler, or Time Ruler. Simply click + drag along the timeline of either. It should be pointed out that the Overview does not display the arrangement true to scale. This means that it does not physically take up the same screen space as the track display does. Think of it as a bird's eye view or approximation – like viewing a holiday parade from a sky cam or blimp. In other words, be careful where you click along the Overview or you might end up very far from your desired location. Take notice of the Zooming Hot Spot to help narrow in on your location.

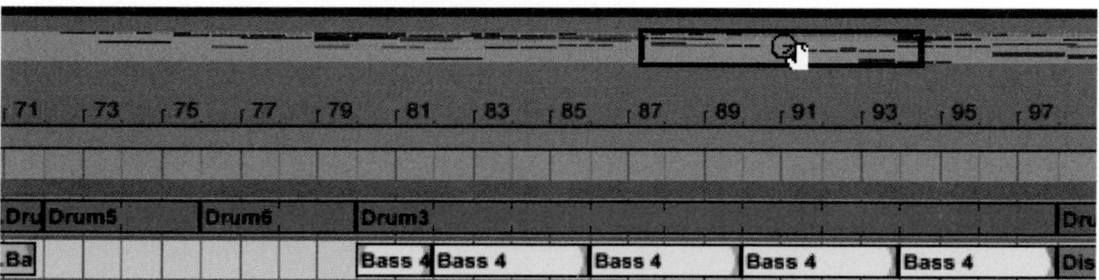

Figure 5.6 Zooming Hot Spot is the black outlined square within the Overview. Use this to focus on a specific area of an arrangement.

Depending on how you like to work, you can also have the display autoscroll during playback with *Follow* activated or stay focused on the current visible song position. Follow is located on the control bar transport section. Click the Follow button to activate scrolling with playback (yellow = on).

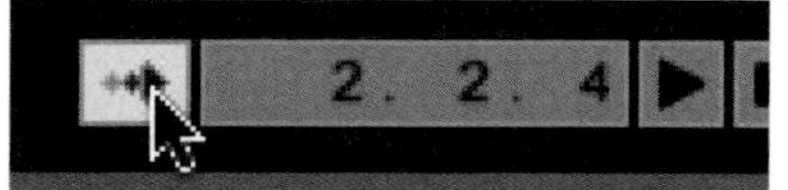

Figure 5.7 Follow On/Off button.

As mentioned, you can scroll through the arrangement via the Time Ruler. A hand-shaped scroll tool will appear when the mouse pointer is placed over the Time Ruler. The hand tool is for scrolling only. You may also access the hand scroll tool anywhere in the track display by holding [⌘+alt] + dragging. In this way, you can also allow for vertical scrolling. This is probably the most efficient method for navigating through the arrangement since you can begin scrolling from anywhere within the display.

5.3.2 Transport

The main playback controls for the Arrangement View are located in the Transport section of the Control Bar. From left to right, you will find: Arrangement Position, Play, Stop, and the Global Record button.

Figure 5.8 Transport Controls and Arrangement Position display field.

The Arrangement Position display field indicates the current playback start location of the Arrangement View when playback is stopped. While playback is running, it displays the current arrangement position by following playback. The exact position is displayed as [bars. beats. sixteenths.]. From this display field, you can manually drag or type number values to position playback at any location. This will move the Arrangement Insert Marker to your desired playback start location. You can also achieve the same result by clicking your mouse pointer in the track display on or around the location you wish to start playback. After an insertion is made, press play.

The Play, Stop, and Global Record buttons are fairly self-explanatory. They function just like any other DAW. One quick note is that double-clicking the Stop button will return the Insert Marker to the beginning of the set "1.1.1." As an alternative to using the Play button or the spacebar, you can start playback anywhere along the timeline by clicking in the Scrub Area.

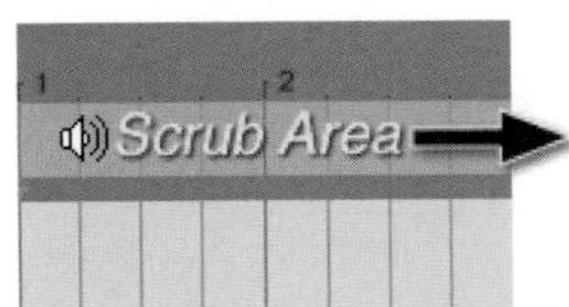

Figure 5.9 The Scrub Area is for starting/jumping playback to a clicked location or to "scrub" audio.

This area is located between the Beat Time Ruler and the track display. It can also be identified by the speaker icon that appears at the end of the mouse pointer when hovering your mouse over the specific scrub areas. Click anywhere in the Scrub Area to jump to and start playback. This can be done while the Arrangement is stopped or already in playback. If in playback, Live will jump to the new position based on the Global Quantization setting. To take advantage of this feature, make sure that "Permanent Scrub Areas" is on in the Preferences>Look/Feel tab. If disabled, you must hold the Shift key while clicking in this area to affect playback.

5.3.3 Locators

At any point along the timeline, you can add a Locator to mark sections, start points, hits, or anything else, you might want to identify. Some DAWs refer to these as "markers" or "memory locations." They make your arrangement very clear and easy to navigate. Locators can be added during playback or when the arrangement is stopped. They appear along the Scrub Area as a triangle connected to a vertical line that extends vertically across all tracks.

To add a Locator, select "Add Locator" from the Create Menu at the top of the main Live screen. It will prompt you to create a name for the Locator. If you wish to change it later, select "Rename" from the Edit Menu ([⌘+R]Mac/[Ctrl+R] PC).

Figure 5.10 Locators are identified by the triangles and connected to a vertical line that extends the entire track display. Locator controls are directly above the track names next to the Scrub Area.

You can also add a locator from the Set/Delete Locator button located on the far right of the Scrub Area above the track names. This button provides the fastest way to create and delete locators. Once a locator has been selected, the Set button changes to "Del" for delete. The arrows to either side of the Set button are for navigating to and from locators (next/previous locator).

When adding locators during playback or recording, they will be inserted according to the Global Quantization setting. When not in playback, they are inserted wherever the Insert Marker is or where your selection begins. Locators make navigating and launching sections of an arrangement as simple as a mouse click. A single click on a locator will move the Insert Marker to that locator. Double-clicking will start playback immediately when the arrangement is not in playback. Jumping or launching from markers during playback or recording is also subject to the Global Quantization. Locators can be triggered via MIDI or computer keys per custom assignments. Remote controls are assigned through Live's MIDI/Key Map Modes ▶Scene 12.

To move locators, you can drag them manually or use the arrow keys on the keyboard. They move along the grid based on the Snap to Grid (Marker Snap) setting similar to quantize settings.

Figure 5.11 Marker Snap (current spacing between grid lines).

Snap to Grid and other grid settings are accessed from the Options Menu or track display contextual menu (ctrl + click or right click). Both menus offer a choice of grid values for snapping and inserting made while working in the Arrangement. The current Marker Snap setting, or the spacing between grid lines, is displayed in the lower right corner of the track display. There are two setting types: Adaptive Grid and Fixed Grid. These values are based on common beat values/subdivisions and are used to show the number of beats (grid lines) per bar. For example,

"1/2" means half note subdivision with one grid line and "1/4" equals quarter note division with three grid lines, etc.

The Adaptive Grid concept is governed by zoom levels and labeled as sizes such as "narrow" or "wide." As you zoom, the grid will adjust the resolution of the grid lines relative to the zoom level. For example, with a wide grid you may only snap at every two bars, but as you zoom in, it may snap at every two beats.

The Fixed Grid is independent of zoom. It remains set to the value you have chosen. To speed up your workflow, you should use the keyboard shortcuts for setting and adjusting the Marker Snap settings ([⌘+1–5] Mac/[Ctrl+1–5] PC).

5.4 The Arrangement View

Working in the Arrangement View is a totally acceptable way of using Live whether it is for composing or producing. Remember that Live is more than a stage performing tool; it's a full-blown DAW. The Arrangement is most commonly used to edit audio/MIDI files and recordings, manage an overall sequenced song or performance arrangement, and a place for drawing and editing envelope automation. These are all part of creating and producing the final product, a finished and well-produced song or piece of music. Feel free to load up the "New in Live 8/Suite Demo" or "Tour of Live/Suite Demo" located in Library>Lessons>Sets folder. Pick a Set and double-click it to load it up. This will be useful for putting Live concepts to practice as we explain them.

5.4.1 Launching

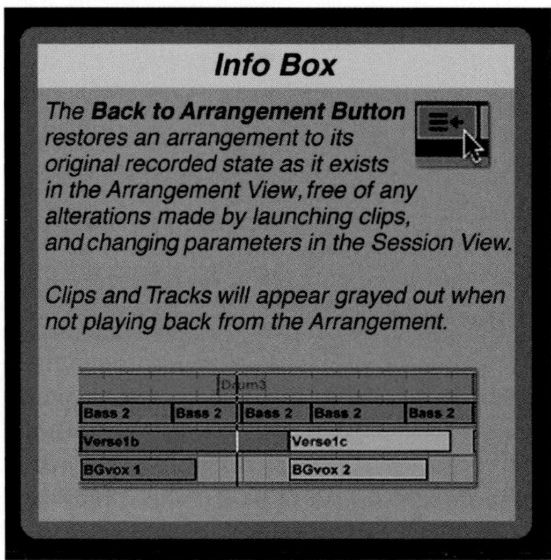

Info Box
The Back to Arrangement Button restores an arrangement to its original recorded state as it exists in the Arrangement View, free of any alterations made by launching clips, and changing parameters in the Session View.
Clips and Tracks will appear grayed out when not playing back from the Arrangement.
Bass 2
Verse1b
Verse1c
BGvox 1
BGvox 2

There are four basic ways to playback an arrangement and Arrangement clips: (1) click the Play button from the transport/control bar, (2) press the Space Bar, (3) double-click a Locator, (4) click anywhere along the Scrub Area (when preference enabled). If you have made a selection within the track display, playback will commence from that point when you press the space bar. When launching from a locator during playback or record, playback will jump to the locator position on the next global quantization value.

While you are launching around, verify that you are listening to the playback of your arrangement. It is very common for Session clips to sneak into playback by accident. Simply ensure that the Back to Arrangement button is not illuminated red. If so, click it to revert to the stored (previously recorded) arrangement.

Figure 5.12 Back to Arrangement button.

This happens anytime you launch Session clips or adjust any parameter that has automation written into the Arrangement, indicating that some or all Arrangement tracks are bypassed and not playing back from the stored arrangement. It is important to note that Session clips have priority over Arrangement clips/tracks. In this case, the affected arrangement tracks will be grayed out in the Arrangement View. More on this concept in clip ▶ 6.7.2.

5.4.2 Looping

There are two distinct types of looping that you will use when working with the Arrangement View: looping playback and looping clips. In general with looping playback, you will make a timespan selection that you can listen to over and over again. You'll find this is useful when creating, editing, or mixing music as you can repeatedly listen to your selection when editing, and temporarily use the technique for loop recording. That's not to say that there would never be a reason to loop playback on stage. Looping clips, on the other hand, is a part of the arrangement and clip workflow design. In this way, loop-enabled clips are dragged out (extended) to repeat or duplicated many times to function as part a musical element of your arrangement. Drum loops are a prime example of this type of looping. For now, we will focus on looping playback segments of the arrangement for tasks such as editing, mixing, and recording purposes. For an in-depth look at looped clips, launch to ▶ 3.5.1 and ▶ 7.5.9. For looping in the Arrangement, launch to ▶ 14.5.

To create a loop selection, click + drag to highlight a segment of clips in the track display. Once selected, select Loop Selection from the Edit Menu ([⌘+L] Mac/[Ctrl+L] PC). Notice that this will also activate the Loop Switch located on the Control Bar to the "On" position. When this switch is "Off," the selection will not loop in playback. The loop length is indicated in the number field box to the right of the Loop Switch represented in [bars.beats.sixteenths.].

The Loop Region is indicated by the Loop Brace, a grey box with triangle bookends, and can be moved by dragging that box horizontally along the Beat Time Ruler. It can also be shortened or lengthened as desired. The concept here is really no different than any other DAW. If you navigate away from your loop

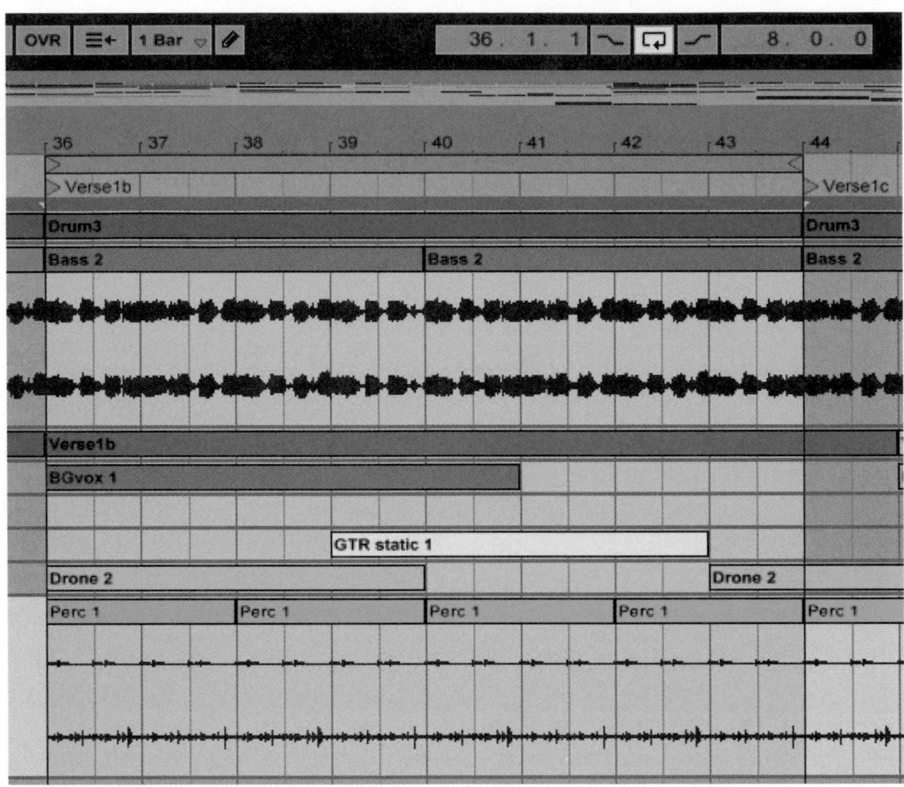

Figure 5.13 Selection in the Arrangement View set to loop in playback.

selection to edit something else, you can easily reselect it quickly by selecting Select Loop from the Edit Menu. If the case you need to loop the entire arrangement, select all ([⌘+A] Mac/[Ctrl+A] PC) and choose Loop Selection as described. Loop Selection is also accessible from the selection's contextual menu. Ctrl + click or right click after highlighting (selecting) the desired region. There you will also find additional features applicable to your arrangement.

5.4.3 Selecting

When it comes time to edit and manipulate your clips in the Arrangement View, you want to feel comfortable with making selections with the mouse pointer and using keyboard shortcuts. The latter will come with time and practice. In the meantime, you can use menu commands and contextual menus (Ctrl + click or right click). For a list of Mac/PC keyboard shortcuts, launch to the ▶ Website.

A single click anywhere in the arrangement background will move the Arrangement Insert Marker to that position. Selections are indicated with yellow highlighting. Click + drag anywhere in a track display to make a selection. To select an entire clip, click on the colored title bar at the top of the clip. You will then see two triangular arrows appear on both sides of the clip. You can also click on the loop brace to make time range selections even when the Arrangement Loop Switch in not "On." This will highlight all clips contained within the Loop Brace's boundaries – start through endpoint. Once you make a selection in the Arrangement, playback will always begin at the beginning of the selection until changed.

Unless a track is unfolded, you will not be able to see the highlighted selection of a clip. It must be unfolded to reveal the highlight. To unfold a track, click the Unfold Track button, a triangular arrow next to the track name by the mixer section. When unfolded, you can make specific selections within a clip's boundaries. Click + drag within the clip waveform/MIDI display area, (below the title bar) to select a specific timespan or event. You may also adjust the height of tracks to unfold or resize to your liking by dragging the dividing line between track names and unfold buttons. Making an insertion or selection is governed by the Snap to Grid settings. This means that the insertion marker will align to the nearest visible grid line. Feel free to adjust the snap settings and width of the grid from the Options Menu or by Ctrl + click or right click in the Arrangement View.

5.4.4 Editing

When it comes to arranging, you will find yourself moving, resizing, splitting, copying, pasting, and merging clips quite often. These are just of few of the features incorporated as standard editing functions necessary to edit your arrangements!

5.4.4.1 Move Clips

To move clips to different tracks or to a different location in an arrangement, simply click + drag from the clip's colored title bar to the desired location. This is no different than any other DAW software.

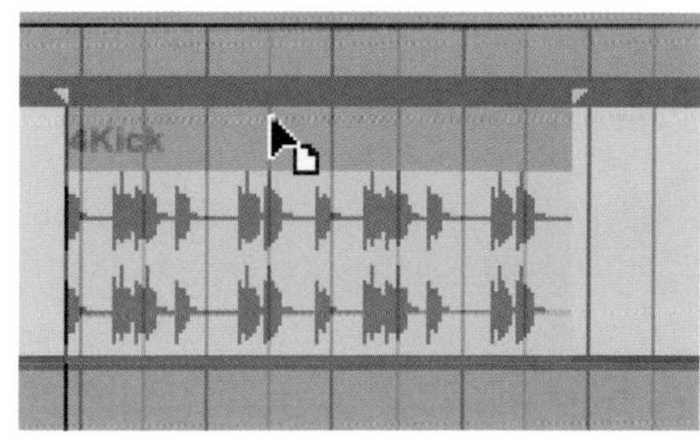

Figure 5.14 Moving clips around the Arrangement View track display.

5.4.4.2 Resizing/Trimming

Place your mouse pointer over the left or right edge of a clip. Once the mouse pointer changes to a bracket, click + drag the edge to your desired location.

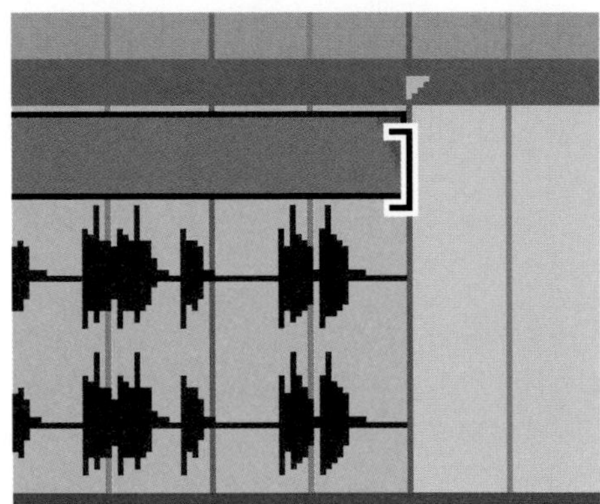

Figure 5.15 Resizing/Trimming a clip.

5.4.4.3 Splitting

To split (separate or divide) a clip into multiple clips, place the Insert Marker at the point where you want to split and select Split from the Edit Menu ([⌘+E] Mac/[Ctrl+E] PC). This will split the clip in two. You can also split a clip by making a selection; therefore, isolating a portion of the clip. Unfold the track, and then make your selection of the waveform or MIDI display. After you split the clip, it will turn into three pieces (three clips). Remember, insertion is based on the snap settings. Alternatively, Live will automatically split clips for you when you drag a clip selection to another location. This speeds up your workflow for certain operations such as splitting out and copying clip selections of isolated percussive hits or drum beats. Once selected, you can drag the isolated beat directly into a sample-based instrument such as Impulse ▶ 15.3 or a Drum Rack ▶Scene 16. Launch to these sections for more.

Figure 5.16 Split a clip from an insertion location.

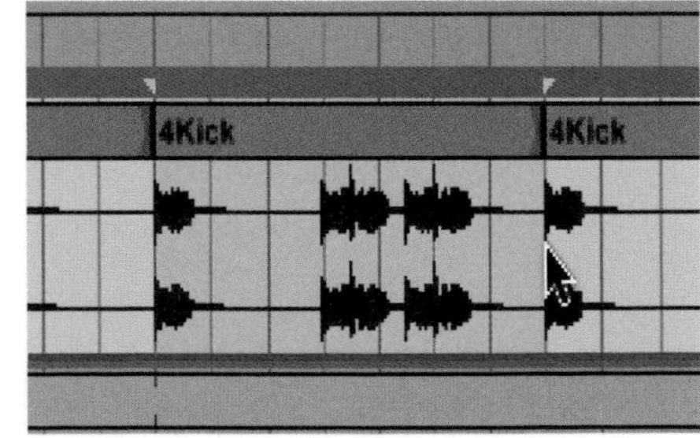

Figure 5.17 Split a clip by making a selection. The result is three clips.

5.4.4.4 Consolidating

In Live, Consolidating means to join multiple clips together as one. Other DAWs use different names for this function.

To consolidate clips, select your desired clips or timespan across clips and select Consolidate from the Edit Menu ([⌘+J] Mac/[Ctrl+J] PC). This will join the clips together creating a new audio or MIDI clip. If you want to include silence before or after the new clip, simply extend your timespan selection to include silence preclip or postclip.

5.4.4.5 Cut, Copy, Paste

These commands follow the standard word processing functions and use the same keyboard shortcuts.

To cut a clip in the Arrangement, make your selection and select Cut from the Edit Menu ([⌘+X] Mac/[Ctrl+X] PC).

To copy, make your selection and select Copy from the Edit Menu ([⌘+C] Mac/[Ctrl+C] PC).

To paste, place the Insertion Marker or make a selection in your desired location and select Paste from the Edit Menu ([⌘+V] Mac/[Ctrl+V] PC). Keep in mind that you need to unfold a track when choosing a discrete selection smaller than the whole entire Arrangement Clip.

5.4.4.6 Duplicate

The Duplicate function creates an exact copy of a clips selection and places it immediately after the selection's endpoint so that they are butted up against each other. In this way, a selection can be repeated along the timeline as a discrete copy. You can select a portion of a clip(s), entire clip(s), or a selection that extends before or after a clip(s) edge.

To duplicate, make your selection and select Duplicate from the Edit Menu ([⌘+D] Mac/[Ctrl+D] PC).

5.4.4.7 Cut/Copy/Paste/Duplicate Time Commands

While navigating through the Edit Menu, you may have noticed Cut/Copy/Paste/Duplicate Time Commands. These commands affect the overall time of the Arrangement, each adding or removing a timespan from it. This means that by adding or removing time, all tracks and clips will be shifted in time or removed from the Arrangement to accommodate the command; therefore, shortening or lengthening the overall duration of the song. To execute one of these Time Commands, all you must do is make a selection on one track and then initiate the command. The rest of the tracks will be displaced automatically.

For example, when duplicating time, make your timespan selection on any track you wish. It doesn't matter which because you're duplicating a timespan not just a single clip. After you've made your selection, select "Duplicate Time" from the Edit Menu. Now your selection will have been duplicated immediately after

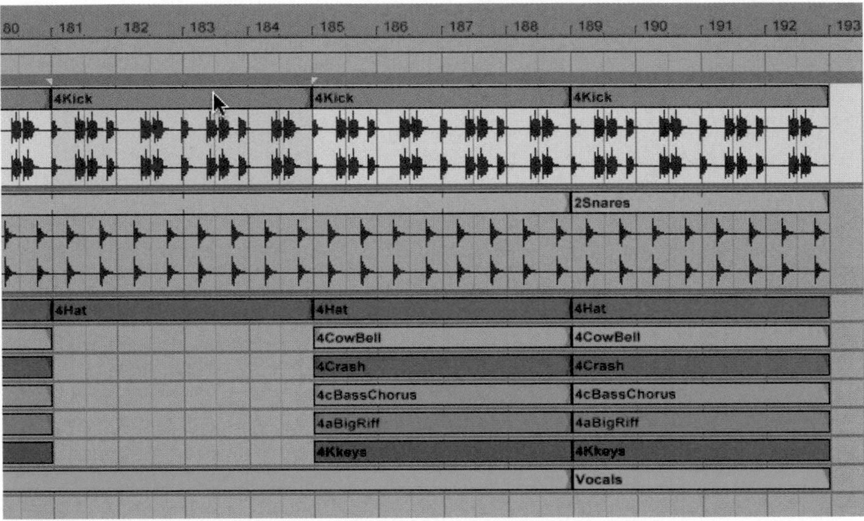

Figure 5.18 To Duplicate Time, make a selection in the track display and then select the command from the Edit Menu.

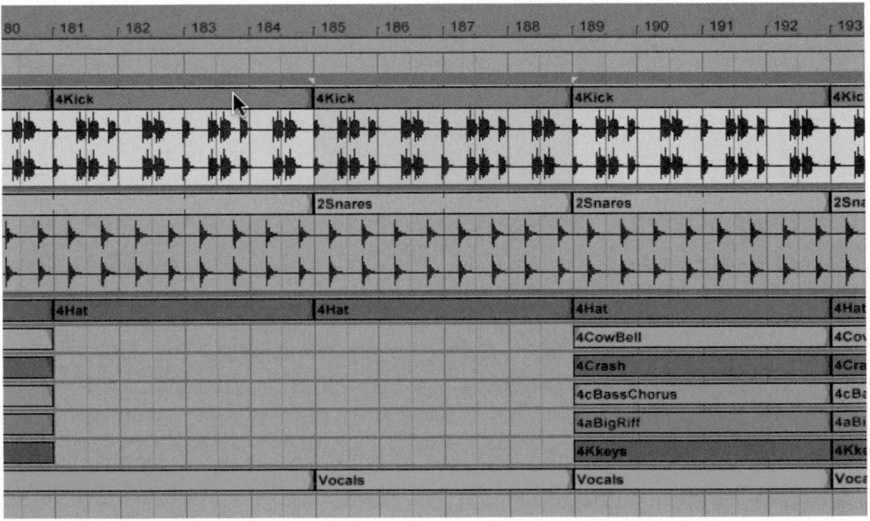

Figure 5.19 Notice that the duplicate selection shuffles all other track clips to the right to make room for the new track clips. This adds "time" to the length of the arrangement relative to the selection duplicated.

the original selection, pushing all other tracks' clips to the right (later along the timeline). If you had used the traditional duplicate command, the duplicate would only be of the selected clip(s) and overwriting any clip(s) in the way while maintaining the current length of the song or arrangement. Take a look at how Duplicate Time works!

Paste Time works the same, except you have the freedom to paste at any location along the timeline. Make sure that you select the same first track – top to bottom – as you did when you copied the original selection. This means that the paste destination should be selected on the same physical track you wish to paste your track clips to avoid pasting clips to the wrong tracks, as you will see in our example.

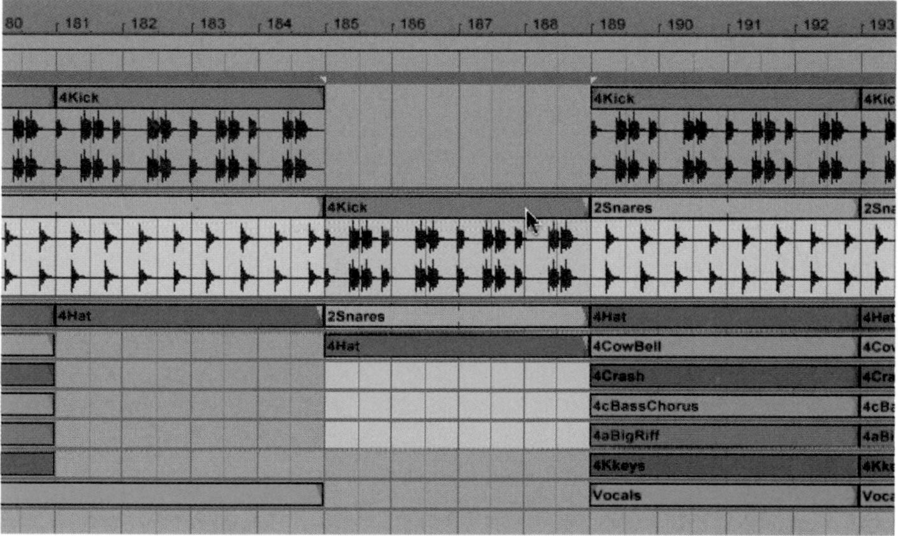

Figure 5.20 A different paste track destination was inadvertently selected; therefore, clips have been pasted to the wrong tracks, offset by one track level.

Obviously, there are times when you may wish to paste clips to different tracks in this manner. Just make sure you do it on purpose, not inadvertently.

5.4.5 Fades and Crossfades

Fades and Crossfades are an essential element of editing audio in any DAW. Fades are used at the edges of audio clips, and Crossfades are used to join two adjacent audio clips. Their main purpose is to eliminate pops and clicks created by edits or cuts made in an audio waveform where the cuts don't exactly cut at a zero crossing point (where no signal is present) or where amplitude levels are significantly different between clips. Controlling volume eliminates these

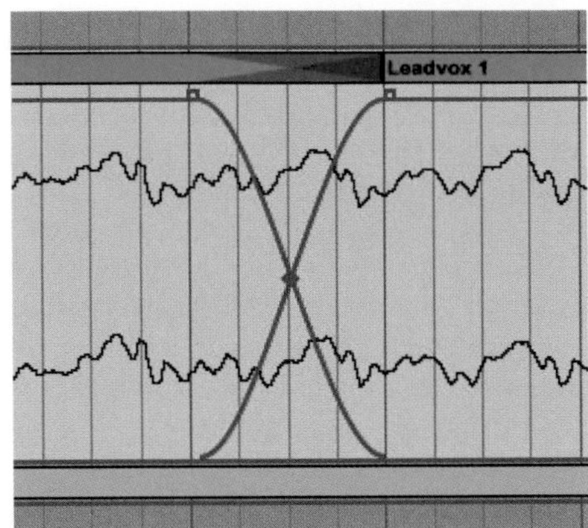

Figure 5.21 Unfolded track showing a Crossfade in the Arrangement View track display.

imperfections hence "volume fades." In this way, silence or blending is used to hide pops and clicks or blend different amplitudes (volume) or spectral content (sound characteristics). It is generally a good idea to use Fades and Crossfades when editing audio content, even if it sounds perfect. For this reason, Live 8 gives you the choice to create fades by default. There are multiple types of fades and crossfade shapes (curves) that determine how quickly a signal is faded in or out. As you become more experienced with using Fades and Crossfades, you will learn that different shapes of fades work better than others for every situation. For now, just worry about what sounds good and gets the job done. You can worry about all the technical audio engineering details on another day.

Fades and Crossfades are viewable from a clip's title bar and from its waveform display. To view them in a track's waveform display, "Fades" must be selected from the Control Chooser menu below the track's name.

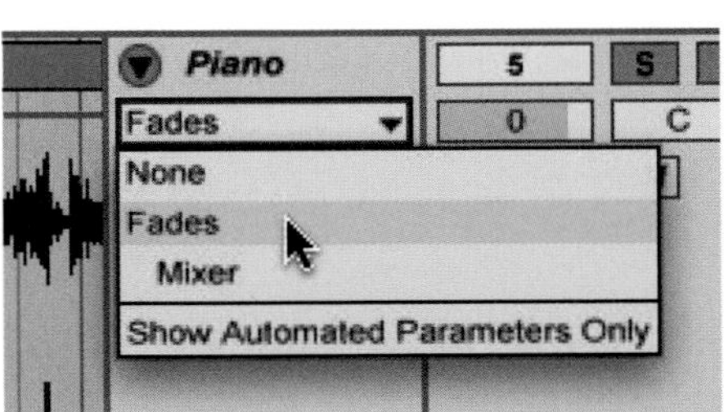

Figure 5.22 To view fades in a track's waveform display, select "Fades" from the track control chooser.

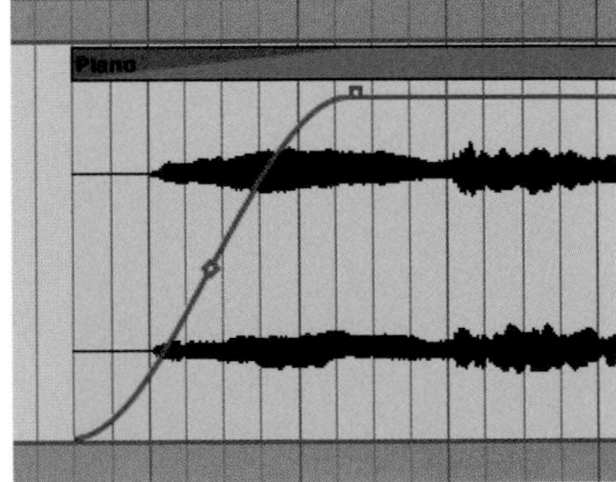

Figure 5.23 Fade in.

The title bar shows a wedge shape indicating that a Fade or Crossfade is present. The waveform display shows an accurate display of the fade location and shape. The waveform itself reflects the volume fade in the display by redrawing itself to match the curve. Obviously, the track must be unfolded to view this, but you must also select Fades from the Fades/Device chooser below the track name. Fades are manipulated from the Fade In/Out Handle (length) and the Fade Curve Handle (slope shape). Click + drag these to the position you desire to change how quick or intense a Fade In/Out happens and how much of the audio signal is affected by its adjacent files, when they are blended together.

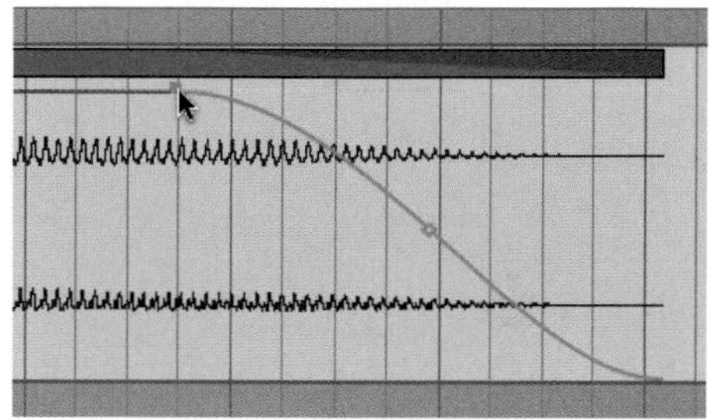

Figure 5.24 Fade out handle.

From the clip contextual menu (Ctrl + click or right click), you can select Show Fades, Reset Fades, and Create Fades/Crossfade. Show Fades allows you to skip the chooser assignment step. Reset Fades reverts a Fade or Crossfade back to its default setting. Create Fades allows you to make a selection spanning from a clip's start or ending edge, or across two clips that are to be crossfaded. After you make a selection, the Create Fade command will do just that, create a Fade or Crossfade. By default – unless set differently in the Record/Warp/Launch Preferences – fades on clips edges and adjacent clips are automatically created with a 4-millisecond fade length. If the Create Fade On Clip Edges option is "Off" in preferences tab, then you can delete fades by selecting the fade handle and pressing the delete key on the computer keyboard. If it is "On," then delete will only reset the Fade to its original position. Keep in mind that Fades and Crossfades are independent of track volumes. Volume fades are clip based.

5.5 Working with automation

Part of taking control of your Live Set includes automating various controls and parameters such as volume, pan, sends, device parameters, and even song tempo. This is really hip! The purpose of automation is to design hands free control over mixer, devices, and other global controls for mixing and programming musical effects. Keep in mind that Arrangement View automation is a linear concept. It is fixed against the timeline; thus only viewable and editable in the Arrangement View. The exception to this is the automation of clip envelopes ▶ 7.8, being that clip properties are independent of the Arrangement View. Automation control is different than Remote control in that automation is the process of moving and changing control parameters over time automatically without a physical human command. These movements are pre-created manually or recorded into your song as a semipermanent element. The same consistent movements are made every time you play the song. That being said, automation is often created via remote control. This gives you the power to not only remote control a device in a real-time performance but also use remote controls to record automation. Think of it like this: automation is executed automatically by Live, based on what you have programmed it to do. Remote Control is the process of physically controlling parameters with something other than the mouse.

5.5.1 Recording automation

In Live, automating a control parameter is as easy as launching Global Record. Click the Global Record button, press play, and start moving faders and knobs with the mouse or a MIDI controller. Yes, it's that easy. Live literally records all your actions while in Global Record mode. This makes automating volumes, sends, track activators, device parameters, and other controls quick and easy. You can automate any parameters by changing or moving them with the mouse, MIDI controller, or even computer keystrokes – anything that can behave as a controller/remote control device.

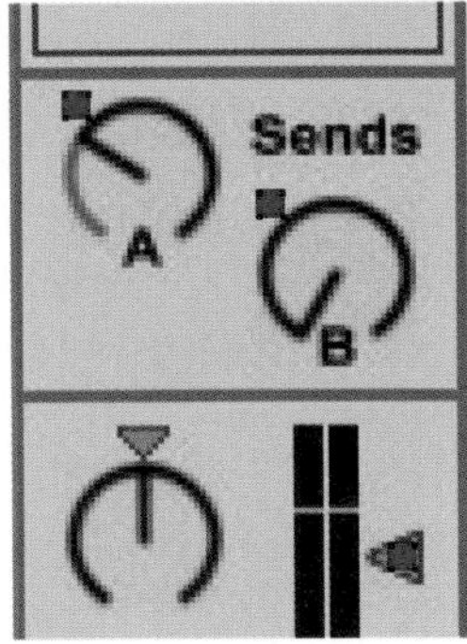

Figure 5.25 A red indicator means that a control or parameter has been automated.

Once a control has been automated, a red indicator will appear next to it to informing you that the control or parameter has been automated. You will see these indicators in the Arrangement View, Session View, and in various other views. Although you can record track automation while viewing the Session or Arrangement, the actual envelopes are located in the Arrangement track display fixed against timeline at the time and place in which they were created. The Session will display automation indicators, but aside from that you will rely solely on the Arrangement for viewing and editing of track automation envelopes. That is unless you make another record pass to write new automation. To quickly delete track automation in the Session View, ctrl + click or right click on the automated control and select Delete Automation from the contextual menu. In the Arrangement View, highlight the envelope in the track display, then select Delete Envelope from the clip's contextual menu – there are other envelope commands here too! Alternatively, you can double-click on any envelope breakpoint to gradually remove automation envelopes.

5.5.2 Automation Lanes

In the Arrangement View, breakpoint envelopes represent the automation of a control or parameter. They indicate the state or value at which the control or device parameter is set at any given time. By default, envelopes will appear as a red horizontal line that extends the length of its track. This is called Automation Lane. When a control has been automated, the envelope will reflect the value changes. Use the Control Chooser Menu to view track envelopes. This menu is located below the track name. Set the Device Chooser first to see the automated device, and then choose the parameter you wish to view. As a shortcut, click on any automated control parameter to bring its envelope into view. If you wish to view multiple automation envelopes at a time or an envelope in a separate lane, click the Add Automation Lane button below the control chooser to the right ⊕. This will add an additional lane containing the current parameter's envelope. Add as many lanes as you need, with each lane set to a specific envelope. Fold lanes to hide them from view using the fold button below the control chooser. To remove a lane from view, click the Remove Automation Lane button, the same button you used to add a Lane ⊖.

5.5.3 Drawing

Recording automation in real-time can bring a more humanistic feel to your music, but the process is not always as precise as you might hope. There are many music genres where precision automation is the only acceptable kind. In that case, envelopes must be as accurate as possible – perfectly aligned to the grid and timeline. In any case, at some point you will find yourself drawing custom envelopes in the track display to correct recorded automation or to create automation.

To draw envelopes, turn on Draw Mode from the control bar. Next, choose the track, control what you want to automate, and select the control parameter from the Device Chooser below. Now with your mouse (pencil tool), draw across the track display – within the envelope editor area – to create your custom envelope. The new envelope will be drawn as stair steps based on the visible grid snap settings. Choose a different grid size to alter this. As an alternative, you can hold down the [alt] key when drawing to override the grid snap. This allows you to draw finer and curvilinear envelopes.

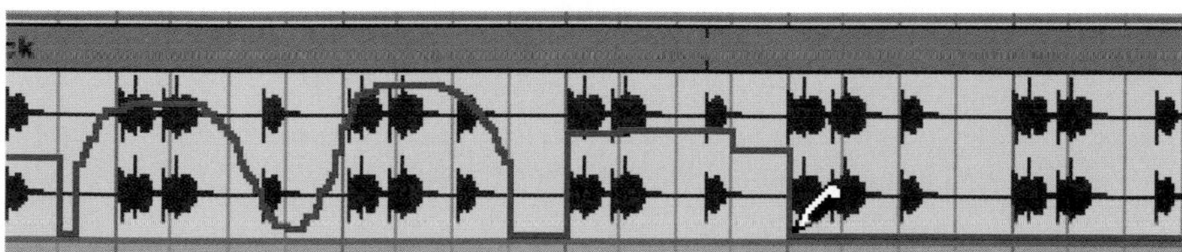

Figure 5.26 Draw custom envelope automation into an Arrangement Clip's waveform or note display area.

5.5.4 Editing

Although Draw Mode can provide high-resolution automation (curves made from very small line segments), it is not always the most efficient or fastest way to work with automation. There are times when you want gradual and smooth diagonal envelopes or fewer line segments, etc. For this reason, many of us choose to create and edit envelopes by working directly with Breakpoints rather than drawing and redrawing them. Breakpoints are the little dots or anchor points along an envelope, fixing it to a value. This is probably the most common way to edit envelopes once they have been recorded or drawn. This will ultimately depend on how you like to work. Take note that breakpoint editing does not snap to the grid.

If you place your mouse pointer directly over an envelope line, you can click + drag it upward and downward to change values. This will affect the entire track unless there are additional breakpoints forming a line segment. To create and automate a segment of an envelope, make a highlighted selection inside the track display with your mouse and then click + drag the envelope up or down in the display. To add breakpoints, double-click anywhere along the envelope line. Double-click on a breakpoint to delete it. If you want to move all or a portion of your automation while maintaining the relative breakpoint relationships, highlight the desired area in the track display and then click + drag on a breakpoint, moving it wherever you want.

Use breakpoints to create automation in segments along a track. They can be dragged in any direction you desire: up or down, forward or backward. Do this to create interesting types of automation shapes such as long volume fades, swells, ramps, shaping, and effectual volume-based stutters, among other things.

5.5.5 Commands

Envelopes can be manipulated in many ways – locked, cut, copied, pasted, duplicated, and deleted. To access these commands, you highlight a selection of the envelope where the specific automation has been created, then ctrl + click or right click. This will bring up a contextual menu where all these commands are located. Each command has an associated keyboard shortcut. Take a moment to familiarize yourself with them to speed up your workflow. Of course, the traditional copy, cut, and paste commands are accessible as a keyboard shortcut or from the Edit Menu, but it should be pointed out that they function differently than the contextual envelope commands. Envelopes cannot be individually copied or duplicated with the traditional copy, cut, and paste commands. Instead, you should use the envelope commands to cut, copy, paste, etc. for isolating envelopes. To lock envelopes, choose the Lock button located to the right of the scrub just above the track names or from the envelope contextual menu. When lock is "On," envelopes are locked to the timeline and remain unaffected by clip editing. When unlocked, envelopes are bound to their clips, following them when they are moved.

5.6 Arranging concepts

By now, it should be pretty obvious that the Arrangement View is very useful for working in a linear fashion. Everything you need to arrange, rearrange, edit, and master is right there in view and fully functional. You can even create your entire arrangement or additional music without limitations right in the Arrangement View. No matter how you like to work, Live can function just like any other DAW in this context – and we haven't even talked about all the effects and mixing capabilities! We'll save that for another Scene, but for now let's look at a few of the create, produce, and perform aspects of the Arrangement View. Keep in mind that any one of these concepts may share similarities with or fall into multiple categories.

5.6.1 Create

As you can imagine, creating in the Arrangement View is fairly straightforward. You can record or import audio and MIDI directly into the tracks just as you would with any other DAW. From a technical standpoint, building arrangements could not be much easier – it's the creative muse or inspiration that is so elusive. Set up your desired track types, microphone, or load up instruments, and off you go. Once your song elements have been laid into place, you can begin to edit and tweak each clip as you see fit. This includes using effects and other manipulations to creatively affect your audio and MIDI events. Live allows you to work, build, and arrange multitrack arrangements (sequenced Sets) full of MIDI and audio without limitation. So, if you prefer to work in the Arrangement View, there is nothing to stop you. As a Film or TV composer, you will have to view your video clips in the linear Arrangement View, which is well suited for scoring music to picture. Launch to the ▶Scene 18 for information on using video with Live 8 and Suite 8. The Arrangement View is a great way to transition from your old DAW to Live, without the burden of learning an entirely new way of composing. Yes it's true that the Arrangement View serves a greater purpose by design, but it will also make you feel right at home while you gradually discover the power of the Session View. For now, there is no reason to "stop the press." Feel confident that you can continue working the way you always have during this period of transition. Just keep in mind that it is never wise to make a major change in your studio – software or hardware – in the middle of an important project. Or at least when it pays the bills. When you've completed the transition to Live, the Arrangement will be a place more for working on edits and mixes rather than full on creation. But that's not to say you won't create now and again.

5.6.2 Produce

Remixing a song, programming beats, or simply overdubbing a recording is all part of producing in the Arrangement View. As mentioned, you can move clips around, duplicate them, and create a variety of loops – grooves, bass lines, rhythmic beds, and hooks. With all the Live 8 and Suite 8's powerful features and devices, you can remix songs just like any other DAW. This is important to understand. Live is a linear production tool. That's the beauty of the Arrangement View. Copy, paste, cut, or lengthen large song sections – verses, hooks, themes, motives, etc. – in unique ways. Automate effects and mixer parameters then copy and paste those too. Finally, master your entire production in the Arrangement View then bounce a final mix. It does it all!

5.6.3 Perform

An arrangement is a ready-made piece of music, a sequenced song that you can take with you onstage. This gives you the option to sing or play a guitar along with the arrangement. Think of Live as your band, or if you're a DJ, it's everything – your tracks, effects, mixer, and so on. Obviously, in the same way, you can perform with one of Live's virtual instruments simultaneously with your arrangement. Going a step further, you could also ReWire Live with another DAW. Use them together to perform with your arrangement in real time or as another sequencer element. No matter how you conceive the relationship, ReWire can open up a whole world of hybrid performance situations. For more on ReWire, launch to ▷ Website.

Those of you who are familiar with Live might be thinking that performing with an arrangement is more limiting than performing from the Session View. That really depends on how you go about performing. Not everyone is a DJ or improviser and Ableton understands that, but there is a performer inside of all us! That is why the Arrangement allows for composers and producers to not only use Live's Arrangement View as a multitrack sequencer and for tracking audio, but also makes it possible for performing. Like we have said, Live is a DAW, meaning that it does it all. No matter who you are, the Arrangement View is a valuable performance tool and its linear work environment will attract DJs, performers, and composers.

In this chapter

6.1 Real-time "launching base" 109
6.2 Layout 110
6.3 Clips 115
6.4 Tracks versus Scenes 120
6.5 Track Status Display 124
6.6 Working in Session View 124
6.7 Sessions into Arrangements 131
6.8 Musical concepts 133

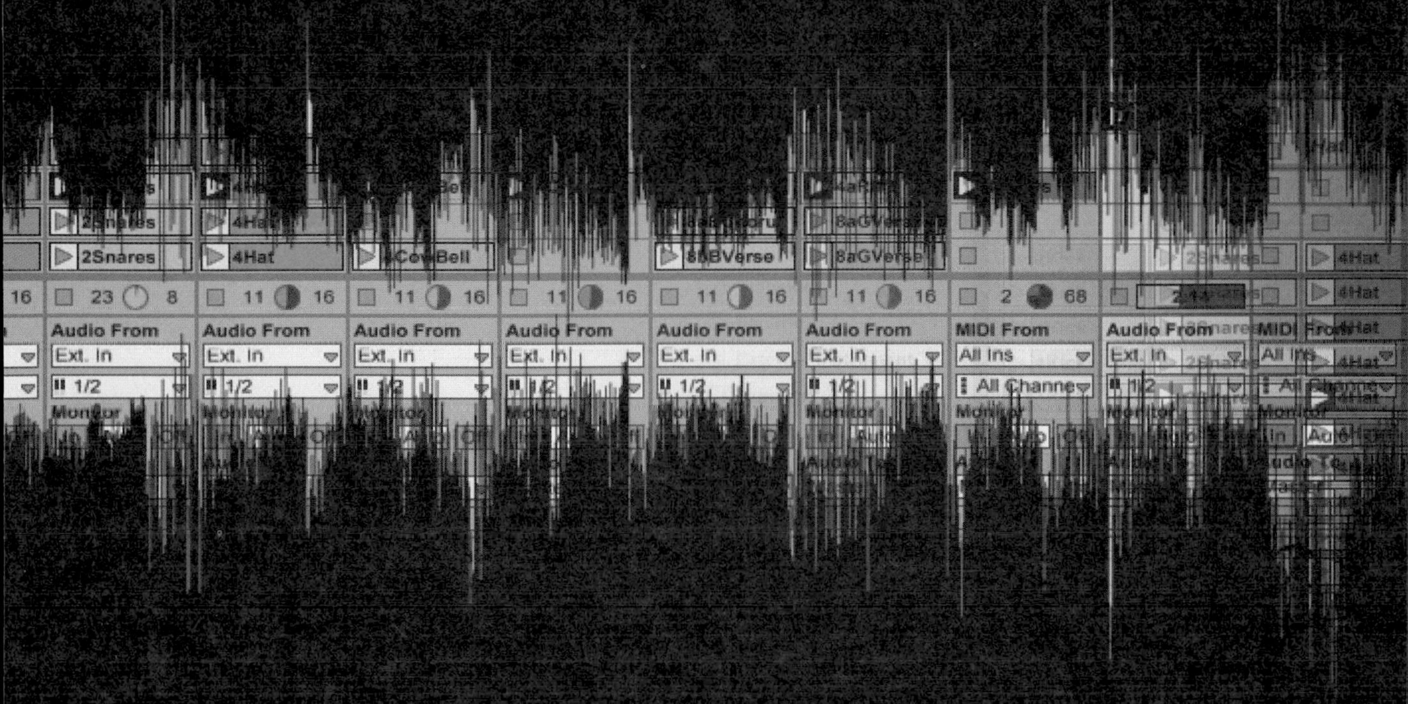

Scene 6

Session View Concepts

Ableton Live allows me to re-create a very intense and personal studio experience on stage. Without Live I couldn't play my music, there simply would be no performance.

Roger O'Donnell, Musician

6.1 Real-time "launching base"

The Session View is both a musical sketchpad and a traditional track mixer, which run effortlessly in tandem. More than that, it's a real-time launching base for all of your musical ideas, motives, beats, riffs, and much, much more. The Session View allows you to be free of a fixed timeline, opening up the possibility of experimentation, development, and improvisation without having to ever stop the creative workflow or Live's playback. This is where you can sketch out your ideas before you know how they should evolve, how long they should remain, and where they should go in your arrangement or production. Create an instrument, build a musical phrase with multiple clips, add layers of tracks, and then launch even more clips all while reacting to the musical moment. Let sounds and grooves inspire how you create. Live is an instrument, whether in the studio or on stage.

Here's an analogy. In many ways, creating in the Session View is like cooking, but with the ability to experiment with flavors and ingredients without ever over-seasoning or burning your food. In Live, you mix it up, add and subtract, and let it simmer for as long as you want. Take your time while clips are playing to choose your next musical idea. The Session View is what sets Live apart from all the other digital audio workstations (DAWs) out there. As we take a look under the hood you will see what makes the Session View so unique and how you can take your music to new levels by understanding its power and real-time flexibility. The Session View offers unlimited creative possibilities when it comes to launching ideas with your own music productions and personal writing flow. As you become more and more comfortable with it, you'll begin to see its infinite creative potential. All you have to do is unleash it. For now, let's stay grounded and stick with some of the more fundamental, yet popular ways to utilize the Session View's workflow and layout.

6.2 Layout

When you first launch Live 8, it will open to the Session View. For many of you this will be like a case of déjà vu. The view will feel oddly familiar, but not quite what you are used to. That is because you are looking at what could be considered a vertical track mixer, but with a ton of unfamiliar Clip Slots stacked above the traditional channel strips. What are these rectangular slots? They're where my inserts usually go! Don't worry, we'll get to all that. In the meantime, looking at the overall layout, the Session View is made up of tracks: Audio, MIDI, Returns, Master, and a Clip/Device Drop Area. Included within tracks are, Clip Slots, Mixer, Mixer Drop Area and Scenes for the Master track. That's a lot of new stuff, so let's break down each element, starting with the basics.

Figure 6.1 Live 8 Session View.

6.2.1 Tracks

The basic default Set for Live will open up with two vertical track columns just to the right of the Live Browser, which are an audio track and MIDI track. There are five track types in Live: Audio, MIDI, Return (aux), Group, and Master. The Audio and MIDI tracks will always be viewed toward the left-hand side of the Session, while the Returns and Master to the general right side with the Master track always the track furthest to the right, labeled Master.

All of these track types function as you would expect. Audio tracks handle audio files and audio recording and are used to playback audio clips. MIDI tracks handle MIDI files and information (note data) and are used to play, trigger, and perform MIDI instruments (Virtual/Software and External MIDI instruments) and MIDI clips. A Return track is essentially an "Aux" in which an audio signal is received from audio and MIDI track sends. Live can handle up to 12 Return tracks. They are best used for hosting "send" effects (Audio Effect Devices). A Send's audio effect could be a reverb device or delay for example. In other DAWs aux tracks are also used for submixing. A submix is the combination of multiple audio signals split off from their specific track's main channel, sent via their Send knobs, and received into a single aux track. The submix is then the result of these various audio signals mixed together at the aux's output. In this way, you can use the aux track to control the output of multiple tracks using one fader. This is common for handling multiple backing vocals or a multichannel drum kit, which can be done with Return tracks. To learn about submixing with Return tracks, launch to Clip ▶ 6.8.1. The Master track serves as the default track where the overall outputs of all tracks are summed. It's your main output and generally from where you listen back to Live's output. Group tracks are a special track type for managing tracks that are grouped together.

Tracks are physically divided into two halves from top to bottom. The top half contains the Track Name (Track Title Bar) followed by Clip Slots. This is where you can drag and drop audio samples and MIDI files to build your musical ideas in the Session View.

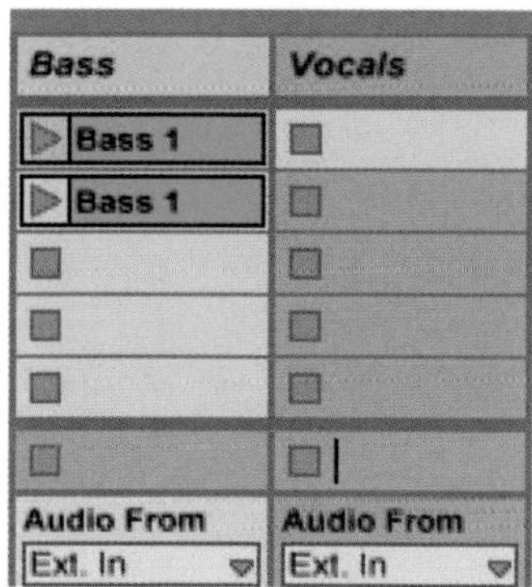

Figure 6.2 Track Name and Clip Slots.

The bottom half of a track column contains the Track Status Display, In/Out Section (Inputs/Outputs), Sends, the Mixer Section, and the optional Crossfader Section. Each component can be hidden from view to maximize screen space as described in the next section. To expand a track's Peak Level and RMS meter, click + drag the dividing line below the Sends in an upward direction.

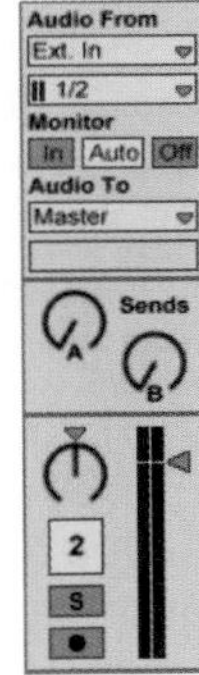

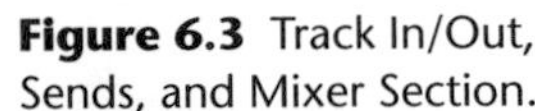

Figure 6.3 Track In/Out, Sends, and Mixer Section.

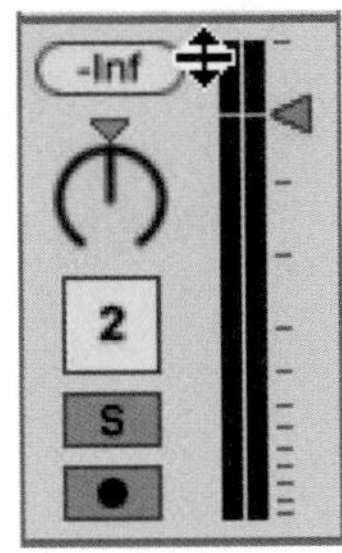

Figure 6.4 Expanded Track Signal Level Output Meter.

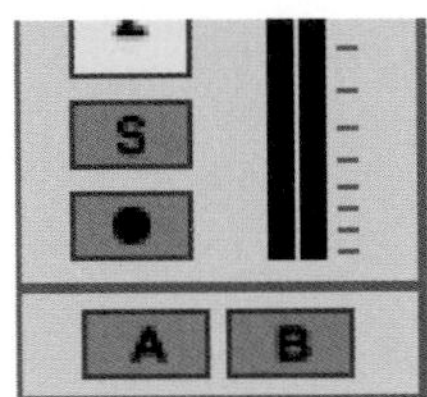

Figure 6.5 Track Crossfader Section.

In addition to expanding the Track Meter section, it can also be resized horizontally to increase or decrease its width. As an added bonus, feel free to customize track colors. Right click or ctrl + click on a track Title Bar and choose any color you like.

6.2.2 Session View Mixer

Hot Tip
The Preview/Cue Volume knob at the bottom of the Master Track acts as a volume control for both the Browser Preview and the Metronome volume.
Solo
75bpm_Swing.wav
Raw
Drag the knob up or down to raise and lower volume.

The Session View Mixer does exactly what it says; it mixes the Audio and MIDI signals of all of your tracks using traditional vertical faders. The mix you create here is the same mix that will be heard in the Arrangement View, meaning that all mixer parameters (faders, pan, sends, etc.) are linked between views. Remember, they share the same signal path; it is the actual clips (contents of the tracks) that are different. In other words, the Session View and Arrangement View share the same audio and MIDI input and output signals but are completely different in clip and sequencing functionality. Note that the Session View Mixer is not viewable in the Arrangement View!

The general layout of the Session View Mixer consists of multiple components that can be shown or hidden from view. Customize your setup in the View Menu or from the yellow buttons to the bottom right of the Master track.

The Mixer Section consists of: Pan, Volume fader, Track Activator button (deactivated = off/bypass), Solo/Cue button, and the Arm Session Recording Switch (also referred to as Arm button, record-enable, or Track Record button). Return tracks do not have Arm buttons. The Master track does not have a Track Activator button. Instead it has a unique Solo/Cue button and Preview/Cue Volume knob. Let's take one moment to explain the "Cue." Just above this knob is the Solo/Cue button. When this is set to solo tracks solo buttons behave as expected. When set to "Cue" soloing a tracks reroutes its track output temporarily to the Cue Out. This is only available when Cue Out is set to a different audio output then the Master Out. This is useful for "cueing" up tracks privately in headphones DJ style.

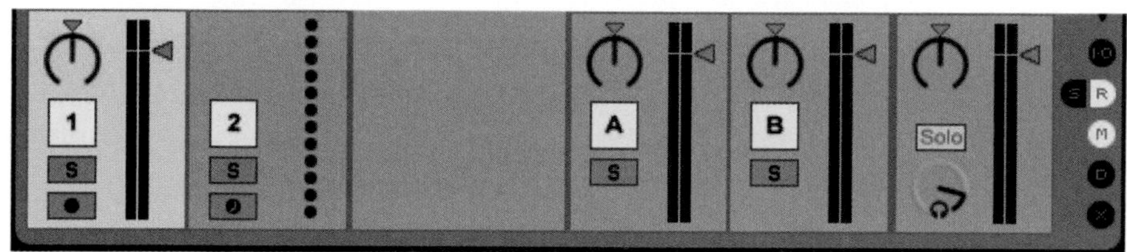

Figure 6.6 Session View Mixer.

The Mixer can be resized to your custom needs. As you saw in our description of tracks, dragging the dividing line at the top of the Mixer expands the height of the Mixer Section and subsequently the Peak Level Meters. This adds tick level marks, a Track Volume numeric scalar field, and Peak Level indicator buttons. Increasing a track's width in this state will add a decibel scale alongside the meter's tick marks.

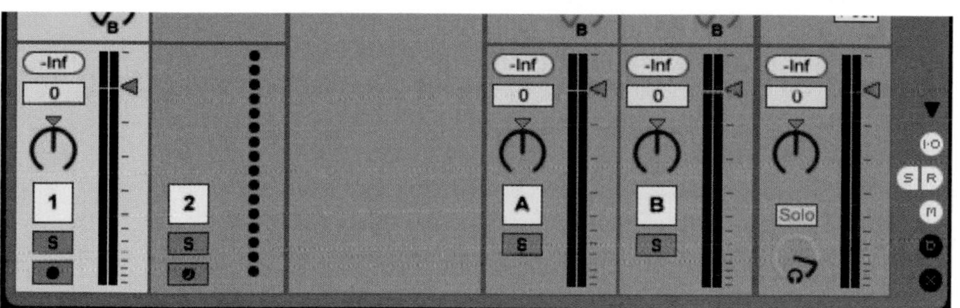

Figure 6.7 Expanded SessionView Mixer.

The Sends Section consists of the traditional bus send knobs labeled A, B, C,... and so on. The Master track does not have send knobs, rather it has Pre/Post Toggle (pre-fader/post-fader) for setting the way in which Return tracks tap into each track audio signal.

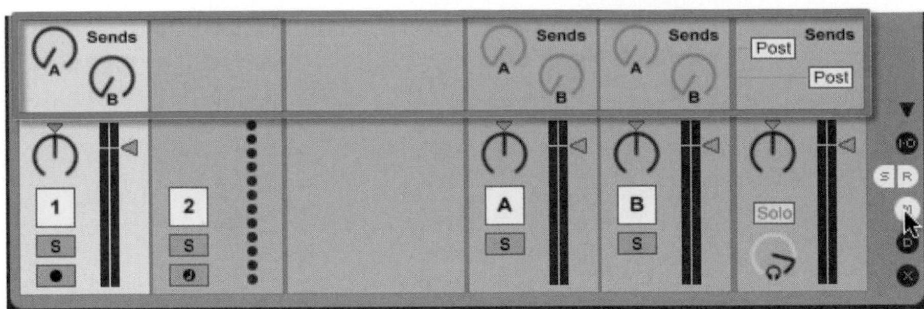

Figure 6.8 Session Sends Section.

The In/Out Section consists of the following: Input Type, Input Channel, Monitor Section, Output Type, and Output Channel. Returns and Master track only have output options.

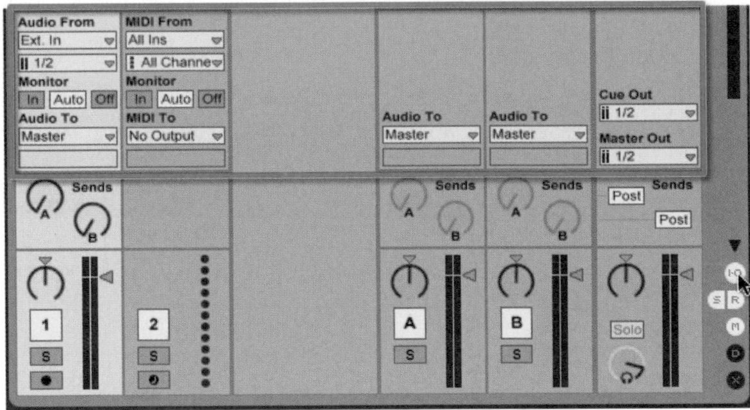

Figure 6.9 Session In/Out Section.

Notice that Cue Out is available when the In/Out Section is in View. Remember to set this to a secondary output (if supported by your audio interface), i.e. 3/4, to privately cue track outputs that are in cue mode.

6.2.3 Drop Areas

This Clip/Device Drop Area is defined as the empty space between the audio/MIDI tracks and the Master track above the Mixer Section. You can create entirely new MIDI or Audio tracks by dragging MIDI/audio files/samples, MIDI/audio effects, and MIDI/virtual instruments. Whichever you drop, Live will automatically create the appropriate track type.

The Mixer Drop Area is the empty space to the right of the Mixer section below the Clip/Device Drop Area. Here you can create new tracks by dropping MIDI/audio effects, and instruments.

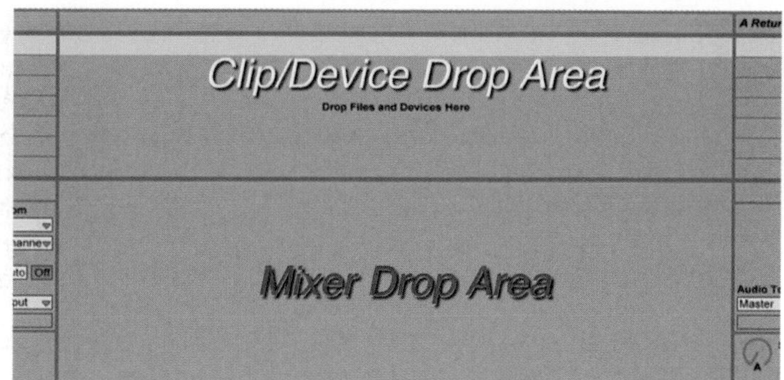

Figure 6.10 Clip/Device Drop Area, Mixer Drop Area.

6.3 Clips

Session clips are the basic music building blocks of Live. They store your audio and MIDI-based musical materials like a container. These clips can be expanded, recorded, and arranged into a basic musical idea or an entire song. Each clip can be customized and manipulated with its own unique set of parameters and information. They are used to build and grow larger musical structures eventually becoming the musical elements to create songs, scores, sound design and effects, jingles, remixes, DJ sets, live stage performances, and interactive performance installations.

6.3.1 Clip Slots/Session Grid

When you first open a new Live Set you will see many empty Clip Slots. They are rectangle-shaped boxes stacked one on top of one another in the vertical track columns. As we have mentioned already, tracks (columns) are divided in two halves, the top half being dedicated to Clip Slots as part of the Session Grid. This is where your audio and MIDI clips will live.

The Session Grid is formed by Clip Slots laid out and aligned in the vertical track *columns* and the horizontal *rows* (Scenes ▶ 6.4 ▷Scene 10). These vertical columns and horizontal rows outline the organization and structure of clips as they are laid out in the Session View's Clip Slots. When a Clip Slot is empty, there will be a small gray square in the left corner of the empty Clip Slot, unless it is record-enabled in which the square will turn into a round record button. These are called Clip Stop buttons. When a clip is loaded into a Clips Slot the clip will automatically assume a color and fill out the Clip Slot, replacing the Stop button with a triangular Launch button.

Now that you have a general idea about what you are looking at, let's open up a Set with some clips and demonstrate how to launch (play), record, and edit clips.

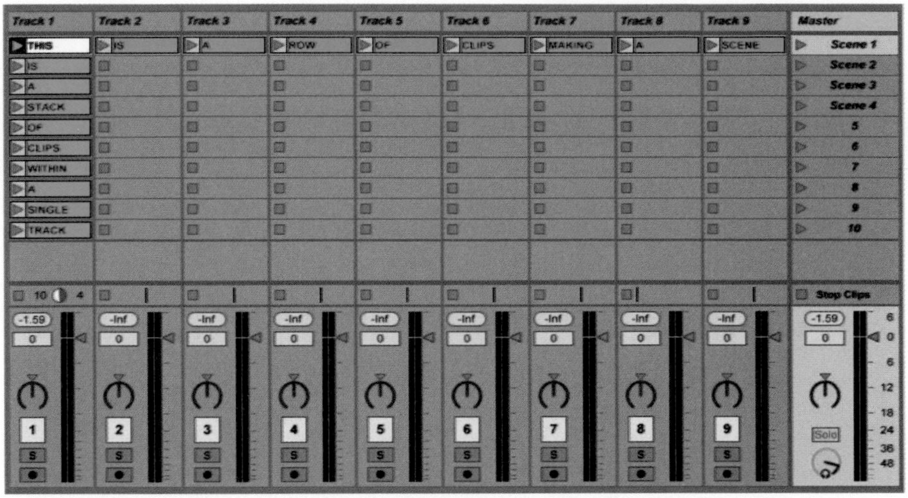

Figure 6.11 The Session Grid: Clip Slots aligned in vertical track columns and horizontal scene rows.

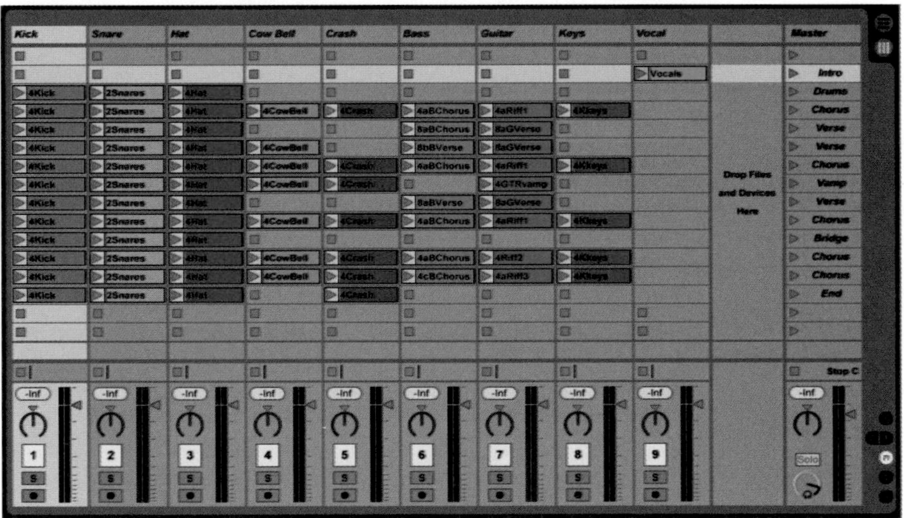

Figure 6.12 Session clips and scenes.

6.3.2 Launching clips

For this section, use the Live Set called **CPP_6-3-2_GlobalQuantize-Off** located on our Website. This way you can try out the concepts we discuss in real time as you read.

Now that we are looking at a Set with filled Clip Slots, notice the triangle to the left of each clip. This is called the Clip Launch button and is how a clip is activated for play back, hence launching clips.

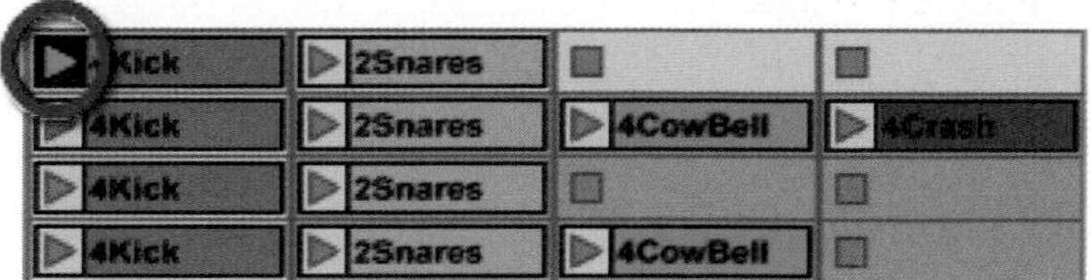

Figure 6.13 Clip Launch button activates a clip's playback.

To launch a clip, simply click on its Launch button with your mouse or select a clip with the mouse and then press the Return (Mac) or Enter (PC) key on your computer keyboard. Once you do this, you can use the spacebar to stop and start your set globally, which will also pause and/or restart your clips.

You can also navigate from Clip Slot to Clip Slot using the computer keyboard's left/right, up/down arrow keys. This is a great practice to get used to early on in your exposure to Live's Session View. A selected clip will be highlighted yellow around its border edge and the play triangle. When a clip is playing, its Launch button will be illuminated solid green. Clips can also be launched via custom computer keys and/or a MIDI controller using Live's Key Map and MIDI Map function. Launch to ▶Scene 12 for more mapping and control.

Rules of the clip: *Important safety rules to remember!*

Clips can be launched at any time in whatever order you choose, but only one clip per track can play at the same time. In other words, you can play as many clips as you want simultaneously as long as they are not on the same track. The order in which clips are laid out in the session does not dictate or determine their playback order. Only you can determine that. This goes back to the nonlinear concept behind Live's Session View. Let's review these properties one more time because this concept is very important to understand.

Each track can contain multiple clips from top to bottom, but only a single clip can be activated at once per track. If you were to have several clips playing at the same time on a single track you would have a difficult time discerning the individual audio outputs from all those clips and things would be quite chaotic. Using multiple tracks to contain and launch multiple clips together opens up a wide palette of creativity and, thus, makes the Session View a great way to express and create different musical styles and ideas.

Stop here and take some time to practice with the example Set you downloaded from our site. Here you have multiple clips with different colors and titles.

Let's start by activating/launching clip 1 "Groove 1" in Clip Slot 1, then after a moment launch "Groove 2," and so on...

Hey, wait a second! My clips are not lining up with each other and things sound off. What gives?!

Don't worry; this was deliberate on our part. This can all be correct with a click of a button. Live has provided a fantastic way to avoid timing errors when launching clips: Global Quantization! On the Control Bar you will find a dropdown chooser called the Quantization Menu. This can be set to various rhythmic values (resolutions) that determine when clips will begin playback after their Launch button is clicked or activated.

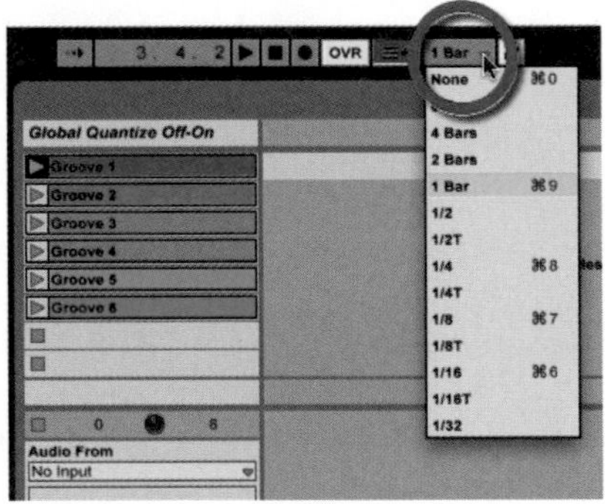

Figure 6.14 Global Quantization set to "1 Bar."

In our example you can see that "1 Bar" has been selected. This means that when your set is already playing, clip launches will commence playback at the beginning of the next bar. For example, if you launch a clip while the Session is in playback, Live will wait until the next quantize value (bar in this case) to launch or relaunch the clip no matter when you launched it. With Global Quantize, Live will force the clip to launch at the appropriate time based on the resolution of the quantization setting, whether it be on the 8-, 16-, or 30-s note. Quantization settings always round to the nearest rhythmic value based on the setting you choose. While a clip is waiting for the correct launch point, its launch button will flash green.

To stop a running clip, click either a Clip Stop button in that track column (generally located in the next Clip Slot just below the currently activated clip) or in the Track Status Display located above that particular track's I/O Section. Stopping clips in this way will only affect the clip on that particular track. All other clips will continue running as will the Arrangement View playback. This is because the overall Arrangement always runs in the background in parallel with the Session so that it can potentially log (Global Record) all of your actions. For more on this, launch to 6.7 or Scene 4.

When the Arrangement is running, the Play button on the Control Bar will be solid green. Its current, or running position is indicated in the Arrangement Position Field on the left side of the Control Bar. Remember that individual Session clips are independent of the overall musical timeline and flow of the Arrangement. The Position Field is there only for reference. It's very important to know that Session clips are not bound by linear order or time. If you want to stop all running clips at once, click on the Stop All Clips button found on the Master track just above the Master track's I/O. Do not confuse the Control Bar Stop button with the Stop Clip button! Using the Control Bar or computer keyboard spacebar to stop playback stops the entire Set, and previously active clips will remain in standby. Thus, they will continue to show the green activator button inside the clip. When you press play/spacebar or launch a clip after this, all clips that were playing automatically begin playing again.

Pressing the Stop Clips button on the Master Section will deactivate ALL currently playing clips completely. Because most other DAWs use start and stop with the transport play button or a spacebar keystroke this can sometimes lead to confusion. Of course, there are many more fantastic ways to take control of clip launching using a MIDI controller or your computer keyboard. Launch to ▶Scene 12 for more details on "mapping" your clips.

6.3.3 Basic editing

Right off the bat you have probably noticed that editing clips in the Session View is quite different than in any other program you have used or are currently using. If you have no previous experience with editing, then this difference won't matter much. For the rest of you, you'll have to relearn some of the simplest functions.

For the most part, moving clips around among Clip Slots and executing the common tasks – selecting, cut, copy, paste, duplicate – are conceptually similar to most other programs. The real difference lies in the actual editing of waveforms and MIDI events, which takes on a whole new approach. First of all, a clip's content cannot physically be edited from a Clip Slot. This happens in the Clip View. Second, clip waveforms cannot actually be physically cut or separated per se, in the Clip View, rather they can be manipulated by selecting where in the waveform a clip can start and end playback. For a more traditional waveform editing you would use the Arrangement View. Detailed information on Clip View is discussed in ▶ Scene 7. For now, just be aware that Clip View will appear across the bottom of the main Live screen when clips are selected (same goes for clips in the Arrangement View). Let's stay focused on the basic editing functions of the Session View.

Use your mouse or computer keyboard arrow keys to navigate and select clips in the Session View Grid. Selections result in highlights by clicking on a clip or shift + click to select multiple clips. Selecting multiple clips allows you to perform global edits on the selected clips, such as copying, pasting, and more general changes to their properties.

Here are the some common Key Commands for editing functions. You will also see these in Live menus and supporting documentation.

Modifiers:

Command = ⌘

Control = ^(ctrl)

Shift = ⇧

Alt/Opt(ion) = ⌥

Contextual Menu = right click or ctrl + click

Commands:

Cut = [⌘+X] Mac/[Ctrl+X] PC

Copy = [⌘+C] Mac/[Ctrl+C] PC

Paste = [⌘+V] Mac/[Ctrl+V] PC

Delete = delete or backspace key

6.4 Tracks versus Scenes

Figure 6.15 Related clips on one track.

We have already determined that tracks are vertical columns in the Session View as opposed to linear horizontal track lanes as in the Arrangement View, or in any other DAW for that matter. We have also established that only one clip can play at a time from the same track. In other words, an unlimited number of clips can exist in one vertical track, but never can more than one clip in the same track play at the same time. Following this principle, clips from the same instrument or audio source that are intended to be played in an alternating fashion would generally be placed on the same track, as they never need to play simultaneously. This follows Live's clip playback rule and represents efficient organization. For example, each rhythmic variation of a drum kit track – MIDI or Audio – would be placed in a new clip on the same track in a vertical direction. Generally, each section or part of a song uses the same drum kit throughout a song, but will vary from time to time, section to section; therefore each variation will have its own clip. These clips are then launched in an alternating fashion among the other drum kit clips that contain the same source material but different beat variations. For our example, you can see that they are all from an Impulse Drum kit.

Ok, but what if you want to play multiple clips at the same time so as to layer the drum kit or any source into a more complex sounding instrument or rhythmic pattern? In that case you would spread the clips across multiple tracks so they can play simultaneously. Of course if it's MIDI you're talking about, then you will need to route each additional MIDI track's MIDI output so that they share the same MIDI instrument source and thus it is generating the sound for all of the tracks you want to play at the same time ▶ 3.4.1. With that said, how do you launch them all at the same time when you have several tracks so that they initiate playback at the same time? A very good question! This is the job of scenes, the horizontal rows in the Session View Grid (highlighted in the white horizontal line). You've already seen them in various examples, but take another look.

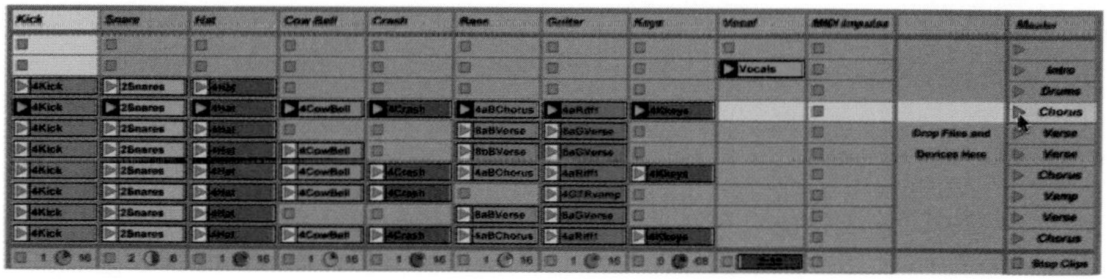

Figure 6.16 Scenes.

When aligned in rows, clips are usually intended to be played together. Scenes make launching multiple clips at the same time (in sync) possible. They are used most commonly for one of two reasons: (1) for launching several clips all at once, layered together as one or (2) for launching a row of clips that make up a song's specific part or section, such as the chorus or bridge in a song. Scenes can also be used in a few other inventive ways, but of the two we've mentioned, the latter will be the most common. We say this because once you get the hang of using scenes, you will want to expand your skills and begin using Group Tracks for launching groups of clips and leave the song sections to scenes. To learn more about Group Tracks, launch to ▶Scene 11.

As you can see, a scene and its enclosed Launch button are located on the Master track. There you will establish a method for titling and listing scenes. In our example, the scenes follow a traditional song form nomenclature, which happens to be a descending sequence from top to bottom.

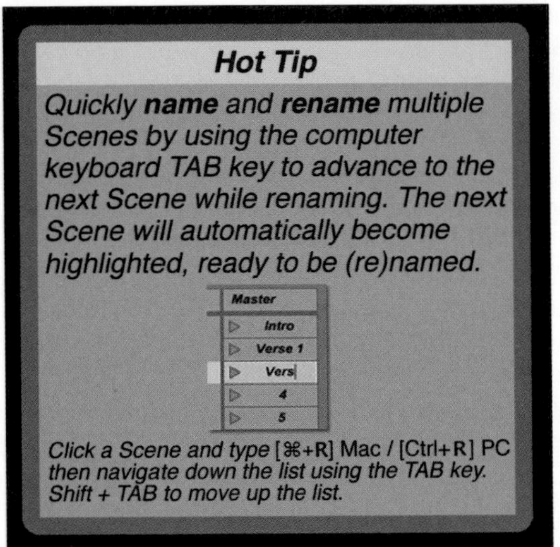

Hot Tip
Quickly **name** and **rename** multiple Scenes by using the computer keyboard TAB key to advance to the next Scene while renaming. The next Scene will automatically become highlighted, ready to be (re)named.
Master
Intro
Verse 1
Vers
4
5
Click a Scene and type [⌘+R] Mac / [Ctrl+R] PC then navigate down the list using the TAB key. Shift + TAB to move up the list.

Remember, their physical ordering does not determine the order you have to play them in. This is just a logical way of organizing a Session considering our brains usually desire some sort of continuity. To rename a scene, select it and choose the Rename command from the Edit Menu (right click or ctrl + click) on the name. The keyboard short-cut is [⌘+R] Mac/[Ctrl+R] PC. In addition, all of the standard editing functions – cut, copy, paste, etc. – are available from the scene contextual menu. Right click or ctrl+click on a scene name. You can also find all of these commands from Live's Menus, and of course the traditional keyboard short-cuts also apply. One very clever feature under this menu is Capture and Insert Scene. Launch to ▶ 10.4 to check this out.

6.4.1 Launching Scenes

Scenes are titled and launched from the Master track as seen on the right of the Session View. To launch playback of all clips in a single row, do one of the following:

1. Click on the triangular Scene Launch button to the left of the scene name.
2. Select the scene and press the Return Key.
3. Use a pre-assigned MIDI or Key Map ▷Scene 12 .

Launching scenes follows the same Global Quantization rules as launching clips as we spoke about earlier. This gives you the ability to launch scenes effortlessly and error free. Just as with individual clips, you can launch them at any time, in any order. Feel free to stop or launch any individual clips while a scene is running. If you want a specific track clip to continue playing unaffected when launching a new scene, you can remove the Clip Stop button(s) below the clip you wish to remain running unaffected as needed. This can be done from the Add/Remove Stop button located in the Edit Menu ([⌘+E] Mac/[Ctrl+E] PC). Take a look at our example. The "4aBChorus" and "4aRiff1" clips continue to play the Chorus Scene while the "4Hat" and "4CowBell" move on to the Bridge scene row upon the launching of the Bridge Scene. Notice the missing Clip Stop buttons on the Bass and Riff track. The "4Crash" and "4Keys" clips stopped upon launch of the new scene because their Clip Stop buttons were triggered with the new Scene Launch.

Figure 6.17 Clip Stop button removed.

Keep in mind that you can always use the Stop Clips button on the Master track to stop scenes and all clips while keeping the Transport playing. This allows you to activate a new scene again in sync with your tempo. All this can be executed in real time without ever stopping the flow. This is an integral part of Live's philosophy, and as you can see by now, nonstop workflow opens up a completely expressive way to work with your music.

6.4.2 Select on Launch

As you become more advanced with launching clips and scenes, you may want to start customizing the way Live responds to your actions, especially how it handles selecting clips after a clip has been launched. Such changes can be made under the Live > Preferences > Record/Warp/Launch.

Take notice of the third option under the Launch section called "Select on Launch." If this is set to "On," a clip or scene will become selected (highlighted) when it is launched, meaning that the session grid is focused on that particular scene's clip. This also causes Clip View at the bottom of the Main Live Screen to change focus (view) from the previous clip onto the newly launched clip, displaying the clip's properties and contents. When launching from scenes, the selected clip is from whichever track column had been selected. If you prefer to maintain focus on the previous clip as it was prior to launching a scene or clip, then set Select on Launch to "Off." This preference works well if you need to concentrate and focus on a particular clip's edit window or parameters while activating and playing other clips and scenes in real time.

To learn about Clip View, launch to ▷ Scene 7. Under the same preference tab, you can also customize how Live responds to Scene launches. This option is called "Select Next Scene on Launch." To learn about scene launch settings, refer to ▶ 10.2.

6.5 Track Status Display

Located at the bottom of a track's Clip Slot section you will find the Track Status Display.

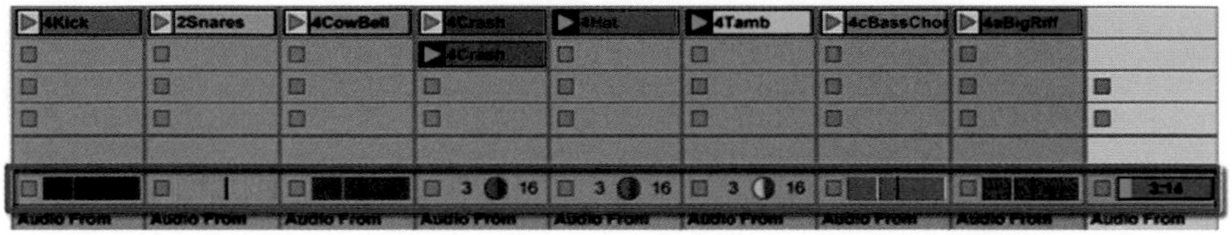

Figure 6.18 Track Status Display.

These keep track of how long a clip has been playing and how many times it has looped. If a track clip is playing back from the Session View, then you will see little circle charts (pie graphs) and numbers. When a track is playing back from the Arrangement View, while Live is being viewed in the Session View, the Track Status will display a condensed version of the Arrangement clip. In our example, the "Crash," "Hat," and "Tamb" tracks are looping and have been launched from the Session View. The last track on the right with removed Clip Stop buttons is a nonlooping or "one-shot clip" that has also been launched from the Session. The rest of the clips – "Kick," "Snares," "CowBell" – are playing back from the Arrangement View. The circle graphs indicate that the clip is looping. The number on the left indicates how many times the clip has looped – three times in this case – and the number on the right indicates how many beats make up the loop's length – 16 beats. The nonlooping track's status is represented by a progress bar (timer) showing how much time (minutes:seconds) is remaining before the clip comes to an end. When recording Session clips, the Track Status Field displays a count of how many bars and beats have passed during the recording.

Be aware that the Track Status Display is also used to display information related to input monitoring 9.1.1.

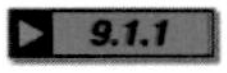

6.6 Working in Session View

Now that you have an idea what clips and scenes are and how to launch them, it's time to import and work with your own. Let's keep it simple and begin with a "New Live Set." Go ahead and open a New Set from the File Menu ([⌘+N] Mac/[Ctrl+N] PC) and make sure that it is empty (no clips). You should now be looking at the Session View and have one audio track and one MIDI track in view. Let's focus on the audio track for now.

For the following exercises, we'll use the drum audio loops found in **CPP_6-6_Working-in-SessionView** located on our Website. You can download these now or use your own drum loops, just be sure to choose four that are similar to each other in timbre and tempo (BPM).

6.6.1 Audio clips

Audio files and clips can be located literally anywhere in your computer filing system. Hopefully you have them somewhat organized! Although you can locate these files and import them into your Session using the Finder/Explorer window, using one of Live's File Browsers is the best and most efficient way. The File Browser system will keep you organized and always ready to work with clips on the fly because they are integrated into Live 8 itself. That being said, let's use the File Browser to locate and drag in our loops for the following exercises.

6.6.1.1 Example A – single track and clips

1. Click on File Browser 1 from the Live Browser on the left of your screen. From the Bookmarks menu select "Desktop" or "All Volumes/Workspace" depending on where your drum audio loops are located.
2. Click + drag **S6-DrumLoop-1** to the first Clip Slot on track 1 "1 Audio". Live will quickly analyze the file. You will see a progress bar in place of the clip during analysis. When this is finished, your Set will have automatically adopted the tempo of the audio loop. Make sure it is set to 120 BPM.
3. Click on the Clip Launch button for DrumLoop-1. You should hear it playback and its Clip View should appear at the bottom of the Main Live Screen. We'll discuss this in more detail later, but for now notice the loop's waveform as it appears in the Sample Display 7.2.
4. Click + drag **S6-DrumLoop-2** to the empty slot below DrumLoop-1 on the same track 1 Audio. Notice that DrumLoop-1 continues to playback uninterrupted.
5. When you are ready, launch DrumLoop-2. Notice that no matter when you click its Launch button it launches in tempo on beat in sync, taking over playback accordingly. (Thank you Quantization Settings!) Our Set is set to "1 Bar" quantization as indicated in the Quantization Menu in the Control Bar next to the Transport buttons. Practice switching back and forth between the two drum loops to get the hang of launching Session clips. Try using the keyboard arrow keys and return key to launch clips as well.
6. Click the Stop button in the empty Clip Slot below DrumLoop-2. This stops the clip, but the Arrangement Position will continue to run. You can also use the arrow keys to navigate to an empty Clip Slot and press return to stop a clip. To stop the Arrangement, click the Stop button on the Control Bar or press the spacebar.

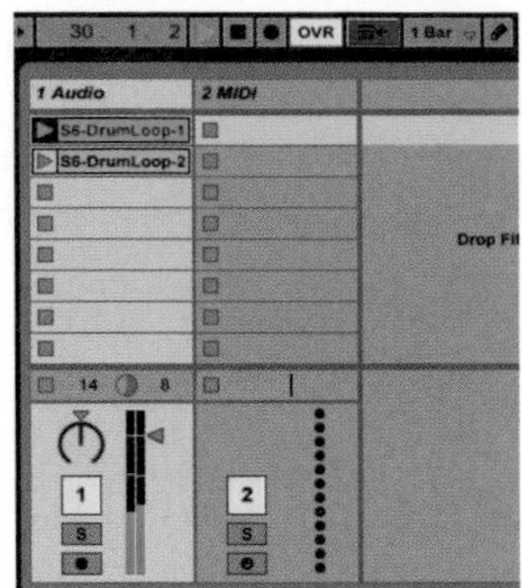

Figure 6.19 Launching clips on the same track with 1 Bar Quantization.

Okay, now that you have the hang of launching clips from the same track, let's switch things up and use multiple tracks.

6.6.1.2 Example B – multiple tracks and clips

1. Select DrumLoop-2 from its Clip Slot and press the delete key on your computer keyboard to remove it.
2. Go back to File Browser 1 and double-click on DrumLoop-2. This will create a new Audio track – "3 Audio."
3. Now let's move the new track's column position. Click + hold track "3 Audio" from its Title Bar then drag it so that it resides between track 1 and 2. It should now be called "2 Audio."

Figure 6.20 Drag a track to a new location from its Title Bar.

4. Launch DrumLoop-1 and let it play for a few cycles, then Launch DrumLoop-2. Feel free to adjust the volume of either loop from the mixer section to balance their output. Keep an eye out for overloading (clipping).
5. Stop DrumLoop-1 using the Clip Stop button below it. You should now only hear DrumLoop-2. Practice stopping and launching each clip so they play

on their own and play together. Once you get the hang of this, we'll add another Loop to the mix on the fly.

6. With DrumLoop-1 and 2 running, double-click on DrumLoop-3 from the File Browser 1. DrumLoop-3 should now appear on a new audio track. Move this track column so that it sits right next to the other two audio tracks like we did for the last track.
7. Launch DrumLoop-3. All three loops should now be playing simultaneously. Check the mix and then add DrumLoop-4. Experiment playing them in different combinations. When you're done, stop all clips form the Master track, then stop the Arrangement.

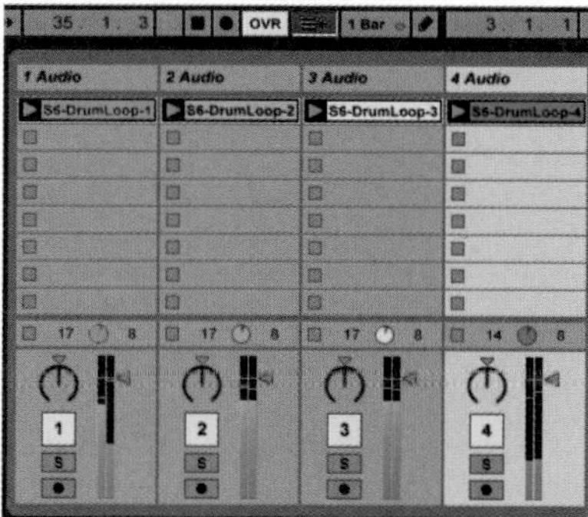

Figure 6.21 Launching multiple clips across tracks.

6.6.1.3 Example C – Scenes

It's time to use some basic editing functions and launch clips with scenes. There are many ways to copy/paste clips – edit menu, keyboard shortcuts, arrow keys, "mousing," and contextual menus – so you will have to decide what works best for you. If you are using a laptop track pad, then keyboard shortcuts with arrow keys will probably work best. If you have a mouse, then you might opt for mouse-based editing. Whenever possible, we use and strongly recommend using keyboard shortcuts.

1. With all of your clips still in place from Example B (four tracks and four loops), select DrumLoop-1 and copy drag [opt/alt+drag] or duplicate ([⌘+D Mac/[Ctrl+D] PC) it to the next Clip Slot below (Clip Slot 2) placing a duplicate there.
   You can also use the arrow keys and copy/paste keyboard commands:
   Copy [⌘+C] Mac/[Ctrl+C] PC;

   Paste [⌘+V] Mac/[Ctrl+V] PC.
2. Repeat this process, pasting Loop-1 in Clip Slot 4 and 6, skipping 3 and 5.

3. Select DrumLoop-2 and copy it into Clip Slots 3, 4, and 5 on its track (Slot 2 will remain empty).
4. Select DrumLoop-3 and copy it into Clip Slots 2, 4, and 5 on its track.
5. Select DrumLoop-4 and copy it into Clip Slots 3, 5, and 6 on its track.
6. Now that the clips are in place, launch each scene (row of clips) successively and listen to the playback.

Figure 6.22 Session clips and scenes.

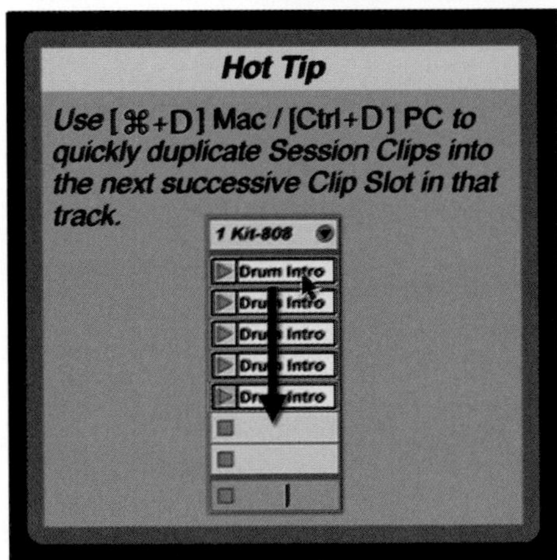

After a moment or two, begin experimenting with launching scenes in a random order. After you get the hang of it, start launching individual clips and observe the behavior of the Session. Alternate between custom clip combinations then back to scene launching and so forth. Feel free to reorder the clips within each scene or create new scenes. While all of your clips are running, experiment with the Mixer Section, deactivating, soloing, panning, and balancing tracks. When you're done, stop your clips and playback.

6.6.2 MIDI clips

As a producer, you will find yourself wearing many hats, some of which will include making beats. Live 8 and Suite 8 make it very easy to make your own beats without having to purchase additional virtual instrument plug-ins. In this section, we will focus on using an Instrument Rack preset to trigger drum samples via MIDI note events (MIDI files/clips). For a closer look at Impulse, launch to Clip 15.3.

Let's begin with a new Live Set. For the following exercise, we will use the MIDI loops from in **CPP_6-6_Working-in-SessionView**. You will also want to download the Live Packs mentioned in Clip 3.4 from Ableton's Website if you haven't already. Once you have a new Set open, follow these steps:

1. Click on the Live Device Browser icon on the left of your screen. Unfold "Instruments" then "Impulse" to view its Instrument Rack presets.
2. Select any Impulse preset and drag it directly to a MIDI track in the Session View or double-click the preset name to insert the instrument and create a new MIDI track.
3. Click on a File Browser and locate **S6-IMP-KitLoop1** then click + drag Kit-Loop1 to the first Clip Slot in your MIDI track. Don't forget to set a Global Quantize value.
4. Click the Clip Launch button for KitLoop1 to hear it playback. Notice that your Impulse drum kit instrument appears in the Track View (Detail View) at the bottom of the Live's main screen. If all you see are knobs and are feeling adventurous, then click the last round button (Show/Hide Devices) to the left of the knobs to unfold the device and take a closer look at the instrument. Listen for a moment then click the Clip Stop button on your track, then click the Stop button on the Control Bar Transport to stop the Arrangement playback.
5. Add **S6-IMP-KitLoop2** to the next Clips Slot beneath KitLoop1.
6. Launch KitLoop1 again, then after a few loop cycles launch KitLoop2.
7. While KitLoop2 is still running, drop in **S6-IMP-KitLoop3** from the File Browser 2 into the next available Clip Slot and then launch it.
8. After a moment or two, start launching back and forth between the three MIDI loops then when ready, press a Clip Stop button on the same MIDI track, then Stop on the Control Bar Transport.

MIDI and Audio clips both function and follow the same concepts in regards to launching, editing, and creating scenes. As you work more and more with Live, you will come to realize that Ableton has done a great job of making Audio and MIDI flow seamlessly, almost to the point that they are one in the same, or at least that is the goal.

6.6.3 Crossfader Section

When in view the "A"/"B" Crossfade Assign Switches are located along the very bottom of each audio track and the Crossfader below the Master track. This area is called the Crossfader Section. When you open Live for the first time, it will not be displayed, but it can be hidden/shown from the View Menu or the Show/Hide Crossfader Section button at the bottom right corner of the main screen visible with the X symbol. The Crossfade Assign Switches ("A"/"B") allow you to physically assign a track's output to the Crossfader. By activating one of the switches, the track's output level (volume) is routed to the Crossfader, which controls all of its assigned outputs in the overall mix. In this way, the volume of any track assigned to a Crossfade Switch is affected by the position of the Crossfader from left to right. In the center is an equal balance. As you slide it right or left, the Crossfader's assigned track outputs will fade and crossfade in/out as the Crossfader's location changes from left to right. All the way to the left mutes all B track's output volume and all the way to the right mutes all "A" tracks output volume. Right click or ctrl + click to access Live's seven different crossfade curves available for the Crossfader.

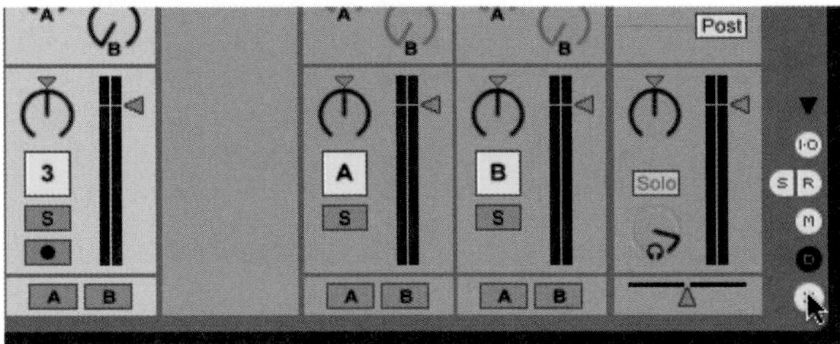

Figure 6.23 Crossfader Section.

Here is a quick example:
Let's say you have one set of tracks assigned to "A," generating what we'll call "Mix 1" and another set of tracks assigned to "B" generating a different mix, "Mix 2." Then with the Crossfader, you could alternate between each mix by sliding the fader from side to side. This fades in the "B" clip's output volume, while the "A" clip's fade out, thus crossfading between mix 1's and mix 2's clip output as your Set's Master Output transfers from monitoring mix 1 to mix 2. Obviously, the mixes will share an equal portion of the Master ouput the closer the Crossfader gets to its middle poisiton (equal balance between the "A" clips and the "B" clips).

The easiest way to practice with the Crossfader is to basically have two audio tracks. Track 1 assigned to switch A and the second track 2 assigned to switch B.

Use your mouse to move your crossfader to the left and right while listening to the results. You'll notice the audio levels rise and fall on each track depending on the position of the Crossfader. After working with this basic method you can assign crossfade switch buttons to multiple tracks. For example, use switch A for all of your vocals and instruments and Switch B for all of your drums and bass tracks. Using the Crossfader with this selection of A and B switches creates some interesting results. Finally, go ahead and activate the Global Record button to record your Crossfader moves directly into the Arrangement View.

6.7 Sessions into Arrangements

Like we keep saying, Live's Session View is a fantastic place to develop and sketch out your musical ideas in real time. Even better, you can record all of your collective ideas, improvisations, and songs sections from the Session View into the Arrangement View all in real time. When we say all, we mean everything. When the Global Record button is engaged and a clip or scene is activated, all of your actions are captured and logged into Arrangement clips along the linear Beat Time Ruler (timeline) of the Arrangement View. This includes the launching of clips, scenes, and changes made to the Mixer, device parameters, and much more. To review this exciting and powerful concept, launch to ▶ Scene 4 Global Record.

6.7.1 Capturing a Session Performance

To record Session clips into the Arrangement View, you will use the Global Record button located in the Control Bar Transport.

Figure 6.24 Global Record button.

Simply click and engage the Global Record button then launch your Session clips or scenes in any order you like. The moment you fire your first clip or scene, Live will begin recording. Select the Tab key or View Selectors to view your newly created Arrangement unfolding and being recorded in real time.

Figure 6.25 View Selectors.

Select the Tab key or View Selectors again to go back to the Session View and continue launching your clips and/or scenes. Press the Global Record button again to stop recording. The Arrangement will continue to run. If you want to stop recording and play back at the same time, or the Arrangement at any time, press the Stop button on the transport or your spacebar. Now you will have an Arrangement of your exact performance awaiting you in the Arrangement View. This will be shown as clips laid out in a linear fashion along the timeline just as you activated and arranged them from the Session View.

6.7.2 Playing back a performance

By toggling over to the Arrangement View, you will see your performance just as you played it, but shown as clips working from left to right in a linear horizontal fashion and free of Launch buttons. Now that you have a captured (recorded) arrangement, play it back using the Transport Play button or spacebar. Playback is now generated from the Arrangement clips rather than the Session clips. At any time you can flip back to the Session View to launch more or record more using the same track clips or using new tracks and clips.

6.7.2.1 Back to Arrangement

As you begin to experiment with launching and recording Session clips over an existing arrangement, you must pay attention to the rules that govern the relationship between the two. Session tracks and Arrangement tracks manage and pass the same audio signals (signal flow); therefore, Session clips and Arrangement clips on the same track cannot playback simultaneously. Whenever a Session clip is launched, it along with its entire Session track takes over playback priority from the Arrangement track. Keep in mind that this also occurs when an automated parameter is manually edited in the Session View at any time, playing or not. You will notice in both the Session View and Arrangement View that the Back to Arrangement button will turn red, indicating that the current playback differs from the stored (captured) Arrangement. This indicates that some of the material is not playing back directly from the Arrangement and that one or several clips are playing back from somewhere within the Session View.

If you toggle over to the Arrangement View, you will see that some or all tracks are grayed out (transparent look to them).

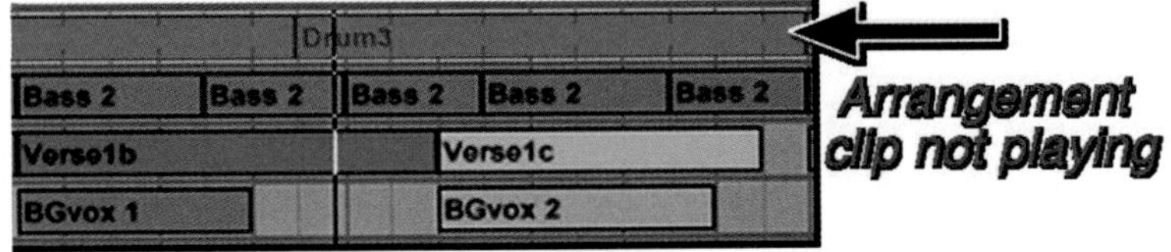

Figure 6.26 Arrangement track/clip not playing.

To revert back to the recorded Arrangement, click on the Back to Arrangement button located on the Control Bar. Live will flip back to the Arrangement playback on the fly in real time so you can keep working. Understand that simply stopping a clip in the Session View does not cause the Arrangement clips/tracks to resume playback or revert to their stored state. The truth of the matter is that it's sometimes easy to forget that you have inadvertently launched a Session View clip while working back and forth between the Session and Arrangement Views. The Back to Arrangement will let you know when this happens.

6.8 Musical concepts

It's safe to say that once you are comfortable in Live, you will spend a whole lot of time creating, producing, and performing in the Session View. Beyond its amazing nonlinear approach to making music with clips, it has the old familiar Track Mixer. This obviously includes the standard vertical tracks and returns. When all is said and done, you'll be working in a hybrid manner, toggling back and forth between the Session View and Arrangement View.

6.8.1 Produce: submixing with Return tracks

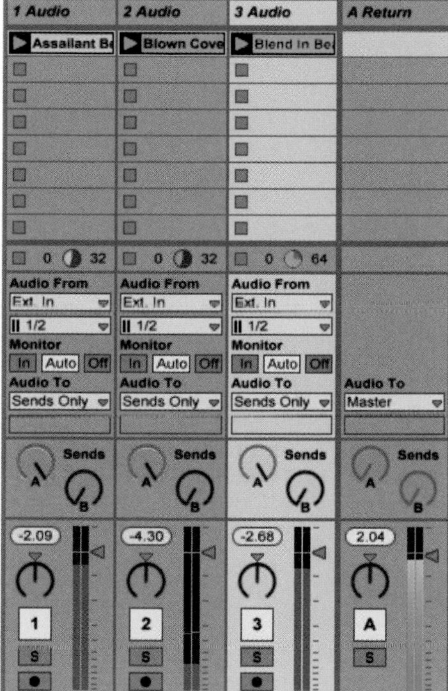

Figure 6.27 Return Track submix.

One interesting way to route audio inside Live is to route your track's Sends to a single or multiple Return tracks for creating "submixes." This is useful for routing a group of tracks to one output where they can be affected and mixed with the same effect and balanced together before going to the Master track. In this scenario, you would use the track send knob to control the level of the track. This allows flexibility in regards to submixing and creating new blends audio from multiple tracks. Let's walk through this.

1. Create three tracks with different audio clips (we'll use four-bar loops) in each track in the first Clip Slot. You should now have a scene row consisting of three separate audio track clips.
2. Bring the Return tracks into view from the View Menu or its show/hide button.
3. For each track, turn Send knob "A" all the way to the right.
4. Set the Output Chooser for each track to "Sends Only."
5. Launch your Scene 1 then add or subtract a precise amount of audio signal to be routed to your Return tracks.

As an alternative, use Group Tracks to achieve similar results. Launch to ▷ Scene 11 .

In this chapter

7.1 Musical building blocks 137
7.2 Clip View 138
7.3 Clip Box 142
7.4 Launch Box 143
7.5 Sample Box 146
7.6 Notes Box 154
7.7 MIDI Note Editor 156
7.8 Envelope Box 162
7.9 Envelope Editor 164
7.10 Musical concepts 170

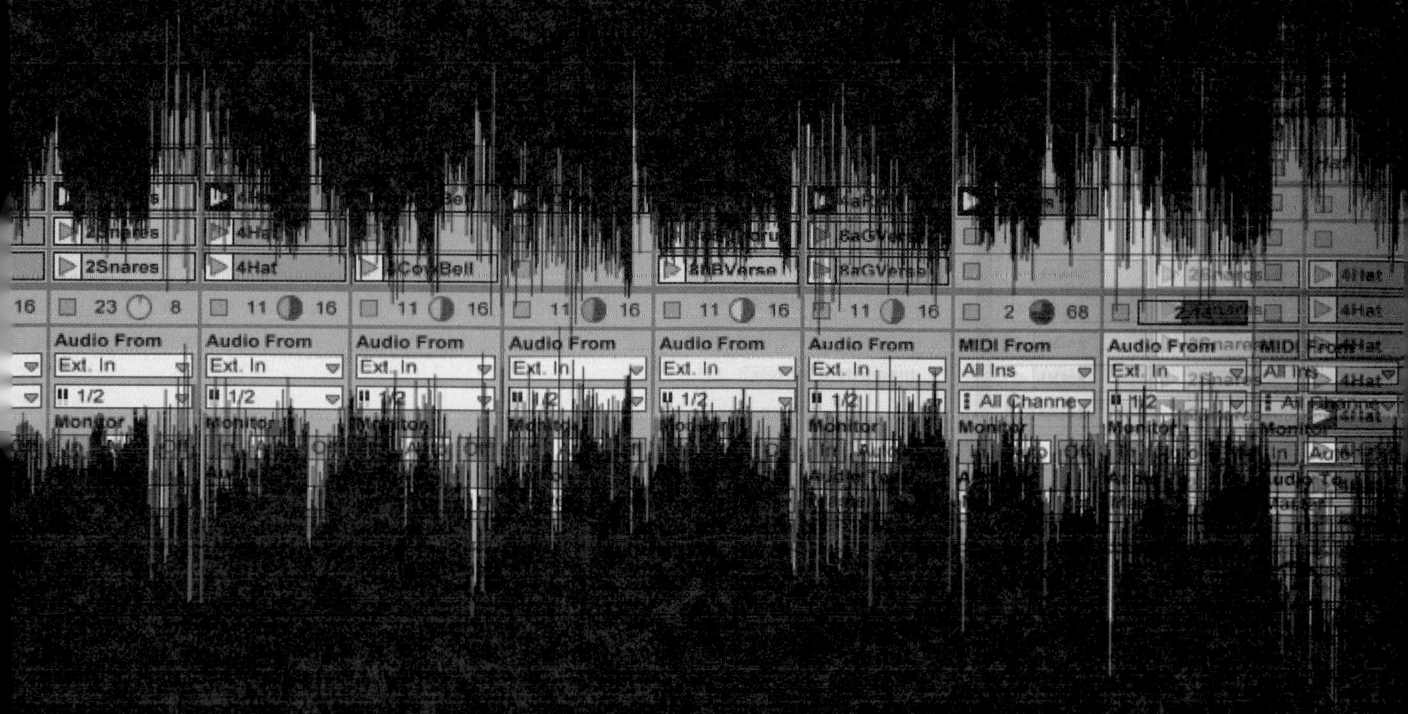

SCENE 7

Clips

I've been using Live as my primary track and soft synth resource with Sugarland and Train for the past 4 years. I can quickly and confidently adjust tempo, arrangement, and routing and not miss a beat. Whether it is the surprise addition of a Mardi Gras Indian dance break or trying an idea on a whim, I get to tell my band mates "sure" rather than "I need a day to reprogram that." As part of my rig, Live 8 functions like an extra instrument rather than an unvarying burden.

Brandon Bush, Keyboardist for Sugarland and Train

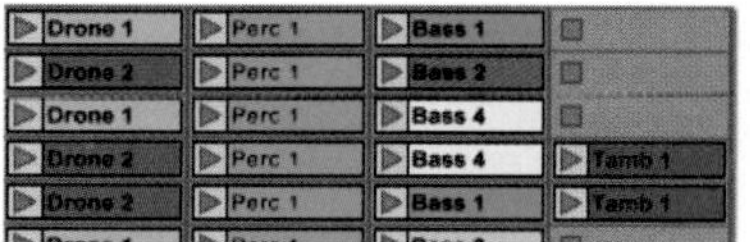

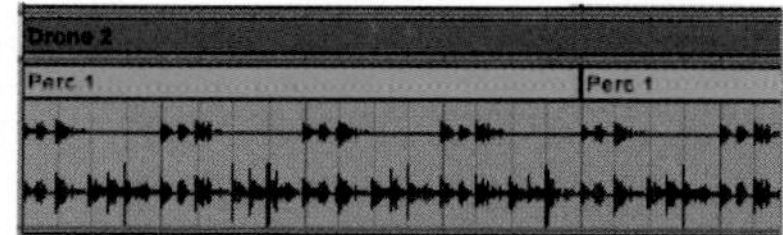

Figure 7.1 Session clips and Arrangement clips.

7.1 Musical building blocks

Clips are the fundamental musical building blocks of Live. Each clip represents a melodic idea or phrase, percussive beat or loop, a grooving bass line, or an entire music track or song. From these clips, Live makes it simple to record, edit, and build larger musical frameworks such as songs, Film/TV scores, jingles, remixes, DJ sets, live stage show performances, interactive music installations, and much more! Whether laid out as Arrangement clips or Session clips, clips store your audio and MIDI-based musical materials. Each includes its own unique set of clip-stored properties and parameters (clip data) that can be viewed and edited while working. This information makes clips flexible and efficient not only in your current Set but also across all your Sets, projects, and as customized clip presets. Launch to 15.2.4 and see how clips can retain device and performance data when stored in the Library or with a Set. Let's take a closer look at these powerful clip properties.

7.2 Clip View

Behind every clip is a wealth of playback information and properties. This information is viewable from the Clip View located at the bottom of the main Live screen. It is easily accessed by double-clicking a clip in the Arrangement or Session View. Notice that the color of the Clip View Title Bars matches the color of the selected clip. Clip View's main purpose is to display Clip Properties when a clip is selected. It displays everything you need to know about a clip, including how it's launched and all of its properties and settings. In addition, MIDI programming/editing and waveform manipulation is made through the Clip View display. There is literally a ton of information, which can be quickly manipulated and individually stored with each clip in real-time on the fly.

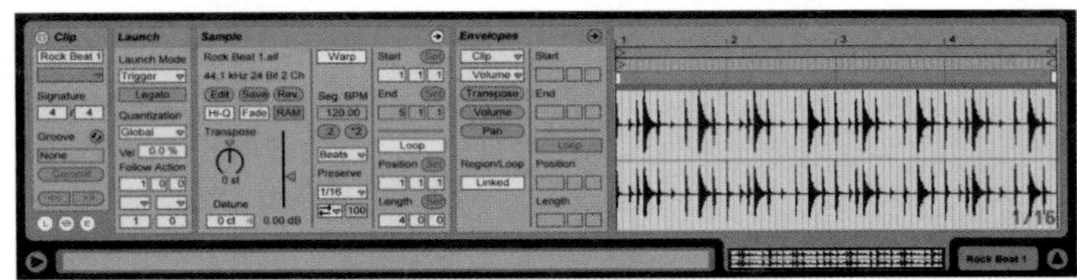

Figure 7.2 Clip View (Detail View).

7.2.1 Sample Display (Editor) and MIDI Note Editor

You have probably already noticed the Sample Display (waveform)/Note Editor (MIDI) on the right side of the Clip View.

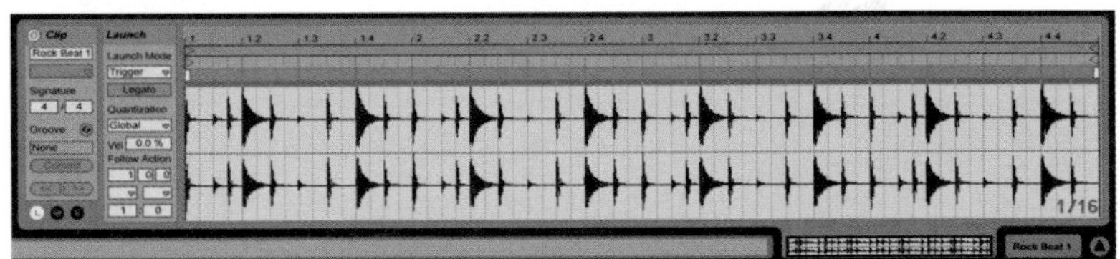

Figure 7.3 Sample Display/Editor.

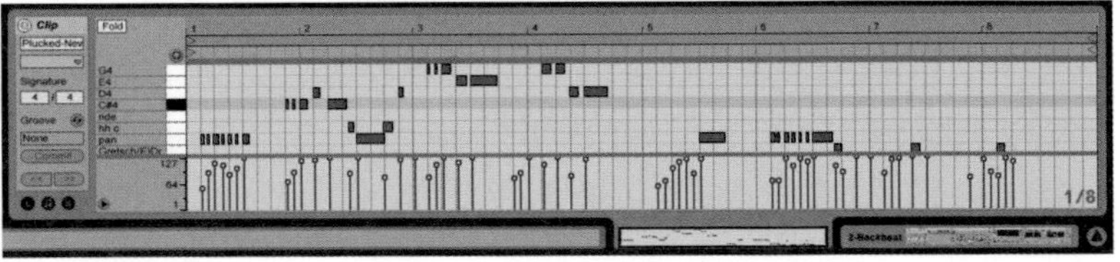

Figure 7.4 MIDI Note Editor.

No matter what type of clip you have selected, the Sample Display (Editor) or MIDI Note Editor will always be shown in Clip View for instant access to a waveform or MIDI content. This display shows an overview of a clip's contents and depending on the type of clip – audio or MIDI – serves as a graphical editor for a waveform's Transients, Warp Markers, MIDI notes, Velocities, and Clip Envelopes as detailed in the sections below. Use your mouse pointer to make selections, manipulations, and zoom in/out on sample or MIDI content within the editor/display. It should be pointed out that the physical editing of sample waveforms in the Sample Editor is quite different than in other music software. There is no cut/scissor tool *per se* in Live 8, but you can in fact crop your samples by setting a loop selection with the Loop Brace, then using the "Crop Sample" command from the contextual menu (right click or ctrl+click). With the Permanent Scrub Areas preference turned "On" (Look/Feel preferences tab), the mouse pointer will show up as either an arrow pointer, magnifying glass, or speaker icon, depending on where in the display you are mousing. Anywhere along the bottom half of the waveform display, you can switch quickly from the speaker (scrub) icon to the magnifier tool. Press+hold the shift key to do this. With Permanent Scrub Areas turned "Off," the scrub icon will not appear, defaulting to the magnifying glass instead. The Editor/Display also shows where the current playback position is located while a clip is running. Use the Follow Switch function in conjunction with playback to allow the display/overview to scroll with playback as the sample "unfolds" along the Time Ruler. Activate Follow Mode from the Control Bar. This is especially helpful when you have zoomed in on a clip's waveform or notes.

Figure 7.5 Follow button.

As you can see in our example, the Sample Display/MIDI Note Editor also consists of navigation and playback elements that impact the function of a clip in the Session and Arrangement Views. Let's break down each section and take a close look at what they do.

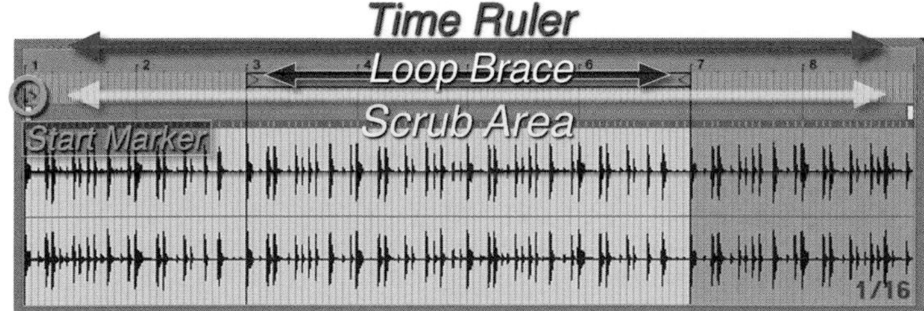

Figure 7.6 Time Ruler (bars.beats.sixteenths), Loop Brace, Scrub Area, Start Marker.

Time Ruler: At the top of the Sample/Note Editor display is the Time Ruler measured in bars.beats.sixteenths. Use this area to align your audio or MIDI and to zoom in/out on the display. A Magnifying Glass icon will appear at the mouse pointer when placing your mouse over the Time Ruler lane. Upon zooming in, the waveform or notes become larger as expected, but when the Adaptive Grid options are enabled to the grid they will also adjust to a finer beat time resolution. Use the bars.beats.sixteenths labels to identify your location.

Loop Brace: Below the Time Ruler is the Loop Brace, which indicates where looping begins and ends in a clip's sample (waveform) or MIDI events. Looping will occur when the loop switch in the Sample Box or Notes Box is clicked and therefore activated (yellow for "On"). At each end of the Loop Brace is the Loop Start/End. In our example, looping has been activated as shown by the slightly darker shaded gray Loop Brace (as opposed to a lighter shade when not looped). To make a quick loop selection, highlight a region of audio or MIDI with your mouse pointer (click+hold+drag), then type [⌘+L] Mac/[Ctrl+L] PC.

Scrub Area: Just below the Loop Brace lane is the Scrub Area, available whenever Permanent Scrub Areas is enabled. When you hover your mouse over this area, your mouse pointer will change into a speaker icon. Clicking along the Scrub Area will cause playback to jump or launch to that position relative to the Global Quantization settings.

Start Marker: Along the Scrub Area, you will see the Start Marker. It is easy to overlook because of its small size and light gray color. It is always located just above the Sample Editor or Note Editor and below the Loop Brace lane. This indicates where a clip's sample or MIDI content will begin playback upon launch. A similar End Marker will appear when the clip's Loop Switch is deactivated ▶ 7.5.13. Hold down the [⌘Mac/[Ctrl] PC key while hovering over the Scrub Area. Then click anywhere along this area – left or right – to instantly set the Start Marker at a new location along the Time Ruler. You'll find this is easier than clicking on the Start Marker and dragging it to a new location.

7.2.2 Clip Overview and Track View Selector

There are a few additional ways to navigate Clip View and Clip Properties. If you haven't noticed already, there are two tabs always located at the bottom right of the Main Live Window. On the left is Clip Overview and on the right is the Track View Selector. These are used to quickly toggle between Clip View and Track View. Since they use the same display area, it's important to know how to navigate between them and to know which one you are looking at. To clarify, Track View shows the devices that are inserted on a track, such as instruments or effects, as opposed to a track's clip contents as shown in Clip View.

We've already mentioned that you can open Clip View directly by double-clicking any clip in the Session or Arrangement View. You can also use the Clip

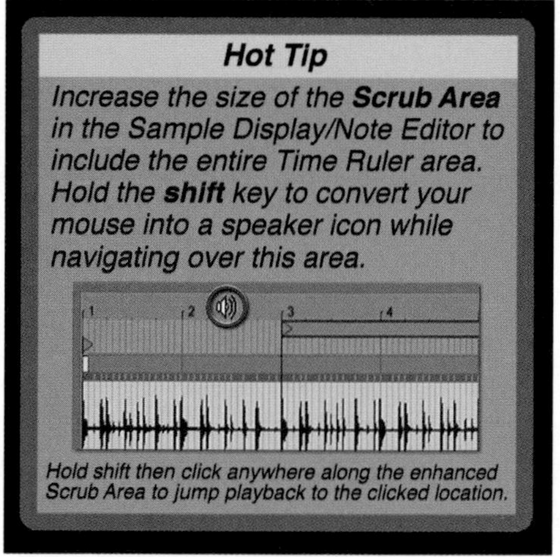

Overview and Track Selector Tabs to quickly toggle between the selected track's Clip View and Track View as seen in our example below. Click directly on either tab or type shift+Tab to toggle between each view. Clip Overview also serves as a Zooming Hot Spot when viewing a clip in Clip View.

Hot Tip
Increase the size of the **Scrub Area** in the Sample Display/Note Editor to include the entire Time Ruler area. Hold the **shift** key to convert your mouse into a speaker icon while navigating over this area.
Hold shift then click anywhere along the enhanced Scrub Area to jump playback to the clicked location.

You should notice that each selector tab displays an overview of its contents, a clip waveform/MIDI notes, and track devices for the selected clip/track.

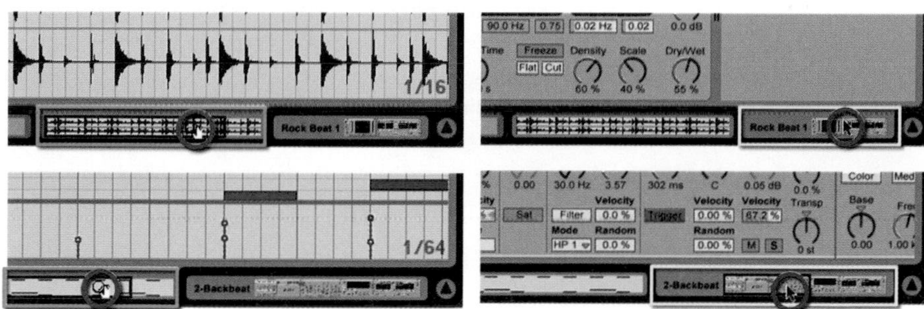

Figure 7.7 Clip Overview. Track View Selector.

Clip Overview is very useful for viewing and navigating the Sample and Note Editor since when you are zoomed in tight, you will not able to see the entire clip contents. In addition, you can put Live in Follow Mode.

7.2.3 Audio versus MIDI clips

Before moving on to dissect the various Clip View Boxes and properties, you should be aware that there are a few differences between properties that are displayed for audio versus MIDI clips. One major difference is that the Sample Box is only displayed for audio clips and the Notes Box is only displayed for MIDI clips. Any further differences will be pointed out as we dive deeper into Clip Properties, just so there is no confusion. Now, take a moment to look at our example showing Clip View for audio and MIDI clips to familiarize yourself with some of the more obvious differences. Notice the difference is in the Sample and Notes Box properties.

7.3 Clip Box

Moving from left to right in Clip View, we begin with the Clip Box. All properties here are the same for both audio and MIDI clips.

7.3.1 Clip Properties

In the upper left corner of the Clip Box is the Clip Activator Switch. This is used to deactivate a clip, muting/bypassing it from playback. This can also accessed via a clip's contextual menu (right click or ctrl+click). A deactivated clip will appear white in color in the Session View and clear and labeled "Clip Deactivated" in the Arrangement View.

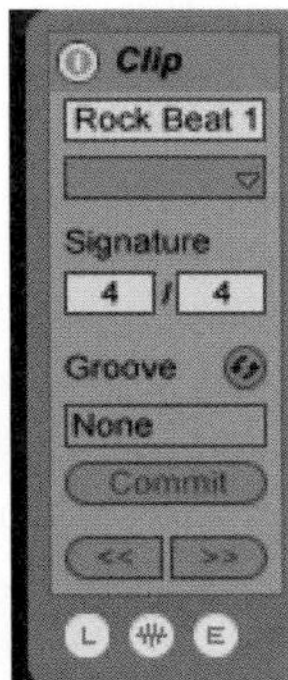

Figure 7.8 Clip Properties.

Name, Color, and Time Signature properties do not affect playback; rather they are for organization and reference only. Type directly inside the Clip Name box to name or rename a clip. This will only affect the clip, not the file on disk. The Clip Color chooser allows you to assign any color to your clips and matching Clip View. Coloring clips is important for organizing your work. Clip Time Signature is there only for your reference. Type directly into the time signature boxes to give your clip a signature label. Again, this does not affect playback.

7.3.2 Groove settings

These properties/settings affect the playback of a clip and are used to alter and adjust its musical timing and "feel" of either audio or MIDI information. Located in the Clip Groove section is Hot-Swap Groove, which is used to quickly select, change, or audition grooves on the fly. The Clip Groove chooser is used to select grooves when loaded and available in the Groove Pool. It is set to "None" (unavailable) until the clip is assigned or grooves become available in the pool. Commit Groove is used to permanently apply the assigned groove to the clip. Note that this is a destructive setting, meaning that when you commit a groove setting to a MIDI clip, its MIDI notes are moved to fit the groove. When applied to audio clips, warp markers are added and adjusted to fit the groove. When applied to MIDI clips, the notes within the MIDI Note Editor move forwards or backwards depending on the timing and quantization strength. To learn about Groove, launch to ▷ Scene 8.

7.3.3 Nudge

Just below the Commit Groove button is Nudge Backward and Nudge Forward. Clicking either of these buttons while a clip is running will skip playback forward or backward, based on the rhythmic value set for Global Quantization settings.

Use this for a stutter or skip effect in real time. You can actually make the clip sound as if it is skipping or stuck on a beat like a CD or record used to.

7.4 Launch Box

Launch Box properties are only displayed when in the Session View. They are dedicated strictly for Session clips and how they behave when launched. This box can be hidden from view by clicking the round yellow Show/Hide Launch Box button just below the Clip Box to the lower left.

7.4.1 Launch Modes

Launch Modes determine how a clip responds when its Clip Launch button is activated (pressed, played, clicked). This is based on the concept of a button having two functional states (positions): on/off (down/up) and press/release. Think of this as click+hold versus click+release or press+hold versus press+release. This concept also applies to MIDI/Key Map launching, note on/note off or key press/key release ▶Scene 12. Obviously, Launch Modes abide by the clip's Clip Quantization chooser setting located in the Launch Box. The best way to familiarize yourself with the modes is to practice using them, eventually determining which works best for you in different scenarios.

Figure 7.9 Launch Modes.

There are four Launch Modes: Trigger, Gate, Toggle, and Repeat.

Trigger: Generally the default mode for launching clips, the one you have most likely been using thus far (Preferences Menu). Down/On position launches a clip. When the launch button is released, nothing else happens (the Up/Off gesture is completely ignored).

Gate: Down/On and Up/Off positions both have an action. Down/On, the clip is launched and Up/Off, the clip is sent a stop command.

Toggle: A clip launch requires a single click/press to start and a single click to stop the clip. This is useful if you are in the practice of suddenly needing to stop clips at random timing points without using your spacebar or selecting the Stop Transport.

Repeat: Launching is just like Trigger Mode. A single click/press launches a clip and release has no action. Down/On+hold causes the clip to continuously repeat back to its clip start position until released Up/Off. The setting and timing resolution for the repeat action can be selected in the Quantization in the

Launch Box. Settings vary from Global to 1/32. This can be both fun and useful when triggering clips from an external surface controller or Pads with velocity sensitivity.

Hot Tip *For a quick description from Ableton, hover your mouse over any Launch Box's Launch Mode chooser menu and their description will appear in Live's Info View. As mentioned before, this is the best way to refresh your memory when working in Live 8.*

7.4.2 Legato Mode

Just below the Launch Mode chooser is the Legato Mode switch. This affects how a clip handles the transition of playback from a previously playing clip in the same track when the new clip is launched. Essentially, Legato Mode allows for efficiently smooth and accurate musical transitions between clips, especially if you want to merge elements from several clips during playback. As a default, when you launch a new clip, it begins playing back from its start point, even when it follows or interrupts another clip when it begins. With Legato enabled, a new clip will take over playback from the current position of the already running clip in the same track. This means that no matter where a clip's waveform or MIDI playback position is, when it is interrupted, the new clip will pick up (sync) at that position of playback. Its exact launch timing is set by the Global Quantize settings as mentioned.

Figure 7.10 Legato Mode switch.

7.4.2.1 Legato Mode On

For example, suppose you have two clips, "Clip A" and "Clip B" on the same track, one right after the other in the next Clip Slot and both of which are four-bar loops. Let's say you launch Clip A and let it cycle through two times. Halfway through a third cycle, you launch Clip B to takeover playback from A, but rather than B starting playback at its start point, it starts at its third bar because that's where Clip A left off when it was interrupted by B. With Legato Mode "Off," B would have instead begun from its clip start point when launched.

7.4.3 Clip Quantization

Live makes launching clips one after another effortless and seamless since clip launches are quantized. In addition to the Global Quantization settings that we have already touched on in Clip ▶ 6.3.2, each Session clip can also be individually assigned to a unique launch quantization amount using the Clip Quantization chooser. Clip launches default to the Global settings but can be set to other rhythmic values. This gives you the freedom to launch clips to the feel or beat flow of its contents. Maybe you want to delay the start or stop of a clip, allowing you to multitask your clip launches and other tasks in the meantime. This is a great way to practice and get familiar with the launch properties of clips in Session View. Get creative and mix and match different quantize values in both the Clip Quantization chooser and the Global Quantize Menu.

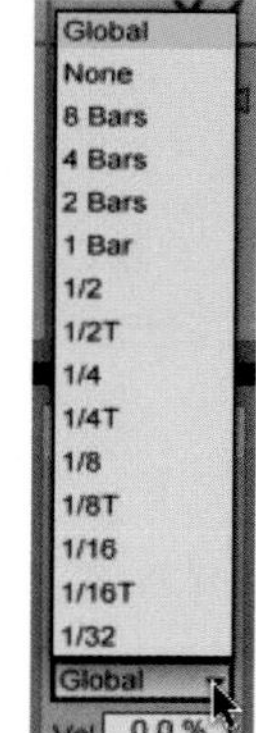

Figure 7.11 Clip Quantization.

7.4.4 Velocity Amount

An interesting clip launch setting is Velocity Amount. This slider determines a clip's sensitivity to MIDI input velocities when launching clips via a MIDI controller. Drag up or down on the slider to adjust the sensitivity. The higher the percentage, the more sensitive a clip is to MIDI input velocities. The sensitivity affect means that low input velocities (soft playing), result in a lowering in volume of the clips playback, and higher input velocities (hard playing) result in louder clip playback. MIDI velocity values range from 0 to 127, the higher values being the hardest or loudest. The percentage of effect is based on 0–100%, with the highest meaning maximum effect and 0 meaning no effect.

Figure 7.12 Velocity Amount.

7.4.5 Follow Action

One of Live's very unique and powerful features is Follow Action. This feature gives you the ability to assign an automated task to a clip or to automatically trigger other clips in a predefined succession or a random order. You can even assign clips to auto stop! The purpose of these actions is to automate clip launches, thus allowing parts of your song or production sequence itself per your instructions. Follow Actions can be assigned to any number of clips. If you want a clip to affect other clips, they must be laid out as a group of clips. That is, two or more clips slotted one after the other on the same track (consecutive Clip Slots). To separate or break up clips, use an empty Clip Slot. Create as many groups as you like and as many Follow Actions as you like and let Live do all the work!

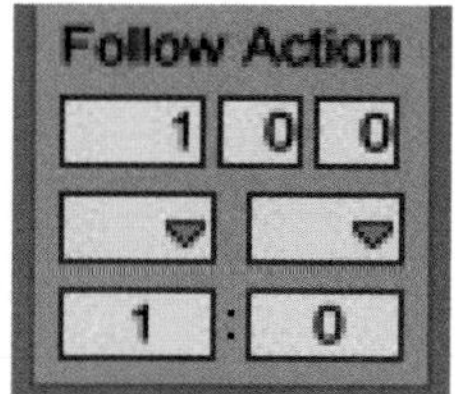

Figure 7.13 Follow Action.

For example, Clip A can be assigned to trigger (launch) Clip B after two bars. Clip B is then assigned to trigger Clip C, which is assigned to stop after four bars. In order for this all to take place, you must have a group of clips. Follow Action assignments are made in the Follow Action section of the Clip Box.

Figure 7.14 Follow Action Group.

The first row of three input fields is used to define the Follow Action Time in (bars.beats.sixteenths). This determines how much time (duration) shall pass before an action takes place. In the second row is the Follow Action A chooser and Follow Action B chooser where you can assign up to two different Follow Actions for a clip. Click on the chooser and all possible Follow Actions will pop-up from the menu.

Figure 7.15 Follow Action A.

Just below the chooser menus in the last row is the Follow Action Chance A and B determinants. Assigning values here will determine the likelihood that either the A or B action will occur when the clip is launched. This means that the each Chance control influences the probability that the other will occur. A setting of zero means an action will never occur and the value of one will result in an action at every launch. The higher the value the less often an action will occur when a clip is launched. When both Chance controls have a value above zero, the probability of actions for either one is affected. There are eight possible Follow Actions, all of which are clearly labeled in the Follow Action chooser menu as you can see in our example. There are tons of possibilities with Follow Actions, limited only by your imagination and chance! For a creative explanation on using Follow Action refer to ▶ 7.10 for an example with downloaded content.

7.5 Sample Box

The Sample Box is "the heart and soul of the clip." It is displayed in Clip View whenever you are working with audio clips. It contains the vital properties concerning the audio sample contained in a clip and is directly correlated to the Sample Display. We'll discuss more about the Sample Display in the next section. For now, let's focus on the properties of the Sample Box.

Figure 7.16 Sample Box.

7.5.1 Sample properties

At the top of the Sample Box is the name of the audio file followed by its file type – sample rate, bit rate, and channels – referenced by the audio clip. Placing your mouse pointer over the file name will cause it to act as a hyperlink to the root folder directory (native file location) displaying the file path in the Status Bar at the bottom of the main Live screen.

Figure 7.17 Status Bar showing sample location.

By clicking on the file link, the file will appear in the Hot-Swap Browser within the Live Browser. There you'll have access to audition the clip, but you can't actually swap it. This is just the browser Ableton has chosen for this function. An additional benefit of seeing your clip in the File Browser is that it can also display the time and date created, and the audio file format. You will probably need to expand (click+drag) the Browser window to the right to view all of this information.

7.5.2 Edit, Save, Reverse

Below the audio file's properties are three buttons: *Edit, Save,* and *Reverse.* All these functions can be activated and used in real time while your Live Set is playing and recording.

Edit: This button opens an external third-party sample editor, as defined in Live's Preferences>File/Folder tab, allowing you the ability to destructively alter the audio file and save it. Click this button to launch the editor application. It will then open up in its native user window.

Save: This button should not be confused with Edit. Instead, it is used to save the selected clip's current settings so that every time it is used after this, the same settings will be applied by default. This information is nondestructive and saved as a referenced analysis file (.asd) along with the audio file. These files streamline Live's use of external audio samples, so they can be used in real time without hesitation.

Reverse: This button does just what it says: reverses an audio sample. When selected, Live creates a new audio file saved to the same location as the original, thus preserving the original audio file (nondestructively).

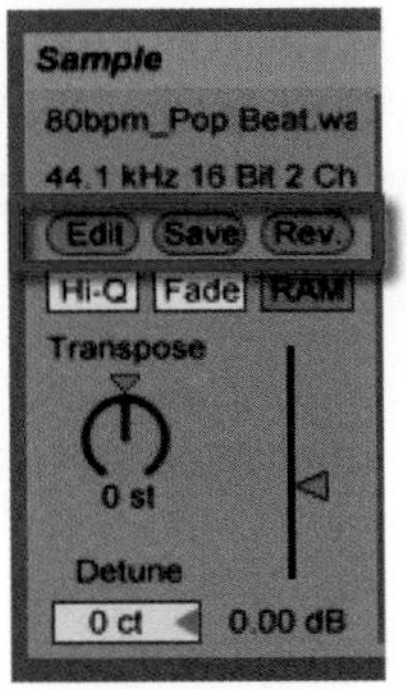

Figure 7.18 Edit, Save, and Reverse.

7.5.3 High-Quality Mode (Hi-Q)

The Hi-Q setting affects how well Live maintains the sonic quality of audio files when you pitch-shift (transpose) them. Unless your computer is struggling with processing power, you should be using this feature. Choosing anything less than the best quality for your audio productions should never be an option or standard especially when using Live.

7.5.4 Fade

To avoid clicks or pops at the start and endpoints of a Session clip, use the Fade feature. This is especially useful for audio clips that have not been cropped at a zero crossing or when looping only a segment of an audio clip. This could be a clip selection that you copied out of a larger clip for example. Fade is only available in Session View. The Arrangement uses its own unique fades and crossfades as you may recall ▶ 5.4.5. Live does a great job at automatically setting fades for you, but take some time to work with Session fades and experiment with them on your assorted clips.

7.5.5 Clip RAM Mode (RAM)

This switch provides the option to load a clip's audio file into RAM rather than reading it directly from the hard disk. Use this when you need to reduce the load on your hard disk. Oftentimes when you have extremely large multitrack Sets, you may experience glitches, pops, or dropouts during playback if your hard disk is too slow or simply becomes too bogged down by having multiple applications open.

7.5.6 Transpose

The Transpose knob is used to pitch-shift a clip's playback by semitones (half steps) raising or lowering its sound. If Warp is active for the clip, transposing will not affect the clip's playback speed, just its pitch. This is ideal when you want to adjust the timbre of a clip but maintain the integrity of the tempo. When deactivated, playback of the audio file contained in the clip will be altered based on the transposition amount selected – just like the old tape playback concept. In addition to simple transposition, it's possible to create some fantastic and original effects for your looped clips using the various Warp Modes and the Preserve settings. We encourage you to try them all out with your clip contents and be sure to use Beats Mode with "Transients" as your resolution for starters! For more on creatively warping audio clips launch to ▶Scene 13.

7.5.7 Detune

To fine-tune a clip by "cents," use Detune. This can help lockup the intonation between two different clips or if you want to create a chorused effect (two or more clips deliberately detuned to so that their pitch offsets create a "beading effect" or phasing effect). When either Detune or Transpose is activated, an orange Reset marker will appear so you can quickly reset to the default setting. Of course, you can also click in the area of the parameter then hit "Delete" on your computer keyboard to remove the setting.

7.5.8 Clip Gain

To change the gain level (volume) of a clip, use the Clip Gain fader. This is independent of the track volume. Instead, it addresses the clip (audio file) itself in a nondestructive manner. The Clip Gain comes in handy if a clip's audio file needs to be balanced with others on the same track or if it was simply a poorly recorded audio file. When you adjust the gain, you will notice the waveform increase in size along with the visible decibel/dB attenuation displayed to the right of the volume slider. Be sure not to add too much or your clip will overload playback.

7.5.9 Warp, Master/Slave

Located in the center section of the Sample Box is the coveted Warp Switch. When Warp is activated for a clip, Live applies time-stretching algorithms to the clip's audio file so that it will follow your Set's tempo and can be manipulated and quantized to create, change, or repair its timing and groove. Grid lines will appear behind the clip's waveform in the Sample Display to represent its alignment to the rhythmically subdivided grid. When Warp is deactivated, the audio clip will play back at its original tempo with playback unlooped. This is also known as a one-shot audio file. You must activate Warp to loop audio clips in the Session View, even if the original tempo is the same as your Set, otherwise, all clip looping features are disabled. You can, on the other hand, deactivate Loop ▶ 7.5.13 so that a

Figure 7.19 Warping properties.

clip does not loop within a Session Clip Slot while Warp remains active. Most of the time your audio samples will always be in warp mode when they are beat/pulse/tempo driven content.

Live automatically assigns a yellow Warp Marker at the beginning and end of an audio file and Pseudo Warp Markers/Transients to each of the files' obvious beats or peaks according to the Warp preferences set in the Live Preferences Menu.

7.5.9.1 Optimum Warping!

Feel free to change warp preference settings in Live's preferences window, including the default Warp Mode that determines which mode samples are imported under – although the default setup is pretty intuitive as is. When you change the Warp Mode default from Beats to Tones, for example, you'll notice an audible change in the sample's sound when playing back the sample's clip. This is because Live is analyzing each transient with a tonal and broader spectrum. Warp Modes are listed and explained in Clip 7.5.11 below. For a more advanced look at Warp Modes and how they apply to different styles of audio samples, launch to 13.2.

A Pseudo Warp Marker is a gray "ghost" marker that appears when mousing over the waveform timeline below the Scrub Area. It's designed to be somewhat of a "helper" in signifying where the optimized transient is located for that particular area of the audio sample. By clicking+dragging on a Pseudo Warp Marker, a yellow Warp Marker will automatically engage unless there is only one preexisting Warp Marker assigned to the waveform.

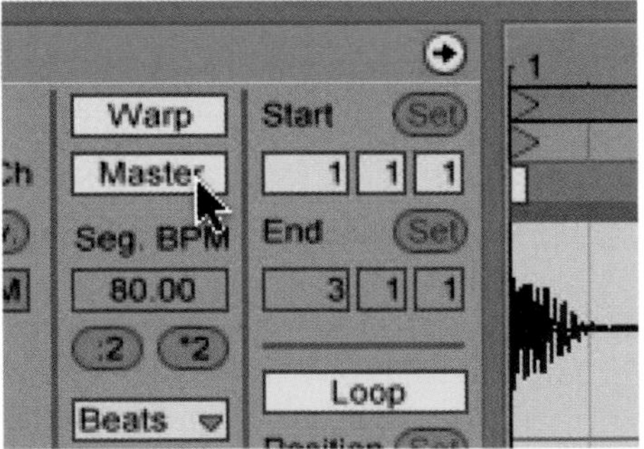

Figure 7.20 Tempo Master/Slave.

By design, all warped clips play in sync with your Set's tempo. Take note, when working with Arrangement clips you can change this sync relationship by activating the Tempo Master/Slave Switch. Setting this to Master essentially overrides and reverses the sync relationship, causing the Set tempo to sync with the specific Arrangement clip's original tempo. Surely you can imagine all the uses for warping audio clips in the Session View. In general, Warp is ideal for tempo/beat matching, looping, and stretching audio material without a noticeable drop in fidelity. For everything you'll ever want to know about Warping, launch to ▷Scene 13.

As you will see, setting the clip to Master overrides and reverses the sync relationship, causing the Set's tempo to sync with the "Master" Arrangement clip's original tempo.

7.5.10 Original Tempo

Live 8's "under the hood technology...the secret sauce!"

For every audio clip, Live automatically analyzes the contained audio file to determine the file's original tempo. It then autowarps it based on the results and is displayed in the Sample Box and labeled as "Seg. BPM" (beats per minute). The displayed tempo may or may not be exactly correct; therefore it is possible to make adjustments by halving or doubling the original tempo. Click the appropriate button to make such changes, which will be reflected by a change in the displayed "BPM." For finer adjustments, you may need to add and/or adjust the audio clip's Warp Markers.

Just below the Original Tempo input field is the Halve Original Tempo and Double Original Tempo. Click on either button to physically expand or compress the audio file by a factor of two to alter its playback speed in relation to the master tempo of your Set. Warp must be selected on the particular clip to execute this function!

7.5.11 Warp Mode chooser

Where would warping in Live 8 be without the Warp Mode chooser? Certainly not as vast and flexible that's for sure. Warp Modes are vital to how Live algorithmically divides and stretches an audio file for warping and tempo synchronization. From the chooser window, there are six Warp Modes, each designed specifically for different genres, audio content, and types of transient/rhythmic content detection. These modes are clearly labeled in the chooser menu, as shown in our example below. Choose a time-stretching algorithm that is best suited for the content of your clip(s). Try out the different modes. You'll be surprised at how well suited one may be over the other for your audio content.

Below the Warp Mode chooser, you will see a variety of control parameters available to fine-tune the time-stretching warping algorithm, whether it be resolutions, transient, grain, or formant-based control.

Figure 7.21 Warp Mode chooser.

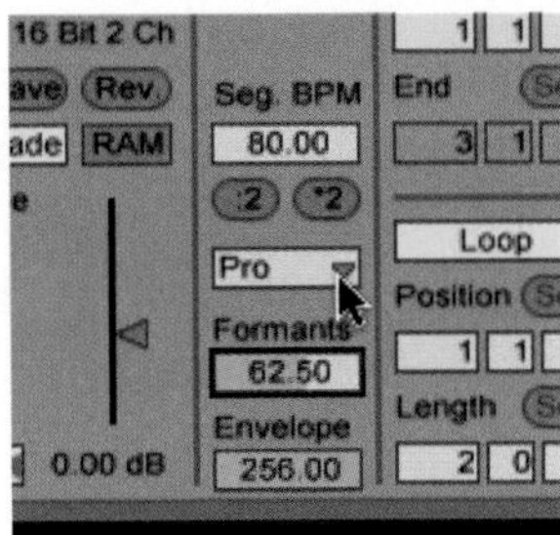

Figure 7.22 Below the Warp Mode chooser are its available parameters.

Each of these parameters is quite advanced and will be discussed in more detail in ▷Scene 13. For now, we'll look at Beats Mode to get a feel how Warp Modes and parameters relate to the Sample Box. Go ahead and select "Beats Mode" for now.

7.5.12 Sample Start/End

Within the Sample Box, you can set an audio file's Sample Start/End point (Start/End Marker as seen in the Sample Display) and Loop Position/Length (Loop Start/End as seen located on the Loop Brace in the Sample Display).

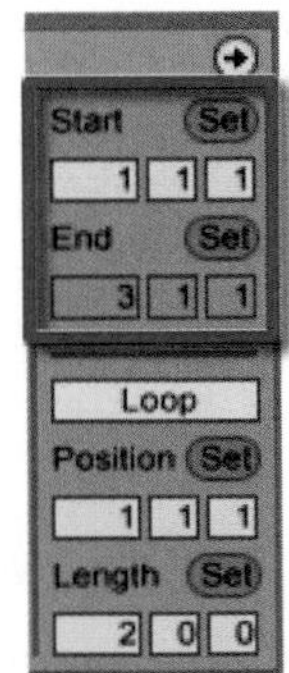

Figure 7.23 Sample Start/End.

The Start point designates where the audio file begins playback when its associated clip is launched in the Session or played back in the Arrangement View. The End point designates where the sample (audio file region) will end. This is only relevant when the file is not set to loop (Loop Switch deactivated).

To set the Sample Start and End, type or click+drag inside the Start/End input field the desired bars.beats.sixteenths value. You can also move them directly in the Sample Display using your mouse. Alternatively, they can be assigned in real time using the Set button. While a clip is running, click on the Set button to place the Start Marker wherever the play position is located at the current moment in time relative to the Global Quantization settings. Use this feature to establish a nonstop workflow environment. You can even map the Start/Stop points to a controller knob, slider, or key map ▶Scene 12. Sample Start/End points also apply to Clip Envelopes, which will be discussed in ▶ 7.8.

7.5.13 Loop Switch and Loop Position/Length

Just as its name suggests, Loop Switch engages looping for a selected clip. This can be for any selected portion of a sample's waveform or the entire sample. When deactivated, neither the sample waveform selection nor the physical clip will loop in the Session, rather, the clip will only play once (one shot).

Loop Position and Loop Length (Loop Start/End as seen located on the Loop Brace in the Sample Display) determine where in the waveform (audio file) a loop selection begins and ends (cycle playback). The Loop Switch must be activated to loop a selection of the sample waveform. Loop Position and Length adhere to the Snap to Grid resolution settings.

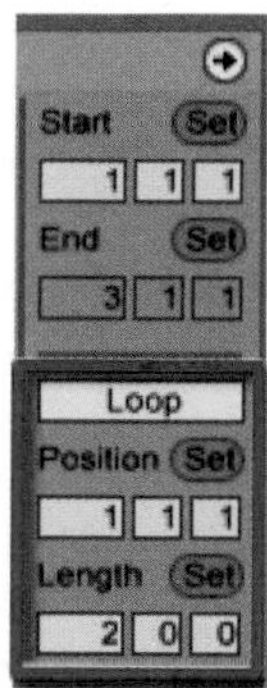

Figure 7.24 Loop Switch, Position, Length.

To set the Loop Position and Loop Length, use the Position and Length input fields, set directly in the Sample Display. Move the Loop Start and Loop End markers using your mouse or the Set button. Notice that when manually moving the Loop Start/End, they will snap to the grid as expected and also to Transient Marks as long as Snap to Grid is enabled. The more you are zoomed in, the more accurate this is. As mentioned earlier, to make a quick loop selection, highlight a region of the waveform with your mouse pointer (click+hold+drag), then type [⌘+L] Mac/[Ctrl+L] PC.

Now let's look at how Sample Start and Loop Positions work together…

For example, a Loop Position of "1" and Length of 2 means that the loop starts at bar 1 and loops for two bars. As we showed earlier, the Loop Brace in the Sample Display indicates where the sample Loop's Start/End points are located. It is very important to understand that the Start/End Markers are independent of Loop Start/End. Loop positions set the looping portion of a clip's audio/MIDI contents while the Start Marker establishes where the file playback will begin regardless of where the Loop Brace (Loop Start/End) is located. In other words, the Start Marker is where playback starts and the Loop Brace outlines where the playback will cycle repeatedly (loop). As our example shows below, they can be separate starting points. We show the playback starting at bar 1 then cycling between bars 3–5, never playing bars 1–2 again, unless the clip is relaunched.

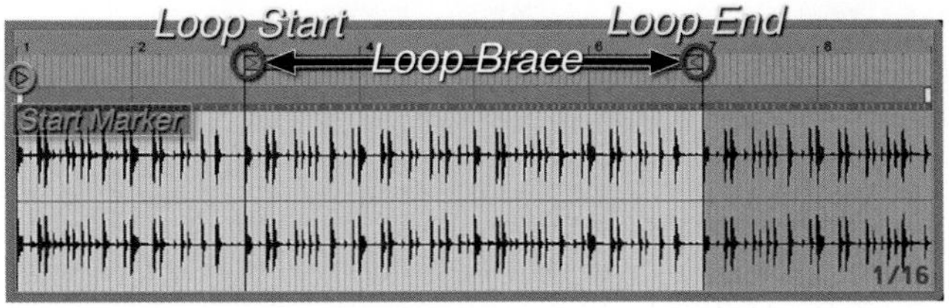

Figure 7.25 Sample Display/Editor: Start Marker, Loop Brace Start/End.

You can practice this concept by selecting a unique Start Marker position along your clip's waveform or notes in the Sample/Note Editor. Then select a different section of the clip to loop using the Loop Brace. Now launch your clip and notice that once playback passes the Loop Start point, it will begin cycling as a loop between the Loop Brace Start/Ends, regardless of where the Start Marker is located. This illustrates that the Start Marker only determines where the clip will start and has no bearing on the loop area.

Take note that Live references the Start Marker with two different names: Start Marker in the Sample/Note Editor and Start Position in the Sample/Note Box input fields.

7.6 Notes Box

The Notes Box is very similar to the Sample Box except that it directly addresses MIDI parameters instead. These properties are only available when MIDI clips have been selected in either the Arrangement View or the Session View.

7.6.1 Original Tempo

The first section you will notice within the Notes Box is "Orig. BPM," which displays the selected MIDI file's original tempo. This information is gathered and used by Live to align the MIDI file to the grid within the Note Editor. This allows the MIDI file to playback in sync with the tempo of your Set. Be aware that if you change the original tempo of the MIDI file, it will play out of sync with the tempo of the Set although the clip itself can still be launched on the downbeat. The clip as a whole remains synced to the current tempo, but the MIDI file note information does not since the clip is a container for the MIDI information. This situation is quite obvious when you look at the Note Editor and see that the MIDI notes are no longer fixed to the grid lines.

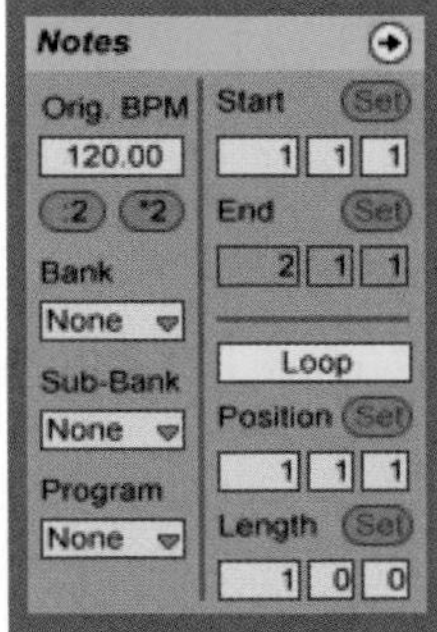

Figure 7.26 Notes Box Original Tempo (Orig. BPM).

Just below the Original Tempo input field is the Halve Original Tempo and Double Original Tempo. Click on either button to physically expand or compress the MIDI notes by a factor of two to alter their playback speed in relation to the master tempo of your Set. This is a very useful way to increase the playback speed of a MIDI file without manually adjusting each MIDI note and rhythm to play twice as fast or as slow. Often, you will import a MIDI file (loop, riff, or song) and come to find that it appears to be programmed at half the speed it should be. Just hit the Double button to automatically conform it. In a more classical sense, this can be used to cut a file's speed in half so as to fit four bars into two (cut time). You could also use this to program in more complex parts and passages at a slower speed and then speed it back up to play it back. This is an old MIDI trick.

7.6.2 MIDI Bank, MIDI Sub-Bank, Program Change

Live handles all External MIDI Bank and Program Change Message information in this section of the Notes Box. Your External MIDI instrument Patch Lists (instruments/sound presets) and Banks (a catalog of instruments grouped as presets) can be navigated and selected from the MIDI Bank, Sub-Bank, and Program Change choosers. They communicate directly with External MIDI devices such as synthesizers, samplers, sound modules, and third-party plug-ins when applicable. The actual bank and program details are dependent on the specific device's protocol, but in theory, you will be able to navigate patches via these menus. If you don't have any, don't feel left out. Current trends favor Virtual Instrument Plug-ins just like the software instruments found in Live 8 and Suite 8. To learn more, launch to External MIDI ▶ 15.9 and Virtual Instruments ▶ 15.7.1.

7.6.3 MIDI File Start/End, Loop Position/Length, Loop Switch

Within Notes Box, you can set the MIDI files Start/End point and Loop Position/Length in the same way as described for the Sample Box.

The Start point designates where the MIDI file begins playback when the associated clip is launched in the Session or played in the Arrangement. The End point designates where the MIDI file will end when it is not looped. Loop Position and Loop Length determine where in the MIDI file a loop selection begins and ends. To loop a selection within a clip, the Loop Switch must be activated.

MIDI file Start, End, and Loop positions can also be assigned in real time with their respective Set buttons. This will instantly drop a position marker to the current playback position, relative to the Global Quantization settings. It should be pointed out that the Set Start and Set End features work well in real time or for setting the positions to a selection made in the Note Editor. Instead, positions can be set from the Note Editor by manually dragging the Start/End or Loop position markers into the desired positions. Be sure to activate the Grid mode switch ([⌘+4] Mac; [Ctrl+4] PC) to help navigate different resolutions of time within the start-end chooser. You can also right+click or ctrl+click within the MIDI Note Editor window to access this feature. If the Grid is deactivated, "Off" is displayed. When activated, a rhythmic value resolution is shown in the bottom right of the Note Editor (Marker Snap). Note that this is present in the Sample Editor as well.

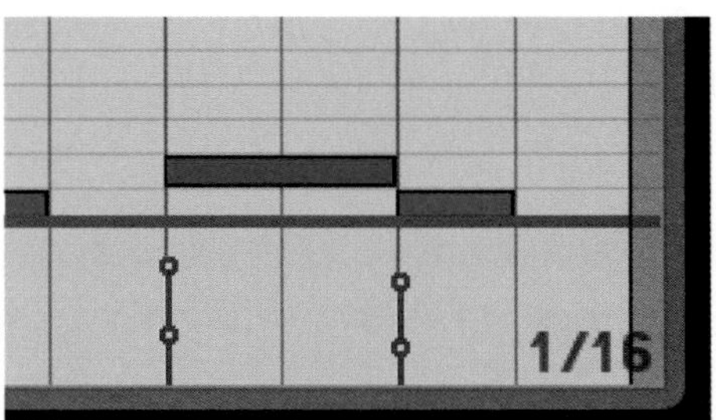

Figure 7.27 Grid/ Marker Snap.

7.7 MIDI Note Editor

MIDI Editing has come a long way since the old days of event list editing. As a matter of fact, with Live's approach to MIDI sequencing, you won't even see an event list editor within the MIDI Note Editor. However, these days pretty much every DAW software packages come with some sort of graphic MIDI editor, often called piano roll. In Live, the MIDI Note Editor is where you will do all your MIDI editing. The Note Editor is available in both the Session and Arrangement Views for each MIDI clip; just double-click on a clip to bring up Clip View to get to the Note Editor. For the following description, we reference a MIDI drum loop assigned to playback with Impulse (Live 8's eight-channel Drum Sample Player) located in the Instruments folder in the Device Browser as we have for previous exercises. Launch to Clip ▶ 15.3.1 for more on Impulse. Let's take a closer look at what is accessible in the MIDI Note Editor.

7.7.1 Basic commands, navigation, and zooming

Many commands in the MIDI Note Editor are the same or very similar to those you will find in any other music production software or word program. Use your mouse (click+hold+drag) to move a note or select multiple notes (shift+click) to move inside the Note Editor. Click on a position in the Note Editor to create an insertion marker. The Insertion Marker is a vertical orange line that determines the point at which notes, time, selections, and pastes will begin. From there, you can *rubber band* (click+hold+drag) notes. Commands such as cut, copy,

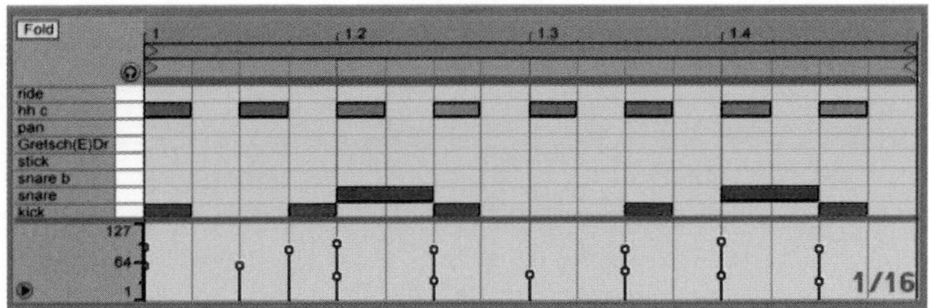

Figure 7.28 MIDI Note Editor.

paste, and duplicate are all available for use within the Note Editor, each using the same key and menu commands we have referenced throughout.

To use the Zoom in and out tool on the Note Editor view, place your mouse pointer over the *Beat Time Ruler* at the top to the Note Editor and click+drag up/down with the magnifier at any position along the top of the timeline area. Once you have zoomed in a bit, you can click+hold+drag left/right to scroll. Feel free to use the Clip Overview/Zooming Hot Spot in the same way. Also, from either edge of the Overview, click+hold+drag inward toward the center of the Overview to zoom and expand the size of the notes within the MIDI Note Editor. For fast zooming, use the +/− on the computer keyboard.

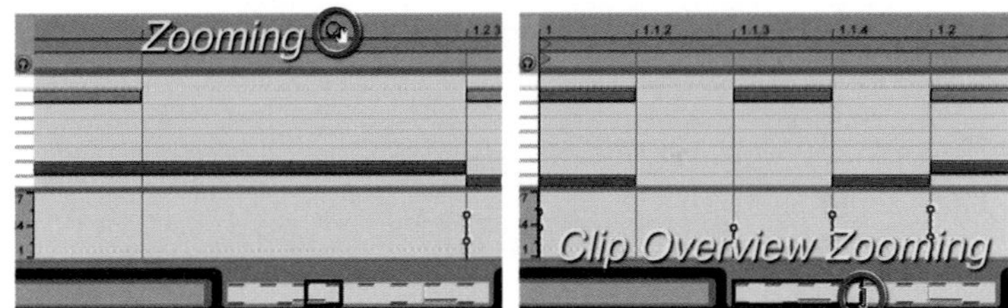

Figure 7.29 Zoom the Note Editor from the Time Ruler or Clip Overview.

7.7.2 Note Ruler

At the left of the MIDI Note Editor is the Note Ruler showing the MIDI key range and note names. In this particular case, it is listing our Impulse's sample names as assigned to those particular keys. Click+drag the Note Ruler to the right and left to zoom in and out on the MIDI notes (making them grow in size for better visibility) and up and down to scroll through the key range. The adjacent white keys are the piano roll. If you want more real estate to work with and visibility of the entire keyboard range, drag the entire Clip View upward to resize the whole window to allow for a larger view of the key range and MIDI notes. Hover your mouse arrow over the top of the black divider line between the Clip View and the Session/Arrange View. Drag upward to resize. Click the Fold button at the upper left corner to hide all unused (no note events) piano key tracks/rows.

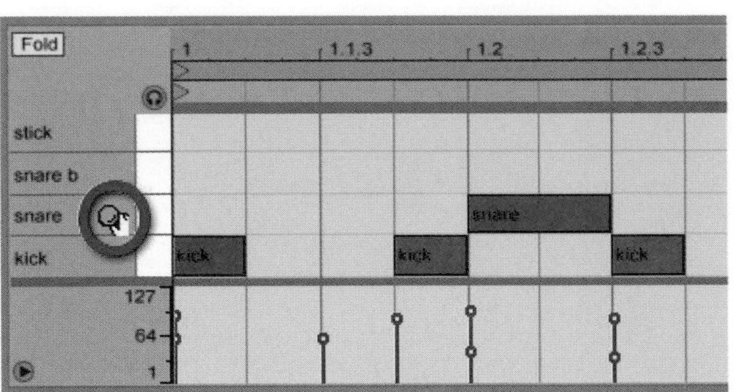

Figure 7.30 Zooming the Note Ruler.

This will hide the Note Ruler names leaving only the piano roll (keyboard) showing to the left. Slide your mouse over the individual notes to reveal specific note names, i.e., C3, D3, E3…

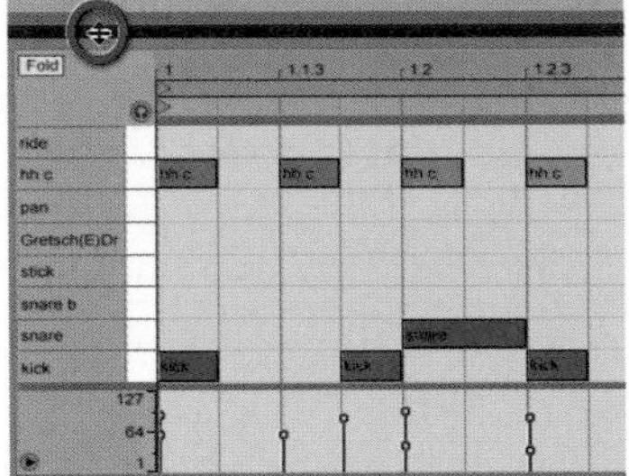

Figure 7.31 Drag the divider line to resize the Note or Sample Editor.

7.7.3 MIDI Velocity Editor

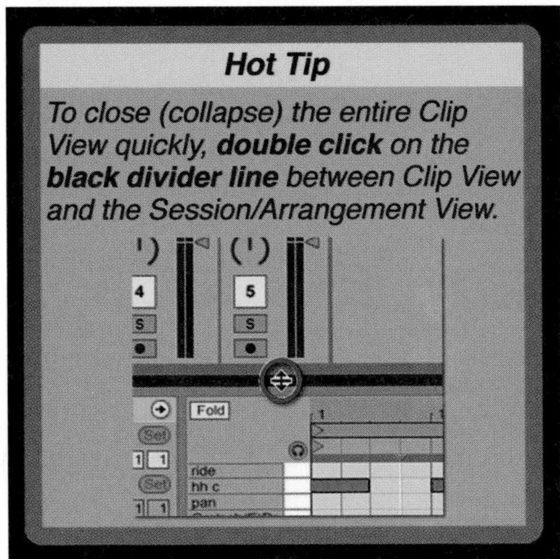

Located directly below the Note Editor is the MIDI Velocity Editor. From here, you can edit the individual velocities for each MIDI note (nothing new if you've used a DAW before). In the lower left-hand corner, you will see a triangle button that is used to hide the Velocity Editor from view if you prefer not to see it. Each vertical Velocity Marker can be dragged up or down to alter a note's velocity. The taller the marker line, the stronger the velocity and greater the velocity value (1–127). As you increase the velocity of a MIDI note, the actual note color will darken and vice versa. For quick access to velocity values while mousing over MIDI note, hold down the [⌘] Mac/[ALT] PC key to reveal a note's velocity value. With this key still held, you can click+drag a MIDI note up or down to change its velocity. Placing your mouse over a Velocity Marker will cause the associated MIDI note to be revealed (highlighted in blue). The opposite is true when mousing over a MIDI note. In this case, the Velocity Marker will be revealed (highlighted in blue). Multiple velocity markers can be selected for mass selection editing. The common commands apply.

If you rather redraw velocity values with the mouse, enable Draw Mode from the Control Bar, then use the pencil in the MIDI Velocity Editor window.

7.7.4 Insert/Edit MIDI notes

Writing and editing MIDI notes in the Note Editor is a very popular way to work in Live or any DAW for that matter as opposed to using a MIDI keyboard controller. This function has been enhanced even further in Live 8. This is especially useful for those of you who do not have access to a MIDI controller or prefer not to enter notes on a keyboard. Not everyone is a keyboard player, but that is no reason not to create music!

Notes can be inserted with the traditional mouse pointer or the pencil tool. These are technically defined as Edit Mode with the mouse and Draw Mode with the pencil. In Edit Mode, all the standard commands apply here, such as copy, cut, paste, etc. Before inserting and editing your MIDI notes, you should activate the MIDI Editor Preview switch located to the left of the timeline.

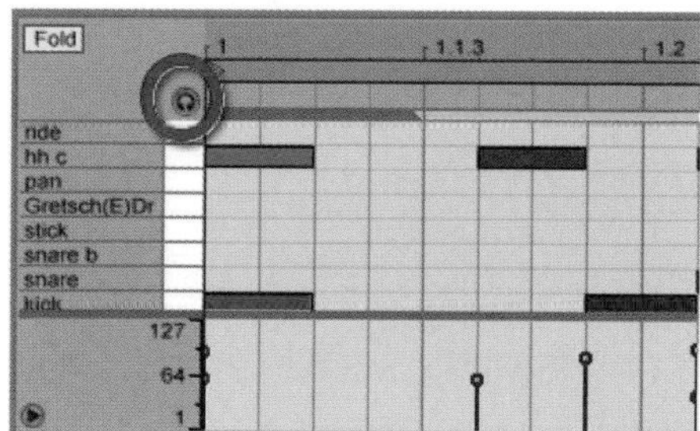

Figure 7.32 MIDI Editor Preview switch.

This allows you to hear a MIDI note as you edit it (dragging and clicking), provided you have an instrument assigned to that track to play them back. Notes can also be drawn or inserted during playback at all times in addition to playing them in during Overdub Recording via a MIDI controller. Just keep in mind that they can only be overdubbed with a controller when Overdub is enabled. For more on Overdub Recording, launch to 9.2.2.

Regardless of the insertion method you choose to use, all note insertions edits are governed by the Grid, Adaptive, or Fixed. With Adaptive Grid, note starts and lengths are automatically snapped (adjusted) to the nearest grid lines upon insertion – wide, medium, narrow, etc. – and changes (adapts) with your zoom level. With Fixed Grid, you can choose actual rhythmic values for snapping to in the grid, which remains constant (fixed) regardless of zoom. Access all these settings from the Options Menu or Note Editor contextual menu. To temporarily suspend note-length snap restrictions set by the grid, hold down the [⌘] Mac/[Alt] PC key while inserting a note, and then drag to extend the note length as long as you desire. To permanently disable the Snap, unselect "Snap to Grid" from the Options Menu ([⌘+4] Mac/[Ctrl+4] PC). To re-enable it, choose a new Marker Snap setting or enable it directly from the Options Menu. With Snap to Grid disabled, you will be able to place notes anywhere you want.

In Edit Mode, adding and removing notes is as easy as double-clicking on an empty spot in the Note Editor to add and double-clicking on a note to delete it. Use either the mouse pointer or the arrow keys to move notes around. Just select note(s) and drag them about, or use the arrow keys. To lengthen or shorten notes, drag them from either edge in the desired direction. If the snap settings begin to hinder your note placements, you can easily change the snap setting from the Note Editor contextual menu (right+click or ctrl+click).

Transpose or move notes either by octave or chromatically using a combination of key commands. To adjust multiple notes by an octave up or down, select or highlight the notes, press and hold shift, and then press the up/down arrow to move the notes up or down the piano roll within the MIDI Note Editor. To move the notes chromatically (one step at a time), select the notes and use the up/down arrow keys (without shift) to move the notes up or down. Move them forward and backward using the left/right arrow keys. In the same vein, shorten and lengthen MIDI notes in the Note Editor by holding the shift key and pressing the left/right arrow keys. These commands adhere to the grid. Using the arrow keys for single or multiple notes will expedite the MIDI note editing process. Activate the MIDI Editor Preview (headphone icon in the upper left corner of above Piano Roll) to audition the notes while moving and editing them.

In Draw Mode, notes are inserted with the Pencil tool. Simply click to insert or click+hold to insert and draw out notes (left or right) within the Note Editor. To delete a note, click on it with the pencil and it will disappear. Turning off "Snap to Grid" will allow you to place notes wherever you want and with customized note lengths upon input. To do this, click+drag the note length with insertion. Changing note lengths is a bit different than in Edit Mode since you can't actually drag an existing length of a note. To make a note longer, set the Grid Snap to the desired length of your note, then click anywhere after the note in the empty space between it and the next grid line. It will then automatically fill out to the next grid line. To shorten its length, choose the grid value you desire, then click just to the right of the grid line where you want the note to end and it should shorten to the appropriate grid line. You can also switch Snap to Grid to "Off," and do this for custom lengths. Either way, the note will be automatically cropped to fit. There are a number of ways to insert notes and make edits. Take some time to practice these methods to get used to them. Over time, you will develop your own fast way of getting the job done, which will enhance your workflow. We encourage you to learn the keyboard commands that will make toggling between each grid mode and inserting and editing quick, easy, and efficient. Beyond that there are many more great time-saving keyboard shortcuts worth learning and using in Live for navigating and editing notes ▶ Website.

7.7.5 MIDI Step Recording

For all of you MIDI aficionados out there, MIDI Step Recording is all about you. That being said, let's define what it is. MIDI Step Recording gives you the ability to use a MIDI keyboard and computer keyboard together to input MIDI notes into the Note Editor. This approach was made very popular by notation software as a speedy way to input notation without having to worry about timing on the keyboard. There is one catch with this, that is, it can only be used with an existing clip. It doesn't matter if it has MIDI notes already; it just needs to be in a Session clip or Arrangement clip. With that, the quickest approach is to do this in the Session View. So, how does it work?

1. First, arm (record-enable) the track you wish to input (record) notes into in the Session View.
2. Double-click in an empty Clip Slot to create an empty clip.
3. Enable MIDI Editor Preview (blue headphone switch)!
4. Using Edit mode, go to the Note Editor and place an Insert Marker at the beginning of the clip (or wherever you want to input notes) by clicking on a grid line inside the Note Editor. The Insert Marker appears as an orange flashing line.
5. Set the grid to "1/8" (just for this exercise). You can right click or ctrl+click on the Marker Snap to change the snap resolution.
6. Once this is set, press and hold the desired key(s)/notes on the MIDI controller that you wish to input, then press the right arrow key on the computer keyboard. Notice that when you pressed the MIDI keys nothing happened but while holding them down and pressing the arrow key, the notes were inserted.
7. Let go of the MIDI note(s). Repeat the last step again and you will see the same note(s) inserted on the next beat or grid line.
8. Do this one more time, but this time, don't let go of the MIDI notes after you press the arrow key. Instead, press the arrow key a couple more times

consecutively. This extends the note lengths (duration) beyond the visible grid to the next grid line and beyond.

To remove a note or shorten its length, while amidst inserting it, keep holding that note's MIDI key down, and then arrow backwards over the note and it will disappear. You can do this to any note to edit its length or delete it all together as long as you don't let go of the notes. If you've let go, then use the mouse to remove notes or set the Insert Marker at the point where you want to rewrite the notes or use the back arrow. When inputting notes, keep in mind that no two identical MIDI notes can occupy the same space. Identical notes will always be overwritten when they overlap.

Making edits and skipping beats becomes quick and easy using only the arrow keys by themselves to navigate from position. Once you find your desired input position, press+hold down the MIDI notes and apply the arrow keys to them. Left/right works the same as up/down, except that the up arrow key will only advance as far as the last MIDI note in the sequence. If you want to move the Insertion Marker around at microincrements, hold down the [⌘] Mac/[Ctrl] PC key while using the arrow keys in any direction. This temporarily bypasses the grid snap, allowing you to insert notes at any finite position.

7.8 Envelope Box

What is an *Envelope*? In the world of DAWs, an Envelope is a line or curve segment-based representation of various parameters – volume, pan, etc. – that are used to shape and manipulate the properties of (in the case) a clip's contents. These lines and curves are made up of adjustable breakpoints that enable them to take on shapes – ramps, bends, etc. – by creating and adjusting multiple points along a waveform or MIDI event. Such parameters and manipulations are linear and evolve over time. Therefore, envelopes can be automated and appended to a clip.

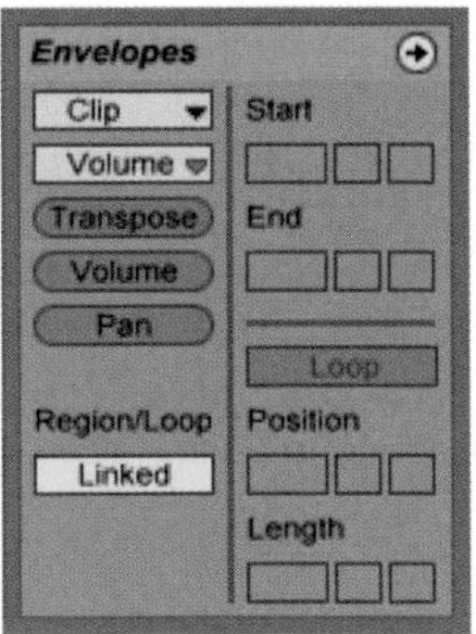

Figure 7.33 Envelope Box.

The Envelope Box provides a unique way of automating clips, device chains, and mixer parameters for both the Arrangement and Session Views. It is used in conjunction with the Envelope Editor, where all clip automation is carried out 7.9. The key point to remember is that any changes to automation of envelopes is strictly clip-specific. This means that automated parameters of a clip such as volume are married to the clip and not to the track, like Arrangement View Breakpoint Envelopes are ▶ 5.5.

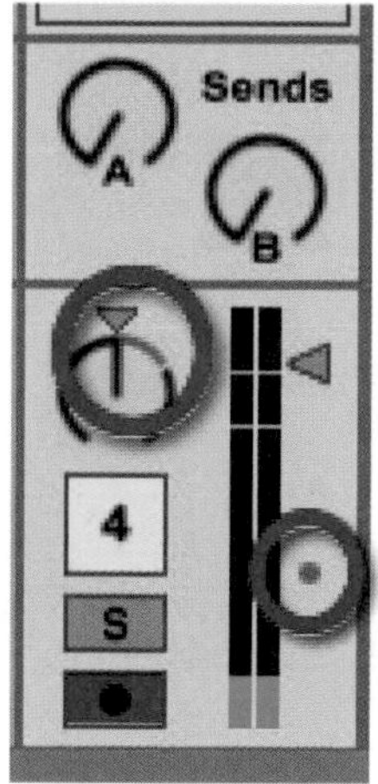

Figure 7.34 Clip Envelope automation is indicated by an orange dot or orange lines.

Clip Envelopes can be set to affect the track mixer, the clip itself, and associated device parameters. While the physical track elements remain unaffected, you will see a small red-orange indicator next to the track mixer setting that has been automated via Clip Envelopes. This is also true of the Arrangement Mixer section, which remains unaffected as well, except that there are no indications of mixer automation in the mixer section. In addition, Arrangement View track automation can also be automated on top of clip automation. For more on track automation, launch to ▶ Scene 5.

Arrangement Clip Envelopes affect the same parameters as Session Clip Envelopes, except they are physically unique to their own view. In other words, Clip Envelopes are specific to an individual clip and can be created independently in their respective views. Such envelopes from Session or Arrangement clips are only carried over to the other view when they are recorded via Global Record (Session to Arrangement) or are physically copied to the other view.

7.8.1 Device, Control, and Quick chooser

To use envelope automation from Clip View, you will choose a device from the Device chooser and the specific parameters of that device from the Control chooser. Whichever device you choose, the associated control parameters will become available in the Control chooser. For audio clips, you will be able to select from Clip, Mixer, or a device in the track. For MIDI clips, you will be able to select between each device in the device chain, including each individual parameter or the Mixer. The three buttons below the Control chooser are called Clip Envelope Quick choosers. They serve as shortcuts to the most common parameter for automating. Clicking on one of the buttons will instantly call up that parameter in the Envelope Editor for editing. When envelope automation has been created, a little red square will appear next to the device and parameter that is affected.

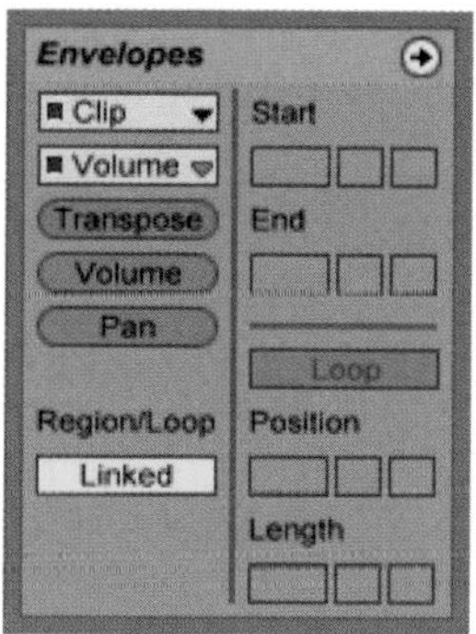

Figure 7.35 Clip Envelope control parameters.

7.9 Envelope Editor

The Envelope Editor is where all the magic and fun begins. It is full of great and powerful automation features. The whole concept is very different than any other DAW you may have worked in and it will take some getting use to. To bring the Envelope Editor in view, click the Envelope's Title Bar or the show/hide envelopes button. You should see some sort of red line through the sample or note display, which is the envelope.

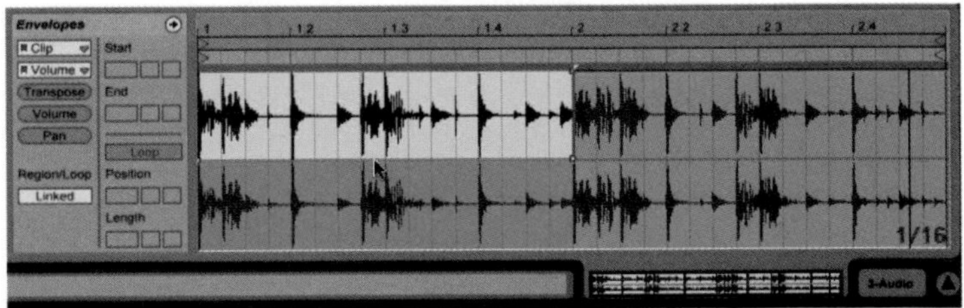

Figure 7.36 Envelope Editor.

7.9.1 Editing Breakpoint Envelopes

Automating clip parameters is easy in Live. Just select a clip to show it in Clip View. With the Envelope Editor in view, select the device and the parameter you wish to automate, then with your mouse click+drag or click+draw the envelope, shaping it to the desired position. Ok, that sounds easy enough, but how do you drag or draw envelope automation? There are two ways to create and edit envelopes, but first let's identify what the envelope looks like. As you can see in our example, the envelope is represented by a red line and pinkish shaded area in the Envelope Editor.

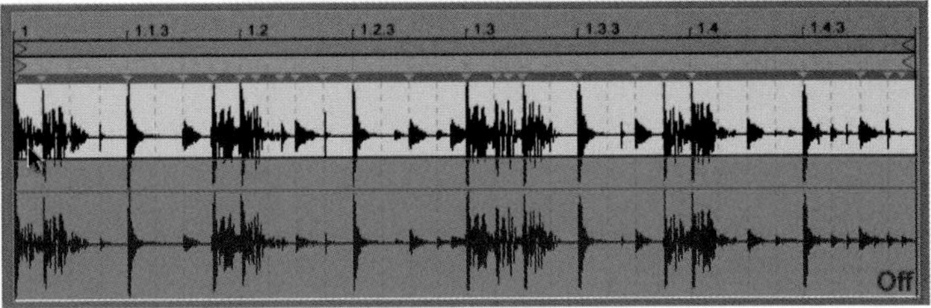

Figure 7.37 Custom Clip Envelope.

The line itself represents the position at which the envelope is set and the actual shape of it. The line is the envelope referred to as a Breakpoint Envelope.

As for working with envelopes, if you place your mouse pointer directly over the envelope line, you can click+drag it up and down. You will notice that this affects the entire clip. To automate only a segment, make a highlighted selection inside the display with your mouse, then click+hold on the line and drag upward/downward in the editor. There you have it! The little circles to either end of the line are called breakpoints. Insert a new breakpoint by double-clicking on the line. To delete a breakpoint, double-click it.

Taking this a step further, you can create ramps in the envelope between breakpoints. Simply click+drag a breakpoint to wherever you want.

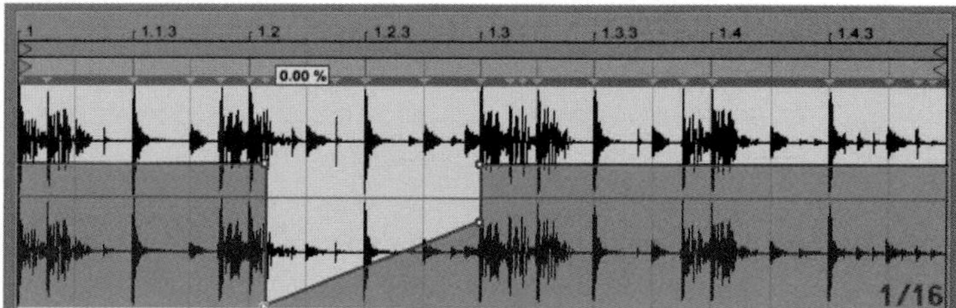

Figure 7.38 Create envelope ramps and other linear line segments.

So far, we've been talking about dragging the envelope around in linear line segments. Draw Mode allows us to literally draw in breakpoint envelopes in either linear or curvilinear shapes without making a selection or adding breakpoints. To do this, enable Draw Mode from the Control Bar, then draw with the pencil within the Envelope Editor whatever shape you like. For curvilinear shapes, turn off the grid in the editor. Right click or ctrl+click in the Envelope Editor to bring up a contextual box, then select "Off" below the Fixed Grid settings to turn off Snap to Grid. Each grid setting dictates the increments at which a breakpoint segment is created (drawn/inserted). To remove an envelope, right click or ctrl+click in the editor and select "Clear Envelope." Once you have created your envelope, you can redraw it as you wish, or switch out of draw mode and use your mouse pointer to drag breakpoints or larger line segments around.

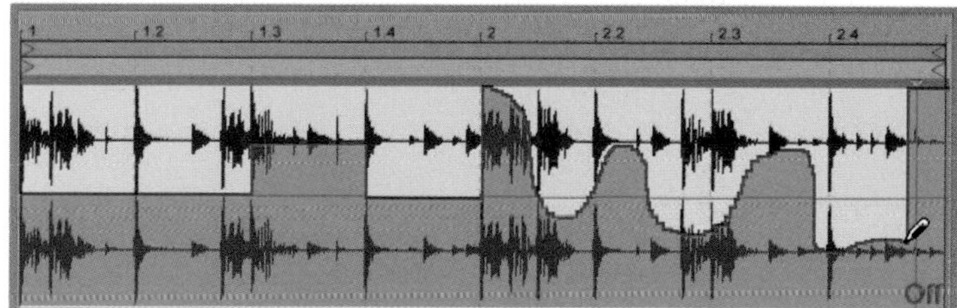

Figure 7.39 Use Draw Mode to create curves.

7.9.2 Link/Unlink Envelope, Start/End Loop Position/Length

When envelopes are linked, they loop with the clip, repeating the same automation over and over infinitely until the clip is stopped – just as you would expect.

When they are unlinked, an envelope becomes independent of a clip's loop cycles and will effect the clip from the envelope Start position until the envelope End position. This is indicated in the envelope Start/End Point Box in the upper right section of the Envelopes Box. This allows an envelope to automate control over a clip for a specific duration or segment. An Envelope can also be set to loop independently of the clip with its own loop length that differs from the length of the clip itself. This is an interesting way to vary or randomize the effect of an envelope on a clip. For example, by using the Clip Envelope Region/Loop "Unlinked," you can literally extend your Clip's Envelope to loop beyond the clip's loop length/cycle to its own unique length. What does this mean? Let's say you have a preexisting drum loop of four bars, but you want it to fade for two bars and then reemerge again at its original volume at bar 1 and cycle (loop) over and over again. Launch to 14.6 for a detailed walk-through of looping with Unlinked Clip Envelopes. Once you grasp how this works, you'll understand the many great possibilities available with Clip Envelopes.

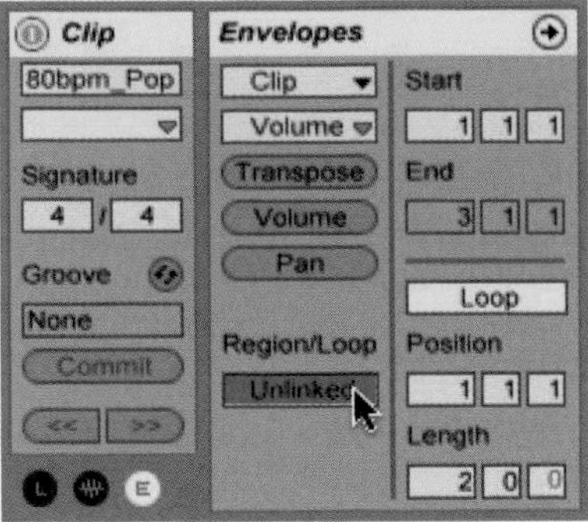

Figure 7.40 Unlink envelopes to create envelope shapes/effects that are independent of a clip's loop length.

7.9.3 Dummy Clips

"Dummy Clips" not only have a uniquely curious name but also happen to be an intriguing and clever trick-like feature in Live. Let's look at the properties of Dummy Clips as a means to control and automate the sound of your audio on each of your tracks in Live. We will be using the Session View with two audio tracks to explain how these Clips work. In this basic setup of Dummy Clips, you will need track 1 containing the audio clips you wish to enhance with your secondary Dummy Clip track. Here we go:

1. Load up your Track 1 with your single or multiple audio clips and make sure they are playing back normally.
2. Create a new audio track (Track 2). Select its input chooser to receive Audio from Track 1 (1-Audio).
3. Turn the Monitor Input of Track 2 from Auto to "IN."
4. Change the Output Type chooser on Track 1 from Master to "Sends Only." Test this by launching a clip on Track 1. With the correct routing assignment, you should be hearing and seeing your audio play through and monitored only on Track 2. The audio levels on Track 1 will show as a blue color signifying they are set to Sends Only.

Figure 7.41 Track 1 with Main Audio Clip, Track 2 set to Input Monitor.

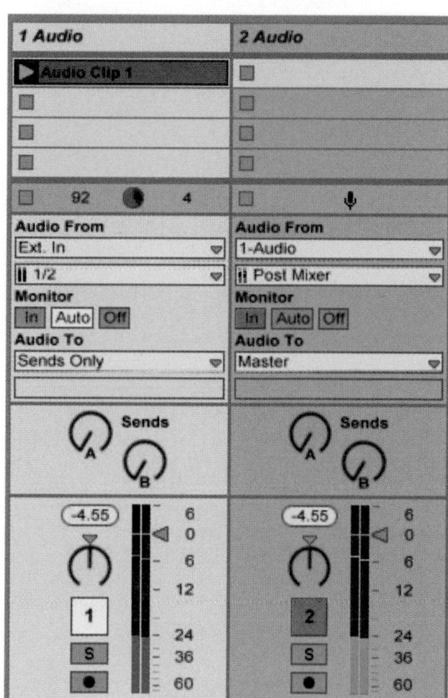

Figure 7.42 Blue signal meter indicates a Sends Only output signal.

5. Now load up two basic audio clips from the Browser into Track 2's first two Clip Slots.
6. Click on the first "Dummy" clip on Track 2 in Clip Slot 1 to open its Clip View and select the Envelope button at the bottom of the Clip Box to open the Envelope Editor.
7. Select "Ping Pong Delay" from your Audio Effects folder in the Device Browser and drag it to Track 2.

8. Press shift+Tab to go back to Clip View, then in the Envelope Box for "Dummy Clip 1," select "Ping Pong Delay" in the Envelope Device chooser menu, then select "Dry/Wet" in the Control chooser menu below it. Drag the Dry/Wet horizontal envelope value downward to 0.00%. (Dragging your cursor over the envelope value and breakpoint horizontal area turns the value line from pink to blue).

Figure 7.43 Drag the Dry/Wet horizontal envelope value downward to 0.00%.

9. Click on "Clip 2" on Track 2 and repeat the same steps, but this time set the Control chooser Dry/Wet envelope automation to 90% (Hold down [⌘] Mac/[Ctrl] PC while clicking and holding on the envelope value line to work in small ⌘ increments).

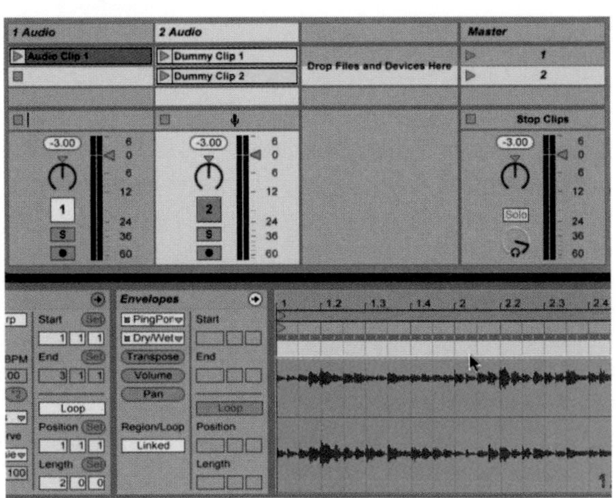

Figure 7.44 Dry/Wet envelope automation to 90%.

10. Launch your main audio "Clip 1" on Track 1 if it's not already playing.
11. While "Clip 1" on Track 1 is playing, launch "Dummy Clip 1" on Track 2.

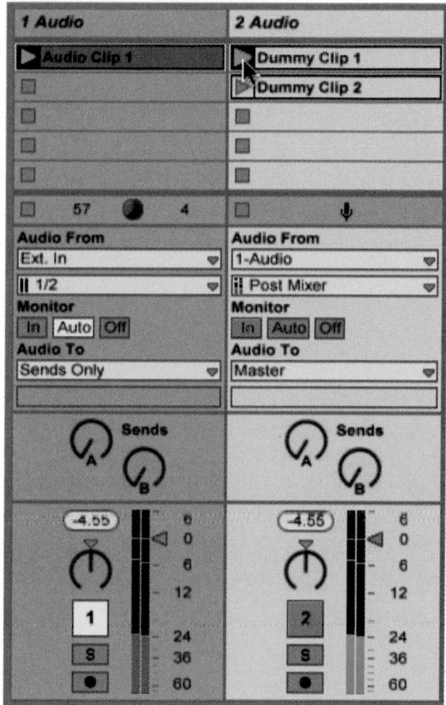

Figure 7.45 Launch "Dummy Clip 1" on Track 2.

12. After a few bars time, launch Dummy Clip 2 just below it on Track 2. You'll now hear the Ping Pong Delay affecting the audio.
13. Click back on "Dummy Clip 1" on Track 2 and you'll hear just the main/dry signal.

To add even more effects and changes to your audio, add another Dummy audio clip to Track 2 and proceed to edit and create new envelope automation parameters for each new clip. It's always a good idea to leave your Dummy Clip 1 in Clip Slot 1 designated and even renamed as a "Dry" Dummy Clip. Expand the Dummy Clip concept once more by adding Audio Effects Racks into the Dummy Clip column. By automating the Audio Effect Rack's various parameters in your Envelope chooser and Sub chooser windows, you can add infinite possibilities to your clip's audio output. Remember that you can always go back to your "dry" Dummy Clip in Clip Slot 1 to achieve a clean, unaffected signal. Also, always make sure that the Monitor section is set to "IN" on your Dummy Clip track.

Figure 7.46 Monitor section is set to "IN" on a Dummy Clip track.

By following the steps above, you'll be creating innovative ways to add effects to your clips in no time. Don't forget that you can always save these new Dummy Clips to your Browser in a custom Dummy Clips folder to recall and load them into your Set when needed. Now that you have completed this exercise, check your Set against our Set. Download **CPP_7-9-3_DummyClips Project** from our Website and see how it works.

Hot Tip *Make sure to keep an eye on the Ableton Website forum for customized dummy clip Live Packs. You can usually find that several of these made available by other Live users. They are great to not only incorporate into your production but also to use as guides and templates for your own Dummy Clip creations.*

7.10 Musical concepts

There are so many ways to create, produce, and perform with clips that we could write a whole other book just on that alone. Saving that for a rainy day, let's focus on some specific ways to enhance your music through Follow Actions.

7.10.1 Create: Rhythmic Loop Creations with Follow Actions

A great way to create unusual and surprising results with Follow Actions is to load up a single track and fill it with 8 to 10 rhythmic style audio beat loops in consecutive Clip Slots. Each clip should be no more than two bars in length for our example. Let's try this together.

1. After you load up your clips, set the tempo to 120 BPM.
2. Randomize each clip's transpose knob by 2–3 steps in alternating directions. Select one clip after the next and make the adjustment to each.
3. Adjust the Loop Brace for each clip to start and end randomly and different from the other clips by shortening the Loop Start/End Loop Brace using a Fixed Grid resolution of 1/4 or 1/2.

4. Now adjust the Follow Actions for all clips at once (multiclip functionality: shift+click) to these settings:
   - Follow Action Time chooser: 0-2-1
   - Follow Actions A: Any Follow Actions B – Other
   - Chance A and B: 0–7

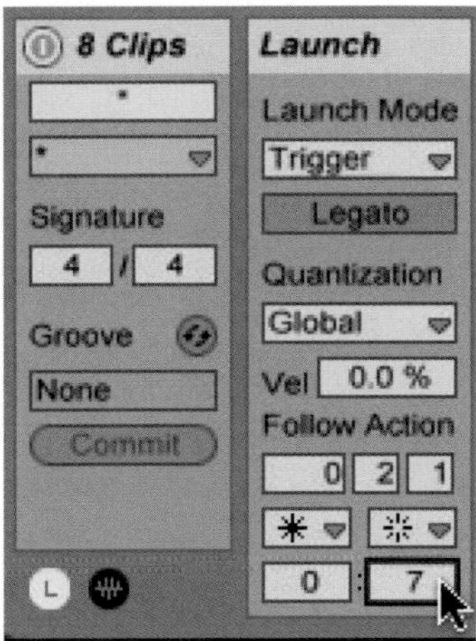

Figure 7.47 Follow Actions being applied to multiselected clips.

   Again, it's important to use shift+click to select all the clips at once (multiclip selections).

5. Launch one of your clips and sit back to see what happens.
6. Now increase the tempo 15–20 BPMs to achieve a completely different sound and vibe with the rhythmic feel of the clips.

This is just an idea, but from here you could add a delay to a Return track as a send effect for your track to access even more possibilities. You can also experiment with reversing one or two of the clips and add even more random results. Now check your Set against ours. Download **CPP_7-10-1_FollowActions Project** from our Website and see what we came up with.

When working with a Follow Action sequence that you generally like the sound of, you might consider resampling the audio from the Follow Action track into a new record-enabled audio track (print to track) 9.5 using the track's Input chooser *Audio From*. You can then create a new and customized audio clip with the resulting and randomized Follow Action parameters you made in the first place. Save this to your Browser for later use. You'll be surprised at what you create. Launch to 9.1 for recording clips.

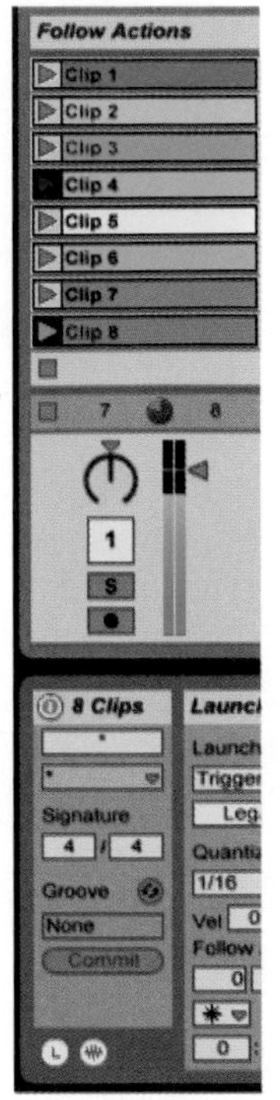

Figure 7.48 Follow Actions in action!

Keep in mind that Follow Actions can be fairly unpredictable depending on how they are set. In some cases, no single Follow Action's "pass" will play exactly like the one before. Therefore, when you do decide to resample the audio output of an unpredictable Follow Action sequence, remember that although it may have the same general rhythmic vibe and feel as it launches from clip to clip, you won't get the exact same pattern each time you launch the Follow Action process. After all, this is one of the basic principles behind using Follow Actions in the first place, randomization and variation. For a more logical and predictable sequence of events, use the "Next" parameter within both the Follow Action's A and B chooser window in order for all your clips to play in order and cycle back around continuously. This is great for setting up a predictably long and precise stage Set or "unmanned" visual performance that requires cues to be in an absolute order.

You will see in our two examples that Follow Action A and B menu's are both in a randomized state "Any" or "Other" and in a more predictable state with the selection as "Next." Notice the other choices included in the chooser menu. By selecting "Any" or "Other," you'll find exciting possibilities in your clip's behaviors.

Figure 7.49 Try Follow Action "Other."

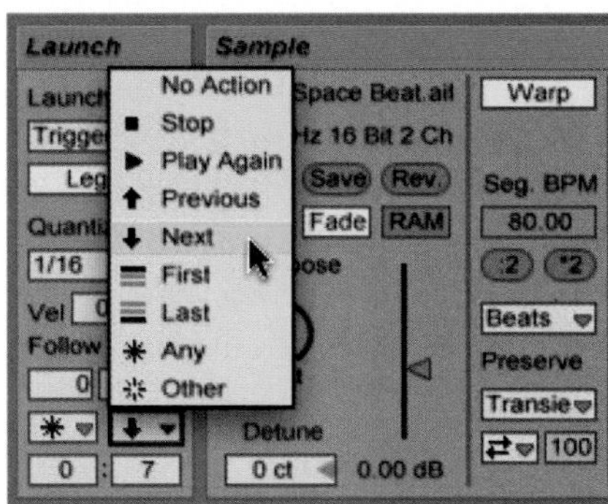

Figure 7.50 Try Follow Action "Next."

Of course, all this can be accomplished by simply aligning your clips in the Session View, launching them in order and globally recording them as a sequence into your Arrangement View. For more on Global Recording, launch to ▶ *Scene 4*.

In this chapter

8.1 Introduction to groove 175

8.2 Grooves 176

8.3 Groove Pool 176

8.4 Commit groove 179

8.5 Extract groove 181

8.6 Musical concepts 182

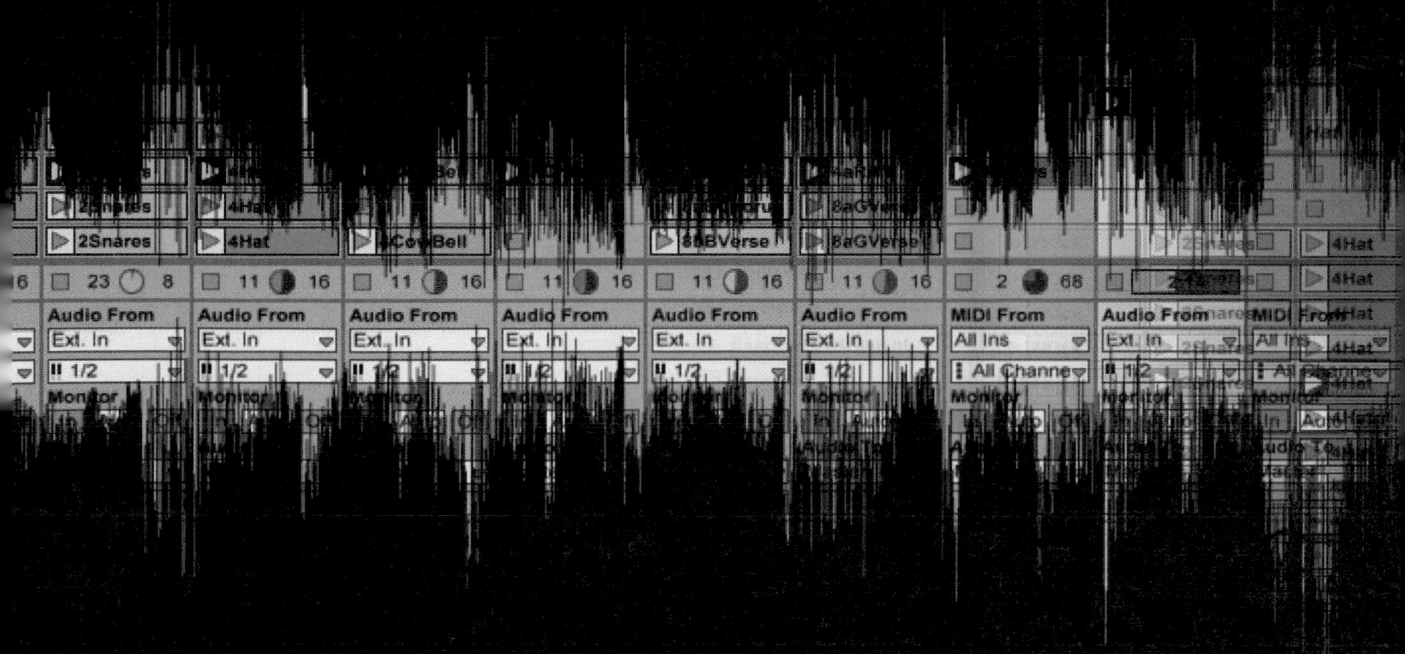

SCENE 8

Groove

Since incorporating Live 8 into my arsenal, I can say it has uniquely changed the way I approach song writing altogether. Once you capture its power, it wraps around you in the way you work. Nothing makes audio liquid like Live.

Barry Ledere, Keyboardist with Brooks & Dunn-Nashville, TN

8.1 Introduction to groove

In a musical sense, a groove is the rhythmic feel of a song or beat, but more than that, it is a real-world phenomena that occurs when a human performs a piece of music live (on stage) or in a recording session. It is the breath of life that makes music come alive, breathing with the beat. Each performer or band generates their own groove when they play and perform, in turn stamping their own music or style with a signature. Their respected fans and admirers often seek after this "signature sound." In an effort to recreate this experience in the digital audio workstation (DAW), Ableton has developed its own groove technology called "grooves" for Live 8. The goal of groove technology is to breathe life into clips when they lack the "authentic feel" of a real performance and/or to conform two different clips so they have the same feel. In other words, to either liven up a "vanilla" beat that doesn't groove and is lacking soul or to reshape the beat to emphasize a different feel than originally performed. For this reason, Live has provided a number of grooves as part of the Live 8 and Suite 8 Library. Applying one of these grooves to your audio or MIDI clips changes its timing and feel, conforming it to the new groove. A prime example of this is the recreation of a swing feel. A true swing feel is not something you can mathematically quantize in the digital realm. It is truly a human experience that doesn't have an exact formula. With Live's groove technology, you can take that real inexplicable human feel and apply it to a straight beat. Even better, extract it from a live performance recording. Who needs the mathematical equations when you can get the groove from the source! We've already mentioned the Clip Groove chooser menu available from the Clip View Clip Box ▶ 7.3.2, but let's take a close look at how grooves and Groove Pool work, and how to Extract and Commit (print) grooves.

8.2 Grooves

Figure 8.1 Library groove presets.

Grooves are located in the Live 8 and Suite 8 Library and are accessed through the File Browser. You'll notice right off the bat that Ableton has done some great research in providing some of the more popular grooves, such as the MPC and Logic grooves, as presets located in the Library>Grooves folder. There are a few different ways to load and work with these grooves. One way to apply a groove to a clip is to drag it directly from the Browser onto a clip. It will instantly alter the clip's feel in real time. Once you've placed the groove on the clip, you can then preview other grooves for that clip using the Hot-Swap Groove button. Hot-Swap Groove is used to quickly select, change, or audition grooves on the fly. Click the Hot-Swap Groove button in the Clip Box then try out other groove presets located in the Grooves folder. This is one way to quickly load and audition the various groove presets. Try them out and see how useful they can be.

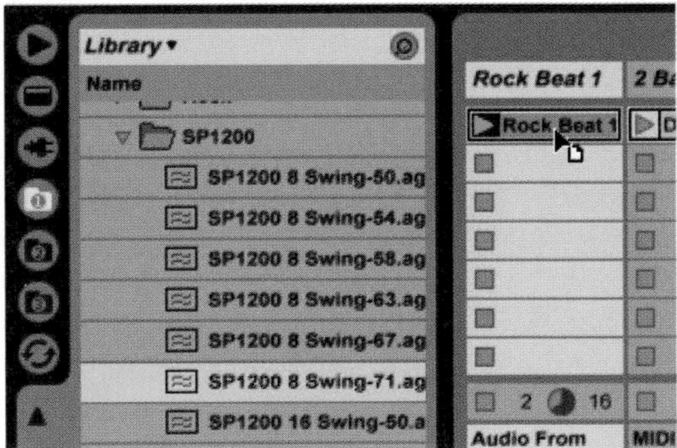

Figure 8.2 Drag and drop a groove from the Library onto a clip.

8.3 Groove Pool

After a groove has been dropped onto a clip, it will be stored in Live 8's Groove Pool located below the File Browser. You can also drag grooves directly into the Groove Pool from the Library. To show the Groove Pool, click the Groove Pool

Selector just above the Info View or choose Open Groove Pool... from the Clip Groove Chooser in the Clip Box. This is where all grooves being used or previewed in your Set will appear. It is also where you view and edit all of the grooves in your current Set. Keep in mind these are not audio files. They are files containing timing information that has been extracted from an audio or MIDI clip. More on extracting grooves in Clip 8.5.

Figure 8.3 Groove Pool.

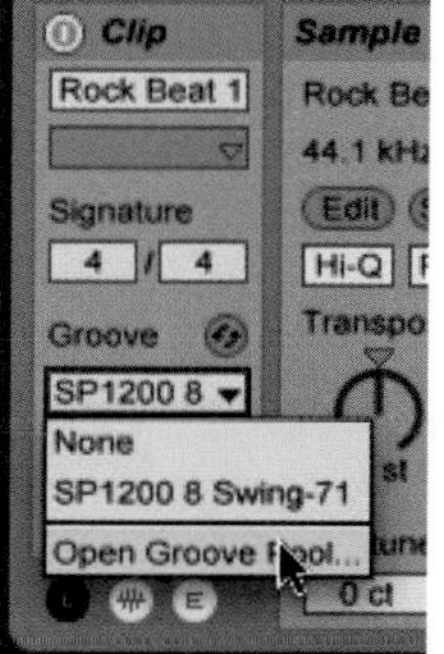

Figure 8.4 Grooves in the Groove Pool are accessible in the Clip Groove chooser.

As an alternative to dropping grooves directly onto clips, you can drag them from the File Browser into the Groove Pool or double-click on one and it will be added. Once they have been added to the Groove Pool they can then be accessed via the Clip Groove Chooser. Use the Hot-Swap Groove button to quickly audition them on a clip. From the Chooser menu, you can select and apply any of the grooves that are stored in the Groove Pool to a selected clip. This integrated list of grooves is always available and even accessible while working and performing in real time. This gives you an incredible amount of flexibility by having the unique feel and timing of various grooves available for your current Set. Grooves in the Groove Pool can be Hot-Swapped with other groove presets in the Live Library or saved as presets to the library. This gives you the ability to Hot-Swap clip grooves with the groove presets stored in the Groove Pool and Hot-Swap grooves in the Groove Pool with groove presets in the Library. Each groove in the pool has its own Hot-Swap button and Save button located next to its name.

Figure 8.5 Save and Hot-Swap grooves to the Library.

8.3.1 Groove Pool parameters

Within the Groove Pool Menu are built-in groove parameters for tailoring a groove's effect on its assigned clips. There are five parameters for each available groove in the Groove Pool: Base (Resolution Base), Quantize, Timing, Random, and Velocity. In addition, the Groove Pool itself has a Global Groove Amount slider that controls the overall effect amount that grooves will have over all clips with a groove assigned. Each clip groove parameter in the Groove Pool determines how a groove affects its assigned clips. When clips are assigned to a groove, the groove's parameters will become active. Here is a description of each parameter.

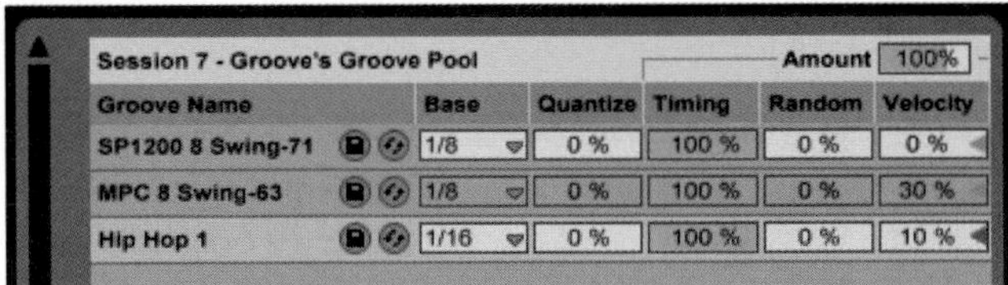

Figure 8.6 Groove Pool parameters.

Base: It sets the subdivision of the beat that the groove and assigned clips are measured against. This is related to the smallest beat value expressed in the groove and clip, such as the lowest common denominator.

Quantize: The amount of quantization Live applies to assigned clips before they are requantized by the groove's feel.

Timing: It shows how strong of an effect the groove will have on all its assigned clips.

Random: The level of fluctuations (randomness or humanization) applied to the timing of assigned clips.

Velocity: It shows how much effect the groove's velocities will have on the velocities in the clip. This can be applied with an inverse effect based on positive or negative values.

Global Groove Amount: It controls the overall percentage of strength of timing, random, and velocity that will be applied for all grooves.

Let's take these for a test drive!

While working with different clips and grooves, take a few minutes to experiment with the Quantize and Timing parameters in the Groove Pool. Once you have a groove loaded in and applied to a clip, adjust the Timing value to 0, and then the Quantization to achieve your desired influence on the current audio clip's "feel" as it plays. If you drag the Quantize percentage all the way up to 100% you will surely start to hear the effect, especially when applying straight 1/16 or 1/8 values to a swing-style audio or MIDI clip. Assuming you have placed the Quantization value in the middle say, 50–60%, gradually start adjusting the Timing value. Timing will utilize the actual audio clip's "feel and groove" to start pushing and pulling the groove toward or away from the Quantized grid location that you adjusted earlier with the Quantize percentage. Heavy stuff, huh?

This is a great way to begin learning how to use grooves. The key is customization. Once you find or create a new groove you like, simply save it in the Groove Library on the fly and keep on working.

8.4 Commit groove

Auditioning and experimenting with grooves on your drum track clips and other clips is a wonderful thing. You can do it all in real time, trying as many grooves as you like. This is possible because it's a nondestructive process, that is, until you click the Commit button. Commit is used to permanently apply a groove assignment to a clip, which writes the groove to the clip by adjusting a MIDI clip's notes or a Sample's Warp Markers. Note that this is a somewhat destructive setting, meaning that when you commit a groove to a MIDI clip, its MIDI notes are moved (quantized) to match the groove. When applied to audio clips, Warp Markers are added and adjusted to quantize the clip to fit the groove. Until you commit a groove, it is only temporary. This means that Live constantly calculates and analyzes the file and groove in real time until you commit it. After you select Commit, "None" will appear in the Clip Groove Chooser signifying that the groove has been permanently written. Of course, you can always use "undo" if you haven't made a million other changes to your Set.

Hot Tip *If you commit a groove to an audio clip then realize that you don't like the feel, what are you going to do? For audio clips, you can always delete all of the Warp Makers added by the groove engine then rewarp the sample back to its original state, or just call up the original clip from the browser. As you have no doubt realized by now, Live allows you many rabbit holes to journey down.*

For MIDI, you can use the quantize feature or manually move the MIDI notes back into position. Here is an example of the results of the Commit Groove feature. In our first figure, you will see the original "straight" clip then a resultant grooved clip to "8th-note swing." Remember, a swing feel drags back the 8th-note "+ beat," compressing its proximity closer to the next downbeat.

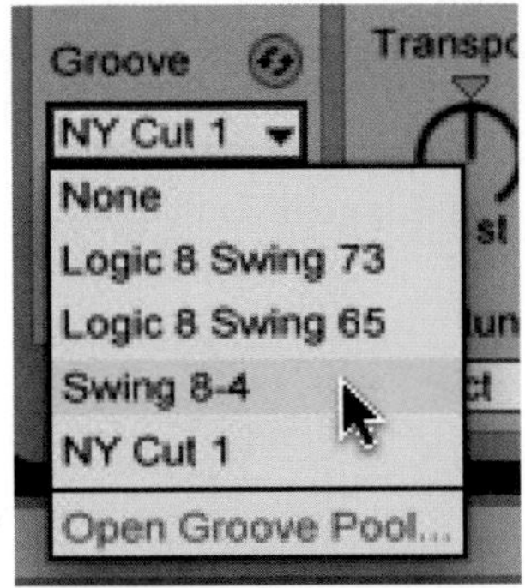

Figure 8.7 Swing groove chosen from the Groove Pool.

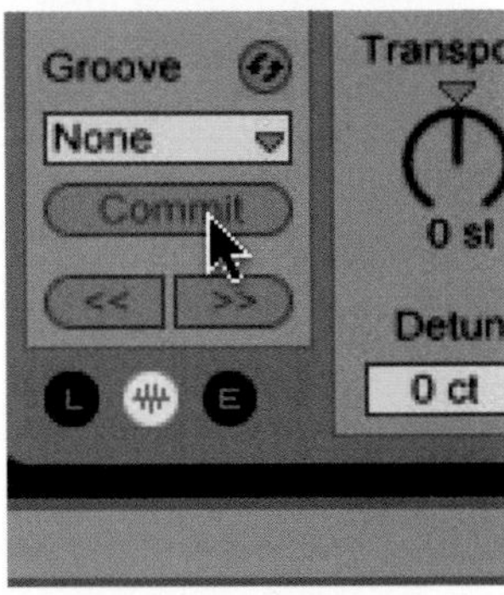

Figure 8.8 Commit groove.

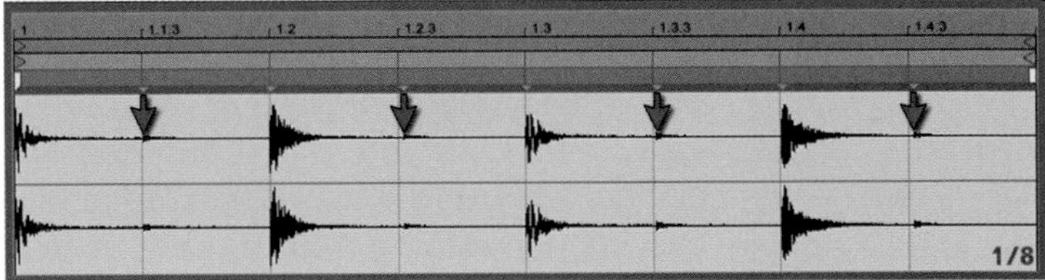

Figure 8.9 Original "straight" feel audio clip.

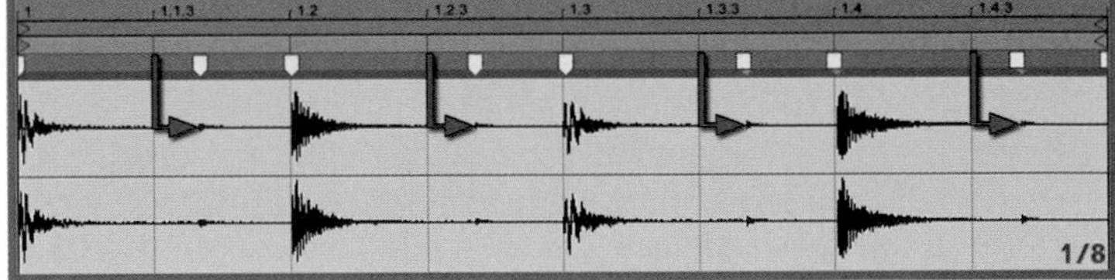

Figure 8.10 Swing groove committed to the audio clip moving the eight-note back.

You can see from our earlier and later example that the eighth notes were assigned a Warp Marker and were slid back towards the next down beat. The similar process happens for MIDI clips.

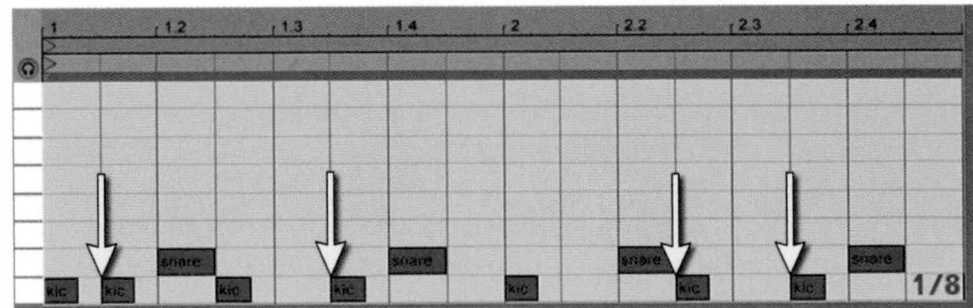

Figure 8.11 Original "straight" feel MIDI clip.

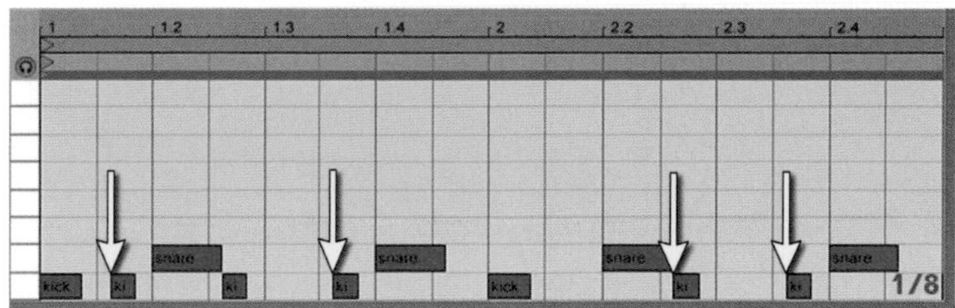

Figure 8.12 Swing groove committed to the MIDI clip moving the eight-note back.

You can see that the MIDI notes have been changed, both position and some lengths. Once again, the eighth-note upbeats have been pushed back.

8.5 Extract groove

Enough with using groove presets, let's use our own! If you have a killer drum groove or loop, Ableton has made extracting and adding it to the Groove Pool very easy. Simply drag any clip or MIDI/audio file directly into the Groove Pool and Live will analyze and extract its groove. It then becomes available in the Groove Pool for applying to other clips. This is very useful for confirming your clips to another song or beat. Alternatively, you can extract grooves by selecting the Extract Groove(s) command from the clip contextual menu. Right + click or ctrl + click on clip and select it.

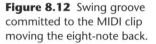

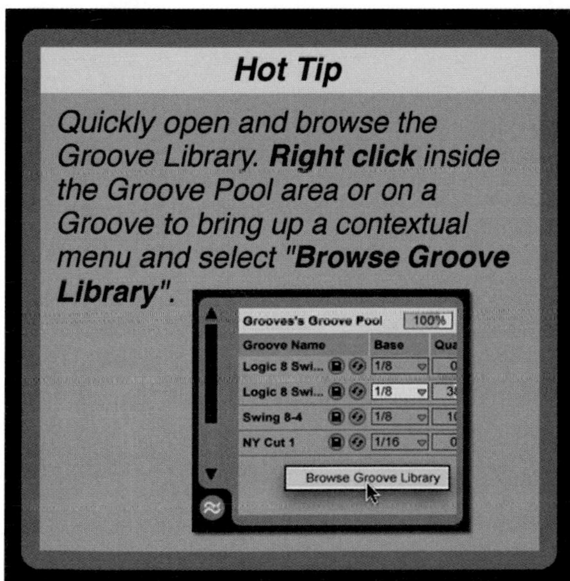

Hot Tip
Quickly open and browse the Groove Library. Right click inside the Groove Pool area or on a Groove to bring up a contextual menu and select "Browse Groove Library".
Grooves's Groove Pool
100%
Groove Name
Base
Logic 8 Swi...
1/8
Swing 8-4
NY Cut 1
1/16
Browse Groove Library

This can become quite addictive as you may find yourself extracting everything you can get your hands on – old vinyl records, vintage sample libraries. You name it. All is "extractable" and that's what makes the groove function so fun and enjoyable to use. Applying a customized extracted groove to a clip that may have been out of pocket or useless in a particular Set can jump start your work 10-fold.

8.6 Musical concepts

Being that a large portion of studio recorded albums are at some point produced "in the box" it's important that they sound as life-like or humanized as possible. Even though your next pop record will be recorded with all live session performers, let's not forget that there are many musical genres where preproduction begins as a mock-up of samples and loops with maybe one or two live recordings. Even the big artist albums or film scores may spend time in the land of MIDI using virtual instruments and samples. To be honest, a large majority of TV, film, video games, and Internet audio content are completely conceived in the DAW without a live performance, some of which never see a live musician. The point is that Live 8 attempts to remedy this by providing the ability to use samples, live performance, and MIDI altogether so that they feel right and organic; not only for sample-to-sample integrated content, but also for grooving with a real studio musician.

8.6.1 Create: grooving your backing tracks

So far, we've shown and discussed many ways to utilize groove functionality in Live 8. Now that you've had some time to see it in action, you'll find it's easy to get caught up grooving with your drum tracks and extracting your favorite drummers. Don't forget, this feature also works great with bass lines and guitar phrases too, not just drum or percussion-based rhythmic loops and recordings! One way to work with guitar and bass phrases is to dial in your drum track clip's clip groove so it feels exactly how you want it to. Once your drum track clips are grooving – Clip Groove in place – select the same groove for your bass and/or guitar track clips. Because your drums are already using the Clip Groove, it's very easy to select it through the Chooser Menu in the Clip Box. Go ahead and do this for each additional instrument you want to conform. Before you launch all of your clips or scenes, you'll want to spend some time working with the timing for your bass and/or guitar clips by dialing in the perfect quantize and timing values in the Groove Pool. Your ears should tell you fairly quickly what is working and what is not working as you massage your bass and guitar clip's groove and feel against your drum clip. As this affects all drum clips as well, there may very well be some compromises made to get the feel to work across the board. If you really want to push things to the limit – for fun of course – try mixing and matching a straight quantized groove template for a bass clip against a swing quantized groove template on the drum clip.

In this chapter

9.1 Recording MIDI clips 185

9.2 MIDI overdub recording 190

9.3 Freezing and converting MIDI clips into audio clips 196

9.4 Recording audio clips 198

9.5 Exporting and printing 207

9.6 Musical concepts 211

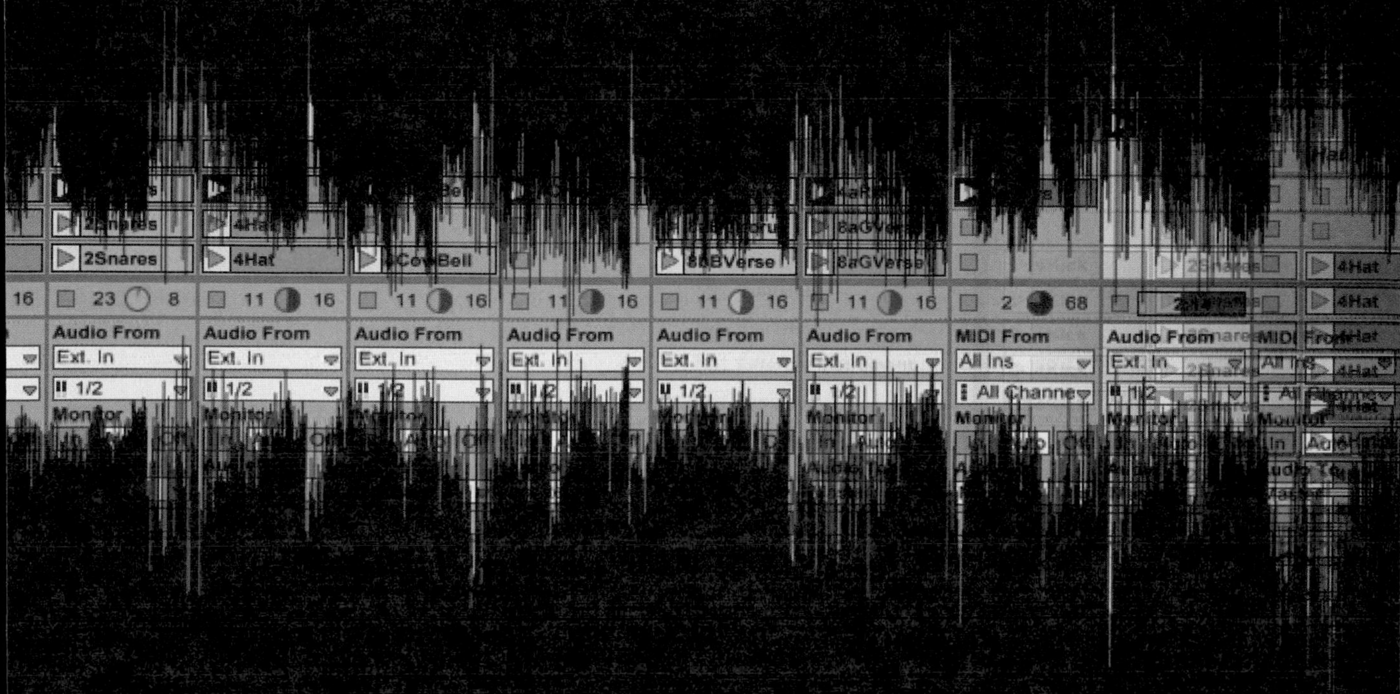

SCENE 9 Recording

Having a wide variety of DAW choices in my studio, I always find myself starting EVERYTHING in Ableton Live 8 first. The ability to use VST & AU plugs is a huge plus for me. I can then change tempo, pitch, harmonics, slice my audio to MIDI, freeze tracks and get real crazy with the new warping features. As a remixer Live 8 makes life so much easier for me.

Mike Acosta aka DJ Michael Trance, Remixer, & Sound Designer

9.1 Recording MIDI clips

After all of this talk about clips, you probably want to know how to record your own MIDI and audio like you have with other digital audio workstations (DAWs). Great news! Recording clips in Live 8 is very easy and quite inspiring. Ableton just loves to make things simple and intuitive, and that's exactly what they've done with the process of recording. To begin, let's take a look at the two ways to record new MIDI clips in Live. You can either start out with an empty *Clip Slot* or an empty *clip*. We'll do both, but first let's go over the empty Clip Slot approach.

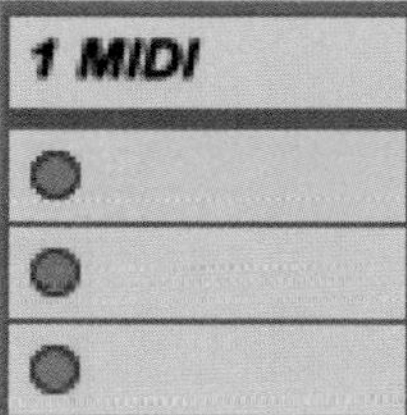

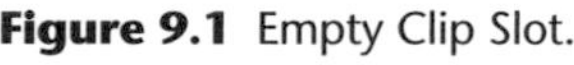
Figure 9.1 Empty Clip Slot.

Figure 9.2 Empty Clip.

9.1.1 Clip Slots and clips

Open Live 8 for this exercise. When you first open Live or create a new Set, you will see empty Clip Slots. Remember, a Clip Slot is where clips are stored in the

Session View. To begin, we need to first set up our Set with a software instrument to create sound.

9.1.1.1 Setup

1. Go to the Live Device Browser and navigate to Instrument>Impulse. With a MIDI track selected, double-click on the "Backbeat Room" preset or drag and drop it onto a MIDI track. Don't forget that the Device Drop area works too. If you don't have this particular preset, then choose a different one or download the related Live Packs from Ableton's Website.
2. Now that you have a drum kit loaded, record-enable your new Impulse "Backbeat Room" MIDI track. Click on the Arm button at the bottom of the track.

Figure 9.3 Record-enable your MIDI track.

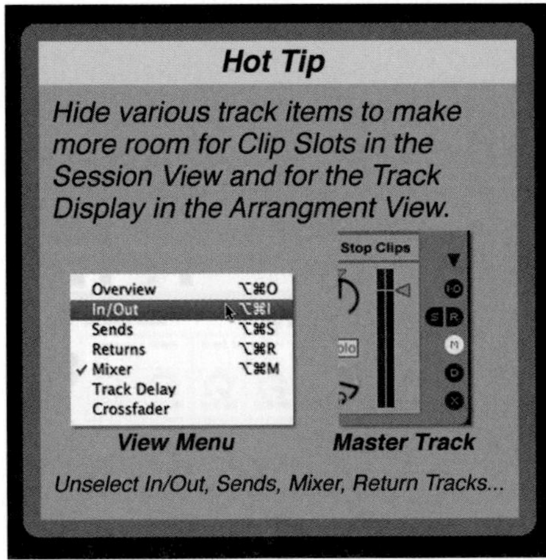

You should notice that all of the Clip Stop buttons in every Clip Slot of the armed track changed to Record buttons.

3. Now test out your MIDI drum kit to make sure that it makes sound when you play a MIDI note. If you have a MIDI keyboard, play the key C3–C4.

If not, use the computer MIDI keyboard, keys a–k (activate the computer MIDI keyboard on the Control Bar).

You should see the MIDI input channel meter lights up yellow, the track level meter lights up green, and should hear a percussive sound. If you are not

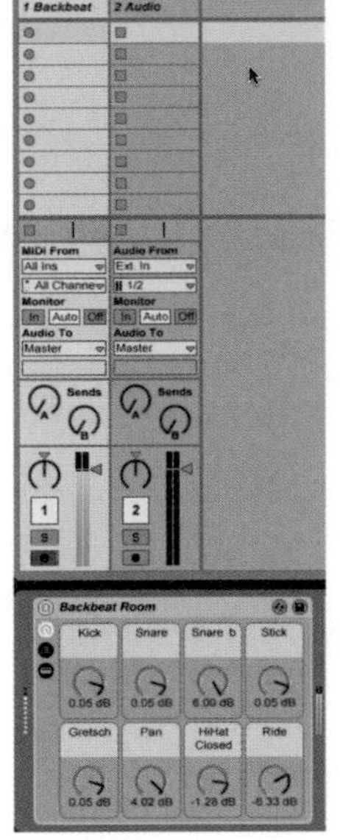

Figure 9.5 The MIDI input channel meter lights up yellow, the track VU meter lights up green, and should hear a percussive sound.

hearing a sound when you play, then make sure that the track is activated, record-enabled, and that monitor is set to "Auto" or "IN." The best way to troubleshoot is to determine where in the signal chain there is a problem e.g. MIDI input or Audio output. Compare your setup to our example.

4. Once your instrument and playback sound is working, we need to set up Record Quantization so we can record MIDI free of timing errors (Edit Menu). Record quantize automatically corrects or alters your MIDI performance relative to the grid as you play. A real-time saver and a necessity for performing on the fly! Let's set it to Eighth-Note Quantization. This means that every note you play will lock to the closest eight-note grid line position. Always choose the smallest resolution (subdivision of the beat) that you plan to play.

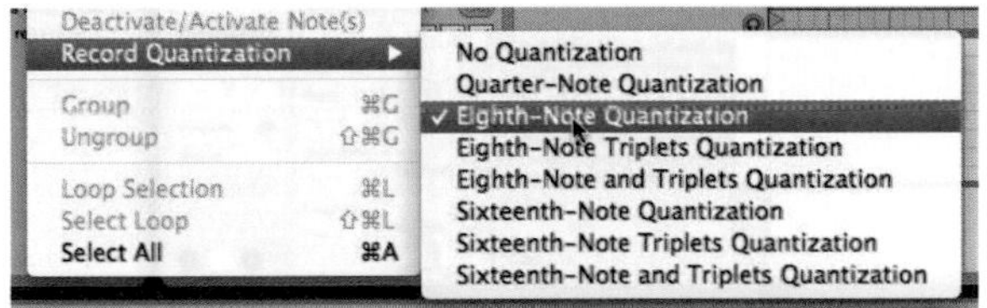

Figure 9.6 Record Quantization settings under the Edit Menu.

5. Activate the Metronome at the top left of the main Live screen and make sure that a count-in is enabled (right click or ctrl + click on the Metronome).

Figure 9.7 Metronome On/Off.

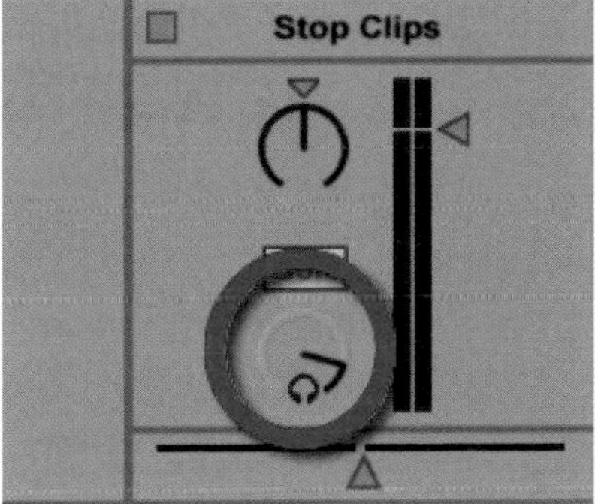

Figure 9.8 Preview/Cue Volume knob controls the Metronome volume.

One-bar is a good setting for this. If you need to adjust its volume, do so from the Preview/Cue Volume knob at the bottom of the Master track.

9.1.1.2 Recording to an empty Clip Slot

1. *Record*: Let's record a four-bar drum loop. Any beat is fine, just keep it simple. Click a Clip Record button and begin playing after the count-in. Record only four bars! (Notice the real-time Clip View display!)

   Hot Tip *You can select your count-in value by right clicking on the Metronome button. Here you'll see a list of values to choose from including one bar.*

2. *Stop Recording*: To stop recording, disengage record on your track by clicking on the Arm button while the clip is still running. You can also click the Control Bar's Stop button or click the Stop All Clips button. Each one has a slightly different effect. For this demo, just disarm the track. This way you will hear it playback immediately – perfect timing too! After it has played through it will start looping at whichever point you disarm the track, most likely a few beats beyond a true four-bar loop, unless you were quick to disarm. If you are even a hair late, Live will loop off beat. If you have the overdub (OVR) button activated, you can click on the clip's Record button to restart recording for adding more MIDI to your newly recorded clip. More on this in Clip 9.2.

   Hot Tip *Make recording even simpler by double-clicking on a MIDI Clip Slot to automatically create an empty one-bar clip template. This allows for easy looping without punching out (disengaging record) at the end of the bar. Duplicate these "templates" multiple times down the same track to fill out consecutive empty Clip Slots. This way you have plenty of ready-to-go one-bar MIDI Clip templates to record into. Easily increase the loop length in the clip's Notes Box Loop Length input field.*

3. *Set Loop*: Adjust the loop length in the Note Editor so that your new clip is a four-bar loop while playback is still running. You can do this by dragging

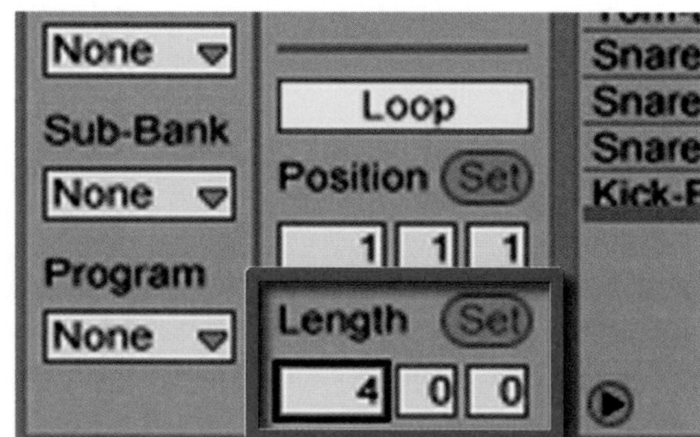

Figure 9.9 Set loop length to four-bars in the Notes Box.

the end of the Loop Brace to bar 5, or set the loop length to "4" bars in the Loop Length input field in real time. For the adventurous users, click the Set button during playback when the Playback Cursor reaches bar 5. Enjoy and listen to your loop as it now plays back in perfect time. Stop whenever you are done listening.

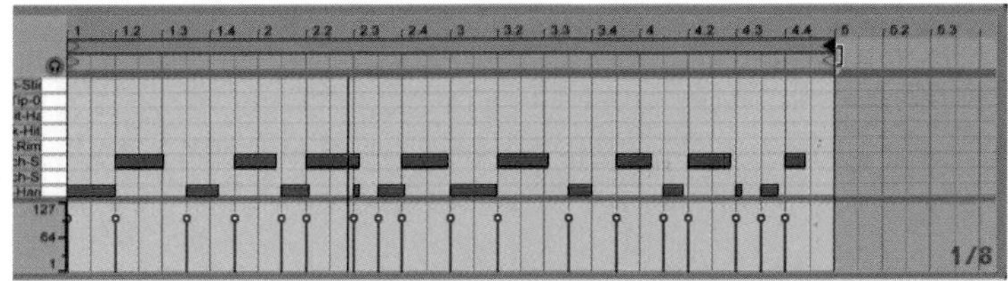

Figure 9.10 Set loop length to four-bars by dragging the Loop End.

Now let's do the same thing, but this time with an empty clip (clip template):

9.1.1.3 Recording to an empty clip (clip template)

1. Double-click in an empty Clip Slot below the one you just recorded in. This will create an empty clip template, which will appear in Clip View.
2. *Set loop*: This time we will preset the loop length. Either drag the loop brace to bar 5 or set the loop length field to "4" bars.
3. *Record*: Arm your track, then click on the Clip Launch button. After the one-bar count in, begin recording/playing your beat up until the end of the fourth bar.
4. *Stop recording*: Although the clip is still recording, you can simply stop playing and listen to the clip playback without disarming the track. Disarm if you want, or simply enjoy the freedom of Live. The Metronome can become annoying, so turn that off. Stop playback when you are done listening (press a Clip Stop button then spacebar).

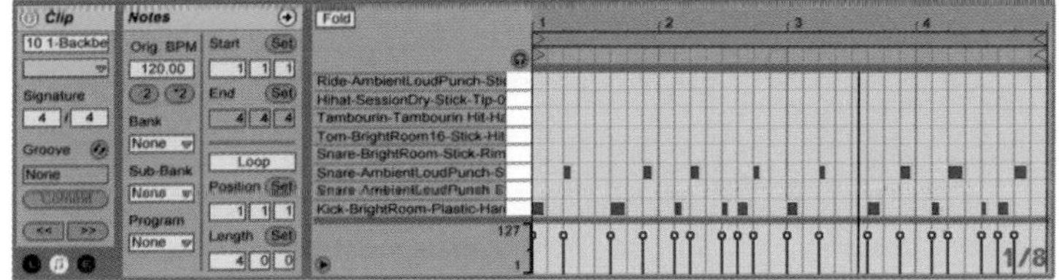

Figure 9.11 Your clip should look similar to our drum kit clip.

9.2 MIDI overdub recording

The OVR Switch [OVR] located on the Control Bar is called Overdub. Its functionality differs when used in either the Session View or Arrangement View. When activated and working in the Session View, you can record multiple takes/-passes of MIDI notes into the same MIDI clip without overwriting the existing notes unless they are the same note. In that case, Live will formulate a consolidated version of the new and the old based on where they overlap. When deactivated, nothing is recorded into a selected clip. In this particular situation, you can only record to empty Clip Slots. This is useful for going in and out of record mode to practice your takes while looping and monitoring input. In this way, consider the following process: with OVR deactivated, arm (record-enable) your MIDI track and then launch it. Practice your take a few times, then when ready, activate OVR to record your MIDI input into the clip. You can use this technique to add to that clip or to overwrite specific notes already in the clip.

Hot Tip *A great way to activate the OVR button is to assign it to a key on your computer keyboard via the Key Map function. Launch to Clip* ▶ 12.3 *for more on this cool feature.*

It's probably safe to say that the Session View is ideal for OVR recording, but often times you need to record a MIDI take directly into the Arrangement View rather than the Session View. Maybe you need to OVR a section in your Arrangement, or a little passage here or there to polish up your work. This is the beauty of the Arrangement View. Not only is it used to capture a performance from the Session View (Global Record), but it's also a traditional linear DAW. When it comes to recording MIDI clips it is very similar to any other DAW. The biggest difference is the way in which multiple takes and overdubbing is handled. In the Arrangement View the OVR function is used in a more traditional sense ▶ 9.2.2 than in the Session View. With OVR activated, it functions the same as already described for the Session View. Notes are added to an existing Arrangement clip, only overwriting identical notes that overlap on the same beat. When OVR is off, each recording is an entirely new take, with each record pass overwriting all of the existing MIDI notes instead of adding to it or doing nothing as in the Session.

9.2.1 Session View loop recording

1. Create a new empty clip (clip template) on your Impulse track. This will insert a new clip and bring up Clip View.
2. *Set Loop*: activate the clip Loop Switch, then set the clip's loop length to extend two bars, ending on bar 3. Use the loop length field in the Notes Box and set it to "2."
3. Turn the Metronome on and set Record Quantization to "Eighth-Note."

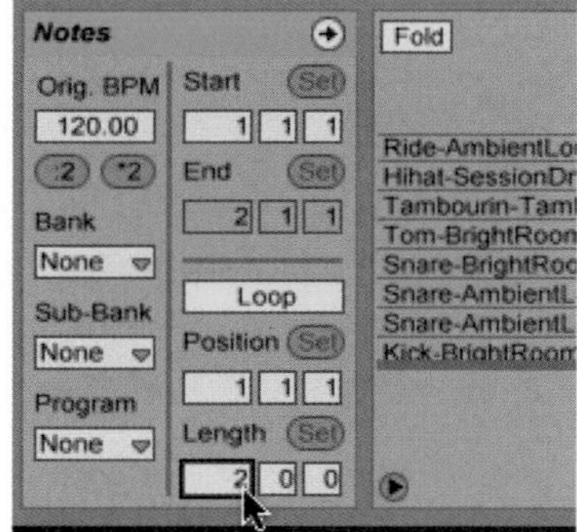

Figure 9.12 Set the loop length field in the Notes Box to two bars.

4. *Record-enable:* Arm Session Recording on your MIDI track and make sure OVR is activated.

   Now you are ready to record a simple "downbeat rhythm" for the kick drum.

5. *Record*: Click the Clip Launch button to begin recording. After the count-in, record the kick drum on every downbeat (8 beats in all). After recording those beats, stop inputting notes and let the clip loop.

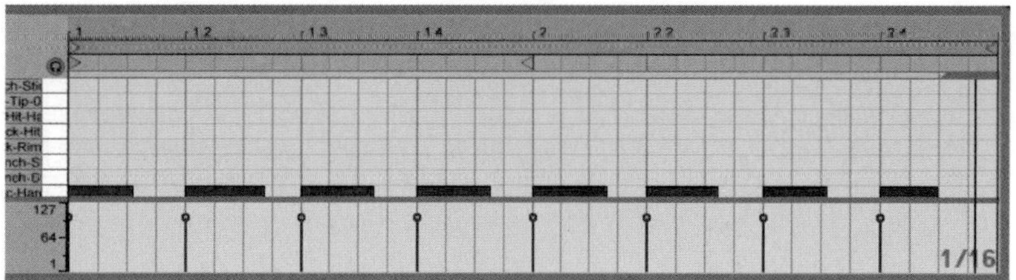

Figure 9.13 Record the kick drum on every downbeat.

Now we are going to add a snare "back-beat" (beats on 2 and 4). Let's play along without actually recording the notes.

6. While clip is still running, deactivate OVR. Begin performing the snare along with the looping clip, playing on beats 2 and 4. After few loop cycles, you should be ready to record the snare.

7. Reactivate OVR and begin recording the snare (four notes in all).

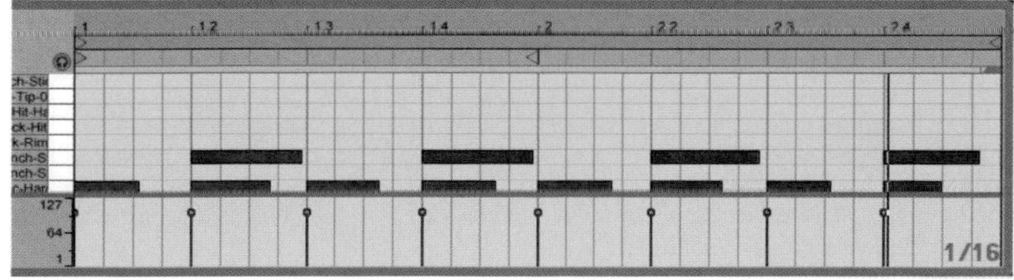

Figure 9.14 Record the snare drum on beats 2 and 4.

Let's make this super simple beat more interesting. While the clips are still looping, we'll add a cymbal ride. You can deactivate OVR if you want to rehearse or just dive right in. It's up to you.

8. Record an eighth-note ride or hat pattern on every eighth-note, e.g. 1 + 2 + 3 + 4 + ... (16 in all). Let this continue looping when you're done inputting the eighth note ride pattern.

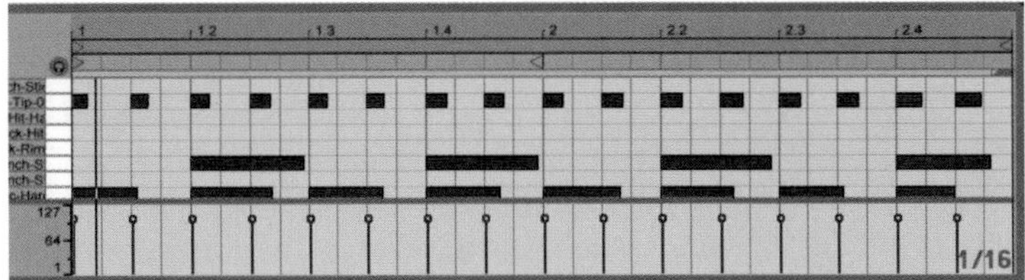

Figure 9.15 Record the eighth-note ride pattern on every eighth-note starting on beat 1.

9. Add a kick drum on the + of 3 in each bar. This is the eighth-note beat right after beat 3 (after the snare hit).

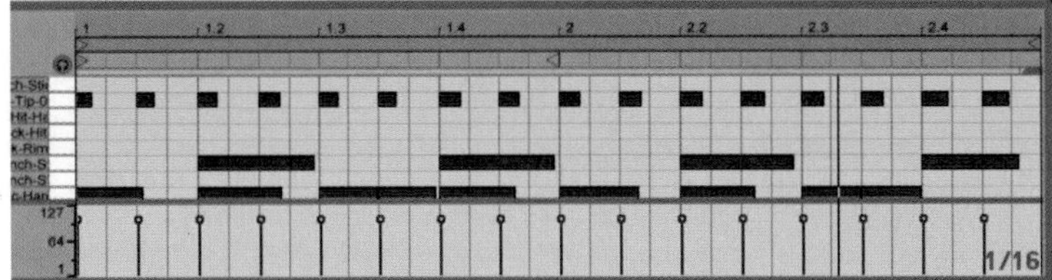

Figure 9.16 Add a kick drum on the + of 3 beat in each bar.

Turn off the Metronome, if you haven't already, and listen to your perfectly quantized drum loop!

9.2.2 Arrangement View loop recording

For this demonstration we will use the same Impulse drum kit.

1. Click the Back to Arrangement button to return playback priority to the Arrangement View.
2. Activate the Loop Switch and set a Loop Brace at bars 5–7 along the Beat Time Ruler (two bars). You can drag the Loop Brace into position or highlight bars 5–7 in the Track Display then type [⌘+L] Mac/[Ctrl+L] PC].

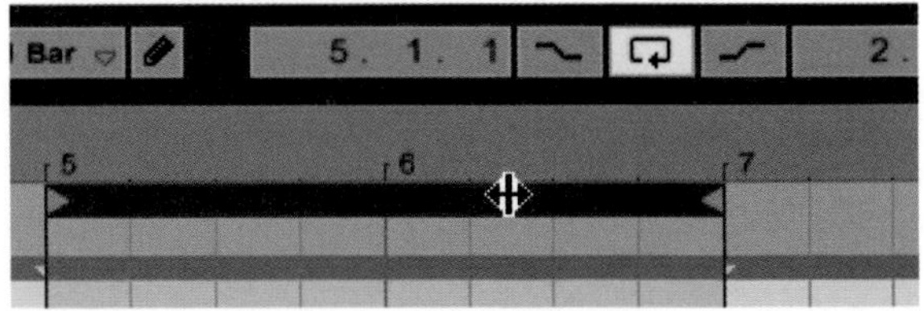

Figure 9.17 Activate the Loop Switch.

3. Turn on the Metronome and set Record Quantization to "Eighth-Note" (Edit Menu).
4. Arm your MIDI track and make sure OVR is activated.
   Now, let's loop record a two-bar drum loop just like before. We'll do this in multiple takes or record passes (loops cycles), kick on the downbeats, snare on 2 and 4, and hat on every eighth (ride-like) note. To keep it stress free, let the loop cycle one time between each record pass.
5. Activate Global Record on the Control Bar then press the Play button. After the count in, begin recording the kick on the first pass, snare on the second, and hat on the third. Only record two bars for each, then listen back to the loop you just created.
6. Click the Control Bar Stop button to stop everything, or disarm the track to keep the Arrangement running.

There you have it. A simple two-bar drum loop created directly in the Arrangement View.

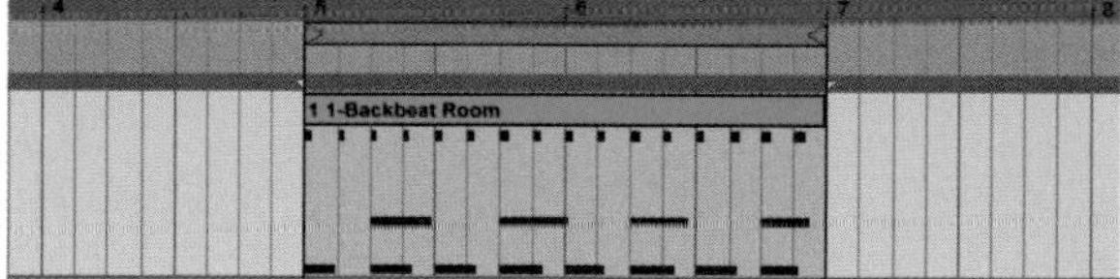

Figure 9.18 Two-bar MIDI drum loop clip.

9.2.3 Takes

Now that we have demonstrated how to loop record with MIDI, let's narrow in on how Live actually handles all of the loop cycles and record passes. On one hand, the OVR recording process in Live is very similar to that of other DAWs, but the way Live handles "takes" or record passes is quite different. With that in mind, click on your Arrangement clip (two-bar loop) you just made and you should see in the Note Editor a very interesting display of the clip. It should look something like our example.

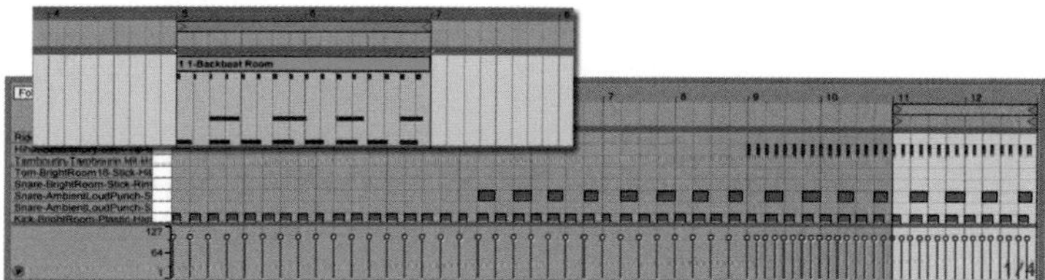

Figure 9.19 Each record pass (take) has been captured into the clip and stored off to the left in the gray area. The white area is the currently active playback area of the clip.

Regardless of the Session or Arrangement, both views use this unique approach to handle "takes."

The entire "take" concept applies to the Session View as well as the Arrangement View. The difference is that in the Session View you govern a clip's starting point (where launch playback begins) and looping region (cycle playback) solely from within the Note Editor. The Start Marker determines where the clip starts, and the Loop Brace determines where the Loop region "Starts" and the loop "Length" (how long the loop region is/cycles for).

From the moment the Arrangement begins to record, it captures every MIDI action/event. If you recall, on the first pass we recorded the "kick," but then let the Arrangement clip cycle through one loop without inputting new notes. Then we did the same with the "snare," followed by the "hat." As you can see, each take (record pass) was captured exactly in that order and what you see in the Arrangement Track Display and in the Clip View Note Display between the Loop Start/End markers is the culmination of our OVR recording. The Note Editor Loop Brace is initially placed where looping was created in the Arrangement, but it has no bearing on the actual looping in the Arrangement Playback until the clip's Loop Switch is activated. At that time the Loop Brace determines the clip's looping selection (more on that in a minute).

Here is what is actually happening with "takes"...

A clip advances with each take (a loop record cycle), logging time as it loops through its loop length whether MIDI note data is input or not until you physically stop the recording process either with the Arm button (disengage Arm Arrangement Recording switch) or by pressing the Control Bar Stop button. Once stopped, your Arrangement clip in the track display remains focused on its final recorded pass, the current state of the last take. The last or final take is essentially a representation of all takes, showing all of the record passes as layers transparently placed on top of each other, as shown on the Track Display and in the Note Editor (white highlighted selection).

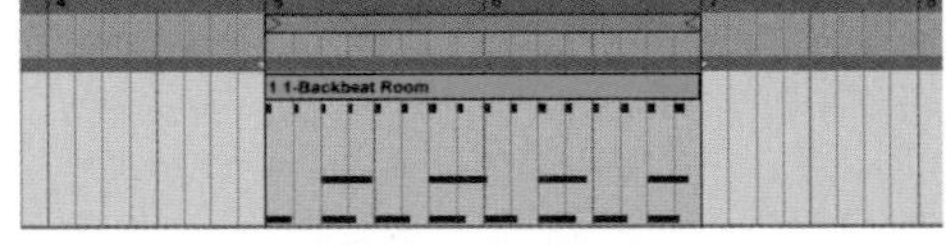

Figure 9.20 Two-bar clip with visible MIDI notes in the Track Display.

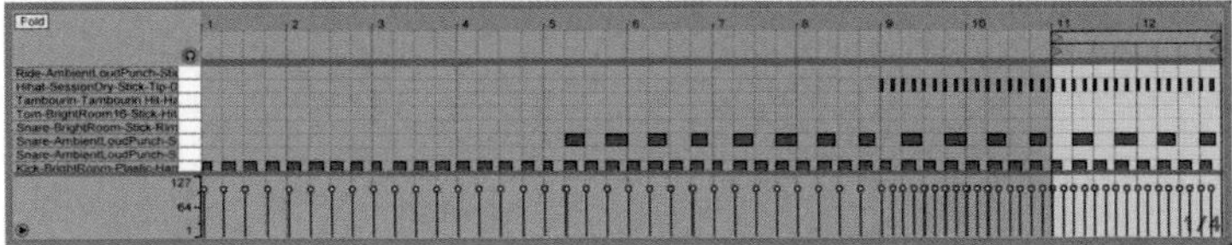

Figure 9.21 Two-bar clip with visible takes on the left and final take on the right in the white.

This is like a birds-eye view, or like the hand-drawn animation technique of layering transparent images (cells) to create a three-dimensional environment. To choose a different selection for playback in the Arrangement View Track Display, you can move the Start or End Markers to encapsulate any segment of the clip from within the Note Editor. These markers indicate what segment will play in the Arrangement. Think of a selection in the MIDI Note Editor as a moment frozen in time that you can choose from to be represented in the Arrangement View Track Display as an Arrangement clip, hence choosing a take. Move a clip's Start Marker in the Note Editor to the left to add and make playable more of your takes in the Arrangement Track Display, as seen in our example. Notice that the clip's length in the Track Display reflects the newly selected Start Marker Position as made in the MIDI Note Editor.

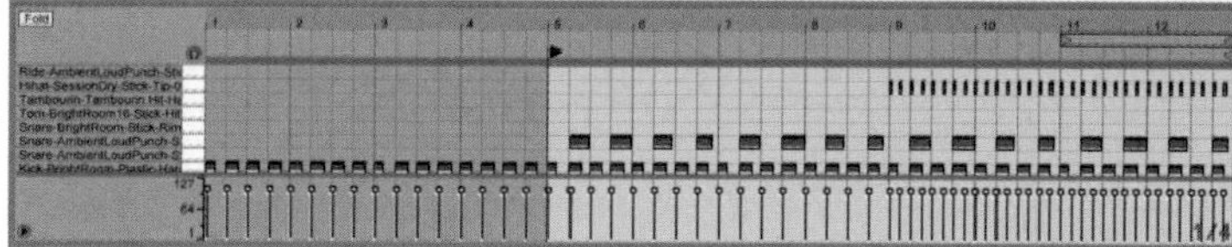

Figure 9.22 The clip's Start Marker Position has been extended to the left six bars in length from within the MIDI Note Editor.

Figure 9.23 Notice that the clip's length in the Track Display reflects the newly selected Start Marker Position as made in the MIDI Note Editor.

Move the End Marker in the Note Editor to adjust the end length of the Arrangement clip in the Track Display. Notice how the Arrangement clip adjusts in the Track Display with the movement of the End Marker in the Note Editor.

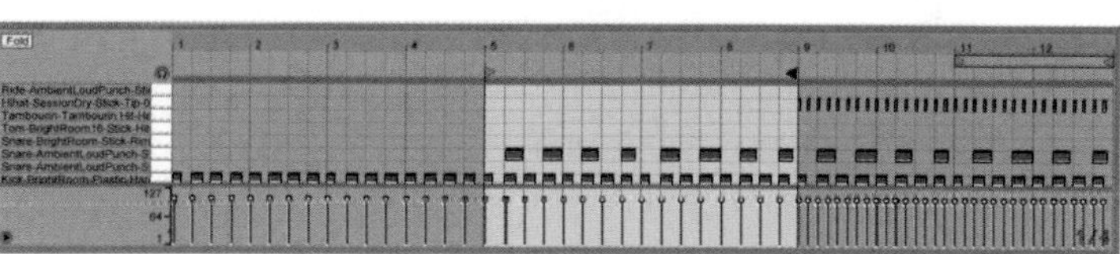

Figure 9.24 Move the End Marker in the Note Editor to adjust the end length of the Arrangement.

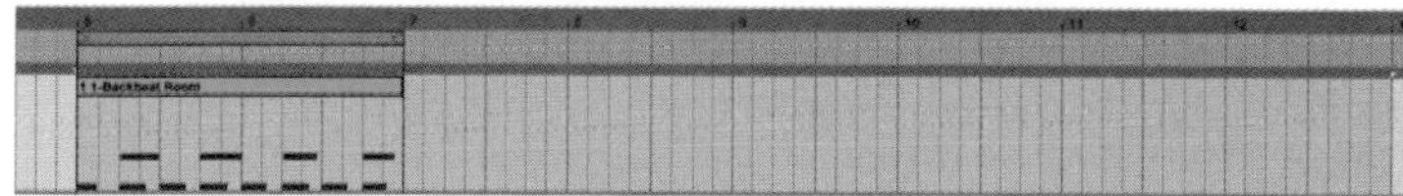

Figure 9.25 The clip adjusts in the Track Display with the movement of the End Marker in the Note Editor.

The last piece to this puzzle is activating the clip Loop Switch. When Loop is activated, the Loop Brace establishes the playback selection within the clip. The Start Marker is then only for assigning the actual start of the clip's content. Nevertheless if the Start Marker is moved outside of the Loop Brace, then it redefines the entire playback selection based on its (the Start Marker's) new relative location to the Loop Brace. See in our example how the Start Marker's location affects the Arrangement clip's playback.

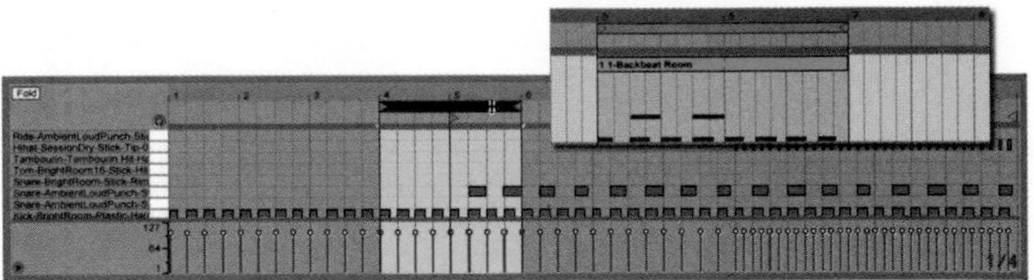

Figure 9.26 The Start Marker's location affects the Arrangement clip's playback.

9.3 Freezing and converting MIDI clips into audio clips

As you may recall from Clip 2.2.3, we talked about using Freeze Track to conserve on CPU and RAM, but there is a whole other aspect to it. Using Track Freeze with MIDI clips allows you to turn them into audio clips in real time. We'll use our last take from the previous example to demonstrate this. This can be done in the Session View or Arrangement View.

In the Arrangement View, select your two-bar MIDI loop that you recorded then right click or ctrl + click to bring up the contextual menu. From there, select "Freeze Track." Once the track is frozen, it will have a blue line through its title bar and the Clip View will be blue.

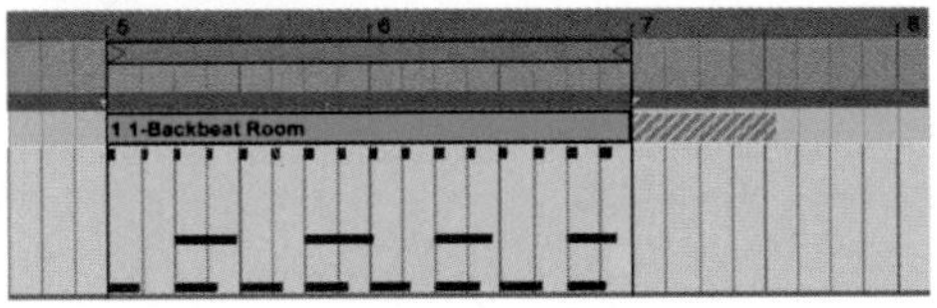

Figure 9.27 Frozen MIDI track.

To convert it to audio is simple. Just click + drag the frozen clip and drop it on an empty audio track. It's a good idea to copy + drag this (alt + drag) to keep the original MIDI in place in case you want to make changes to it. This is what we've done in our example.

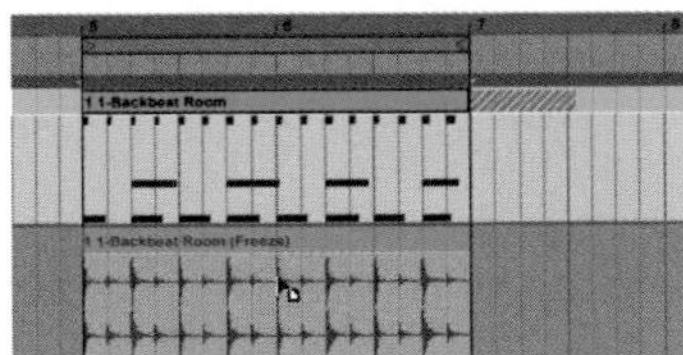

Figure 9.28 Copy + drag the frozen clip and drop it on an empty audio track.

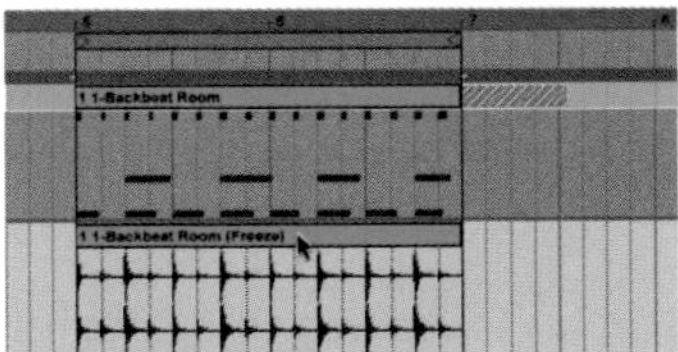

Figure 9.29 Frozen MIDI clip is still in place, and a new audio clip of the MIDI in the track is created below.

The same process works with Session clips too. You can even drag frozen Session clips directly into the Arrangement View.

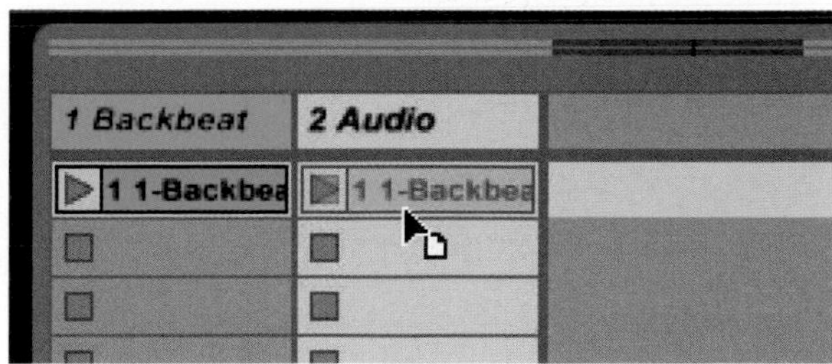

Figure 9.30 Session View: copy drag the frozen clip and drop it on an empty audio track Clip Slot.

Your newly converted MIDI to audio clip is labeled as "Freeze." This feature is so powerful and efficient! Imagine if you just completed a 3-min keyboard part as MIDI for your entire song. Why should you spend the time to print or export an audio stem (bounce) of it? Let Live do it for you right in your Set in real time or offline. It doesn't matter which, it's just the fact that it's a quick, easy, and a very "dummy proof" way to work and convert. You do of course have the option to "flatten" your frozen track too. This is essentially the same process except it converts the MIDI clip, by replacing it with the audio track and clip.

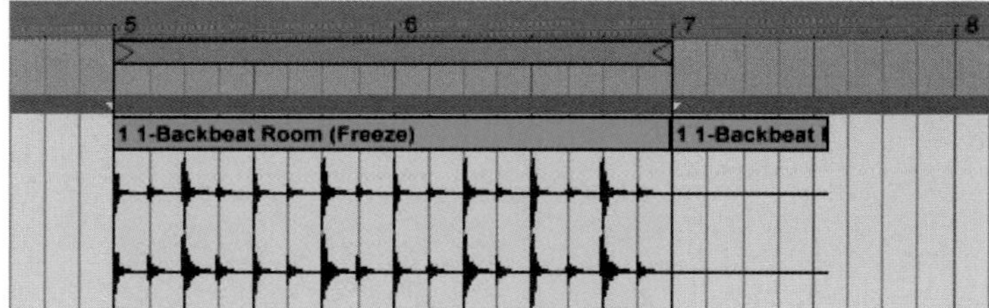

Figure 9.31 Flattened frozen MIDI clip.

Hot Tip *By dragging a frozen MIDI clip to a new Audio track as opposed to flattening you will leave not only the original MIDI track's clip intact, but also the Instrument or Instrument Rack Device in place so you can continue playing and recording on that track. Having your newly converted tracks ready to play at any time keeps the inspiration flowing.*

Before we finish this section, you're probably wondering what the extra clip tail is all about in the frozen clip and in the flattened clip?

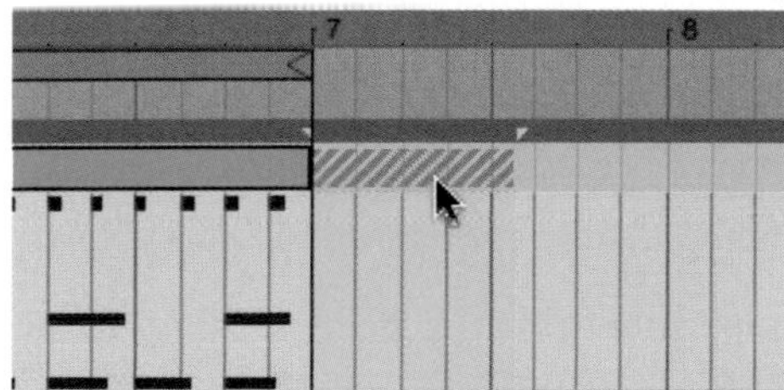

Figure 9.32 Freeze Tail clip.

This little region is called a Freeze Tail Clip. Its purpose is to include any audio carry over from a time-based audio effect, such as delay or reverb, so that integrity of that effect is maintained. Otherwise any lingering effect would be truncated. This feature can be used in many ways, but it is best used for playing back in its entirety along the timeline, because the audio and effects are no longer "virtual." In this way, the Arrangement View is used to playback frozen clips with Freeze Tail Clips. In our previous example, there was not really any audio to carry over, so here is another example with a delay added to the device. After we drag the frozen clip with tail to a new audio track, you will definitely see that it has an effect tail.

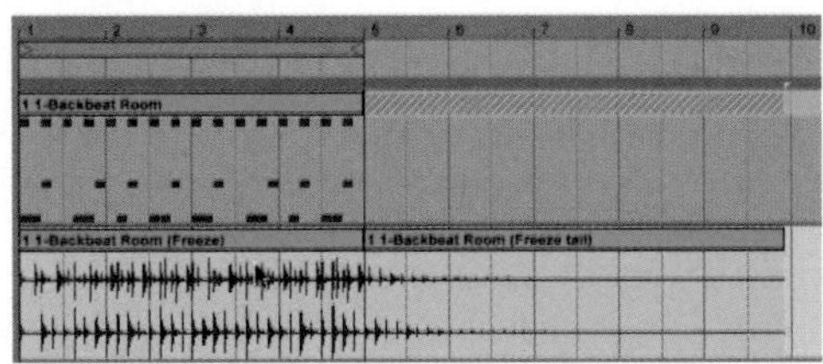

Figure 9.33 MIDI clip converted to audio along with its Freeze Tail.

Hot Tip *Copy a Freeze Tail Clip from a frozen clip and place it anywhere in the Arrangement View or Session View to convert it to audio. You can also try reversing the converted Freeze Tail Clip and place it just before the original clip you removed it from as a transition or build-up to the original clip. An audio effect, such as reverb, is perfect for this, creating that backwards "inhale" effect leading into the main clip.*

9.4 Recording audio clips

If you understand the concept of recording MIDI in Live, then recording audio is fairly simple. The concepts are very similar with only a few differences: (1) You cannot create a new empty audio clip in a Clip Slot by double-clicking. If you double-click in a Clip Slot, nothing will happen. The Empty Clip Slot template approach is used only for recording MIDI to a clip that has been preset with a loop length or Loop Brace selection for overdub recording or for loop recording. (2) In the Session View, you cannot record audio into or over an existing clip. The Clip Slot must be empty. (3) The Overdub record feature is not applicable to audio recording. It only works for merging MIDI notes and takes without overwriting existing notes. This is standard for all DAWs. Instead, you must adopt a "punch in/out" approach for recording audio in the Arrangement View. The new Looper Audio Effect in Live 8, however, is perfect for overdub recording with audio. Launch to ▶ 15.5.4.

Hot Tip *KEY Mapping is great for toggling OVR. Click OVR and select a key on your keyboard to map it to. Then you can punch in/out by tapping the computer key.*

Beyond these differences, recording audio follows the same protocol as MIDI, and once you grasp the overall recording concept, you'll be set to go. Of course, the input source differs, so you must have some sort of audio source to record. As far as recording in Live 8 is concerned, the Arrangement View is just about the same as any other DAW out there. It's the Session View that applies an entirely new and inventive way of recording audio in a nonlinear environment.

9.4.1 Session Clip Slots

When recording audio clips into the Session View, you must start with an empty Clip Slot. An empty clip cannot be inserted. This makes audio recording straight forwarded. To begin, select an existing audio track or insert a new one and set it to the appropriate audio input channel based on your computer's audio setup. As we are going to record a voice, we'll set Audio From input to channel 1 (mono). If you are using a laptop, you can use the built-in mic for the following exercises and demonstrations, just be sure to use headphones or turn your record-enabled track's volume slider down about halfway or more so that there is no audio feedback!

Let's record two bars of vocal counting ("**1**…2…3…4…**2**…2…3…4…"), as this is the simplest way to practice audio recording for the first time. We'll use the Session View for this first exercise.

9.4.1.1 Recording to an empty Clip Slot

1. Turn the Metronome "On" and set the Tempo to 100 BPM.
2. *Record–Enable*: Arm your audio track then test to see that you are receiving signal input by speaking or snapping your fingers into the mic. You should be seeing the track mixers VU meter lights up green with signal. If not, double-check your audio configuration.
3. *Record*: Click a Clip Slot Record button on your audio track and then begin speaking after the count in; only record for two bars.

Figure 9.34 Activate Clip Record button.

4. *Stop recording*: Click the Arm button to disarm your track then press spacebar to stop the song position. You could also stop recording using the Control Bar's Stop button ▶ 5.3.2, or click the Stop All Clips button ▶ 6.3.2. Each one has a slightly different affect.

5. *Set loop*: Double-click the new clip recording to open Clip View. In Sample Editor, adjust the clip's loop length so that it's set to a two-bar loop. You can do this by dragging the end of the Loop Brace to bar 3, or set the Loop Length field from the Sample Box to two bars. For the adventurous users, click the Set button during playback when the playback cursor reaches bar 3.

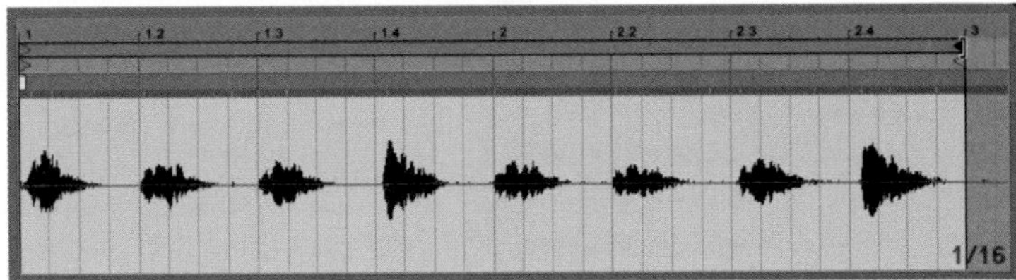

Figure 9.35 Set the Loop Length.

6. *Playback*: Make sure that the Loop Switch in Clip View is activated, and then listen back to your newly recorded audio loop. You'll probably notice that your timing was a bit off, but that's okay for now. Stop when you're done listening.

Now that you have successfully recorded audio into Live's Session View, you'll probably notice that there are some minor timing errors in your recording. Maybe you spoke a bit early on one beat and late on another? That's okay for now, but if you want to fix these timing issues, you'll need to decide how to go about it. First, consider selecting the section of the loop that was performed the best – bar 1 or bar 2 in this case – and then set the clip's Loop Start/End positions to use that specific selection in the Sample Editor. Of course, this still won't solve timing problems that may be obvious. In this case you have two options: (1) move the errant words manually in the Sample Editor using Live's Transients and Warp Markers ▶Scene 13, (2) copy the clip into the Arrangement View track display, edit the audio, and then bring the clip(s) back to the Session View. Our suggestion is that the latter is not as efficient as the former, unless you have some complex edits to make. Then again, if that were the case, you would probably just record them in the Arrangement View in the first place. Simply put, Ableton is all about making recording easy and intuitive for you. Live 8 will also allow you to edit audio just like you have in other software and without a hitch, and to do it even faster with the use of Warp Markers! For all the information you need to adjust timing and quantize audio, launch to Clips ▶ 13.1 and then ▶ 13.3.4.

There will come a time when you want to record multiple tracks at once that make up all or part of a scene, or when you want to simultaneously launch clips and record while recording on another track at the same time. Accomplish this by using Scene Launch buttons to activate record for all of your tracks that are currently record-enabled. This is called Start Recording on Scene Launch. You must turn on the feature in the Live Preferences: Record, Warp, Launch tab. For more on this feature, launch to ▶ 10.2.2. As an alternative, you can use Group Tracks to launch and record multiple tracks at once. Launch to ▶Scene 11.

9.4.2 Arrangement clips

Very few people can record a flawless performance in one take. Most of the times you'll need to OVR some element of the performance, or at least "comp" (compile) some elements together. For this reason, we turn to Live's Arrangement View. Here you can build a vocal comp, for example, and then bring it back into the Session View or you can lay in (record) a live vocal directly to the Arrangement View audio track. Like we said earlier, that is the beauty of the Arrangement View. Not only is it used to capture a performance from the Session View, but it's also a traditional linear DAW.

Let's continue with recording our vocal counting, this time in the Arrangement View directly as Arrangement clips.

9.4.2.1 Recording audio clips

You have a few options at this point. You can set up a Loop Brace along the Beat Time Ruler and Loop Record your vocals, or you can just record them without a Loop Brace, stopping when you reach the end of your desired recording segment. Either way is really no different than any other DAW or approach you may have encountered before. For this demonstration, let's put Live in Loop Mode (Loop Switch – "On") so that we can also choose between takes and comp it all together.

1. Turn the Metronome "On" and record-enable your audio track.
2. *Set Loop*: In the Arrangement View, set the Loop Brace to two bars (bars 3 and 5) using your mouse to drag it into position. Once in position, click on the Loop Brace to select the loop region so you can record starting at bar 3. If you are more comfortable starting your loop recording at bar 1, feel free to adjust and begin there instead.

   Now, let's loop record our two-bar vocal counting just like before, but this time we'll repeat it when the clip loops. All in all, you will record your two-bar

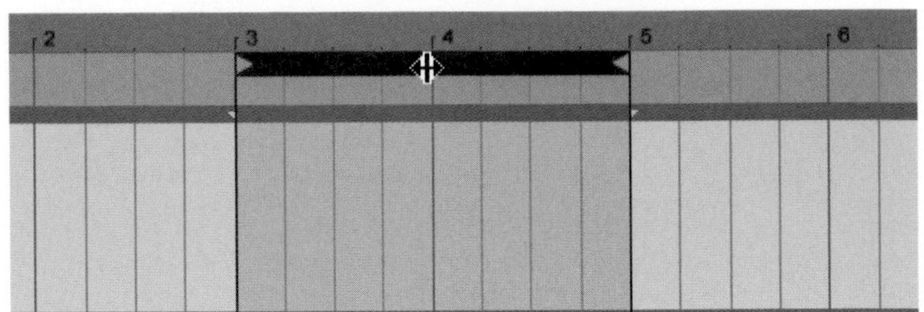

Figure 9.36 Set the Loop Brace in the Arrangement View.

vocal, counting two times through (two passes), starting your vocal counting over again on the second record pass (loop).

3. *Record*: Let's record two bars of vocal counting ("**1**…2…3…4…**2**… 2…3…4…"). Make sure the loop brace is selected, then activate global record on the control bar and hit the play button. After the count-in, begin speaking. Remember to start over immediately following the second bar and repeat the vocal counting.
4. *Stop recording*: Click the Control Bar Stop button to stop when you are done. You should now have a two-bar recording currently showing the second record pass, like our example.

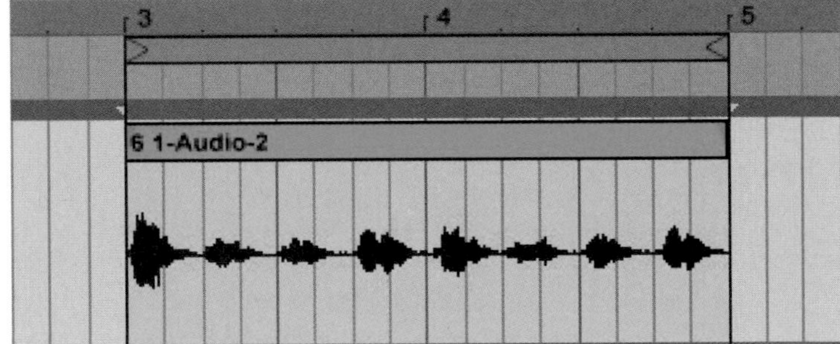

Figure 9.37 Two-bar recording currently showing the second record pass.

When you stop recording, there will be a break in the clip towards the front edge or wherever you stopped recording at. This happens when you stop a record pass mid clip, essentially creating an incomplete take. To remove this, drag and extend the edge of the complete take to the left so that the clip fills out all the way to the loop selection.

5. *Playback*: Go ahead and listen back and see how you did. Click the Play button on the Control Bar or press the spacebar.

Take a look at the clip in the Sample Editor. You will see your entire audio recording including the first and second record pass. The first take will be grayed out, as it is not currently being used in the Arrangement View.

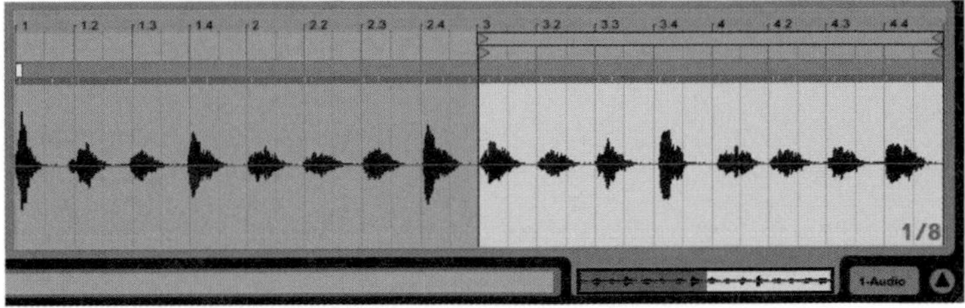

Figure 9.38 The entire audio recording including the first and second record passes with the first take grayed out, as it is not currently being used in the Arrangement View playback.

As always, there will be some timing errors and speaking mistakes. For whatever reason, you may want to use one "take" over the other, choosing the one that you like better. To choose a different take, move the Start/End Markers in the Sample Editor to encompass the take you like. If it's the last take, then you can leave it alone. If it's the first take, then move the clip Start/End Markers to bookend the first two bars. In that case it should look like our example.

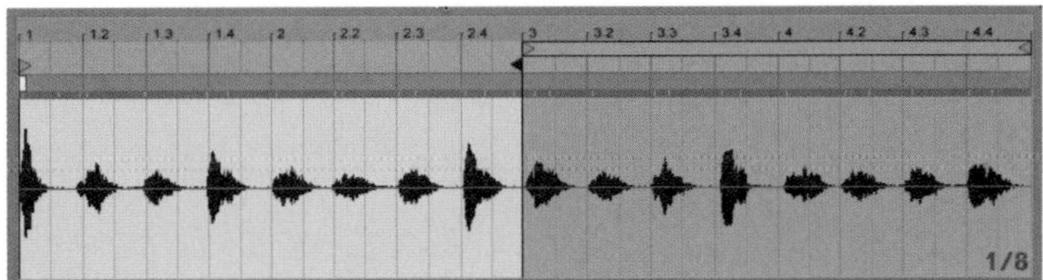

Figure 9.39 Choose the first take. Move the Start/End Markers in the Sample Editor to encompass take 1, bars 1 and 2.

Click opt/alt + drag on the clip's Start Marker to move the Start/End Markers together in one mouse drag motion. Wherever they are moved to, the visible clip region will update automatically in the track display. You can also do this using the Sample Box Start/End input fields.

If you want to loop your clip take, then activate the Clip Loop Switch and drag the Loop Brace to the desired take. As for timing errors, you can fix this with warp markers just like we explained for the Session View.

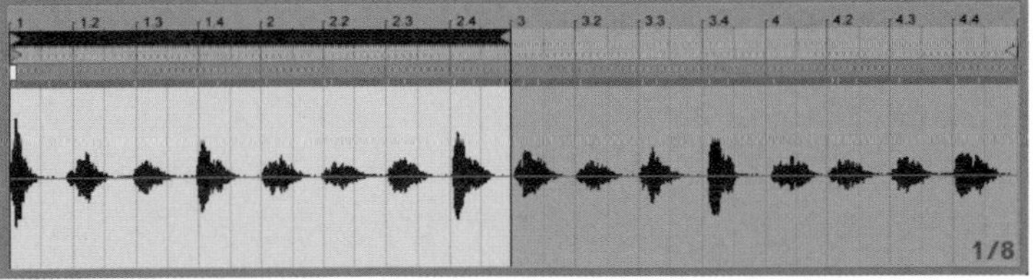

Figure 9.40 To loop your chosen clip take, activate Loop in the Sample Box and move the Loop Brace accordingly.

Finally, as a little treat before moving forward, we would like to again mention that Live 8 and Suite 8 offer a new and hip way to loop record audio on the fly using *Looper*. Another Ableton original, Looper provides some fantastic alternatives for real-time audio loop recording 15.5.3.

9.4.3 Comping

The term "comping" has been a part of the recording and production process for ages. It's the art of compiling multiple vocal takes into one solid, single, and clean vocal track. Here is one way to do this in Live's Arrangement View.

So let's say you like the last "take" (second record pass), but you prefer the way you spoke "3" in the first "take" that you are not using. That's okay; you just need to compile the two together in the Arrangement View. This is why we left room at the front of our recording instead of starting at bar 1.

1. In the Arrangement View track display, extend the edge of the clip out to the left until you see both takes. Now you should have four bars in View consisting of your entire recording in the track display.

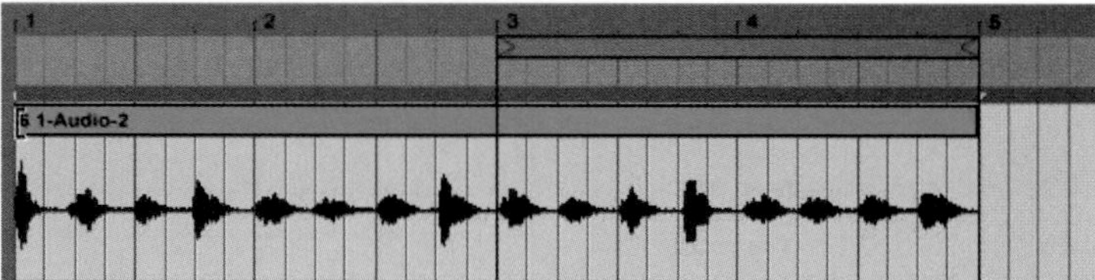

Figure 9.41 All four bars of your entire recording in view.

2. As you like the way you said "3" in the first bar of the first take, for example, we will make a selection around the recorded word in the grid. Set the Grid to a reasonable resolution such as 1/4 or 1/8 so that you can select the audio segment perfectly. Highlight the specific segment with your mouse (click + drag), and then copy it [⌘+C] Mac/[Ctrl+C] PC.

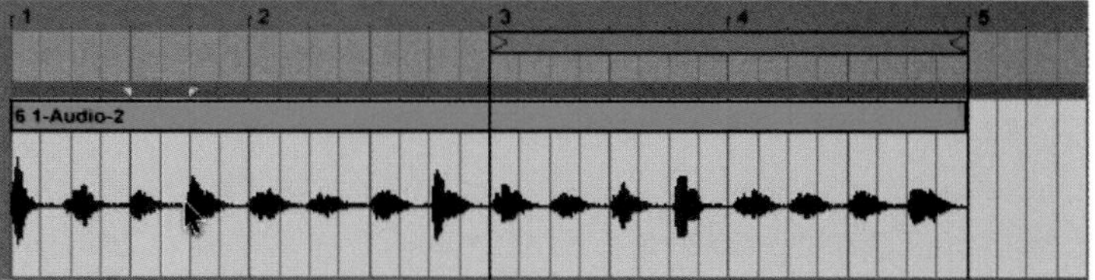

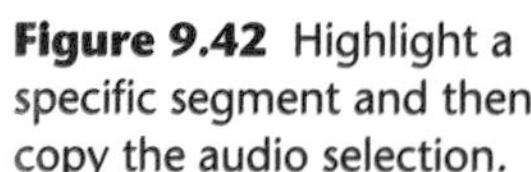

Figure 9.42 Highlight a specific segment and then copy the audio selection.

3. Now highlight "3" from the second bar of the second take and select paste [⌘+V] Mac/[Ctrl+V] PC. Now that it is pasted, Live will automatically add crossfades 5.4.5.

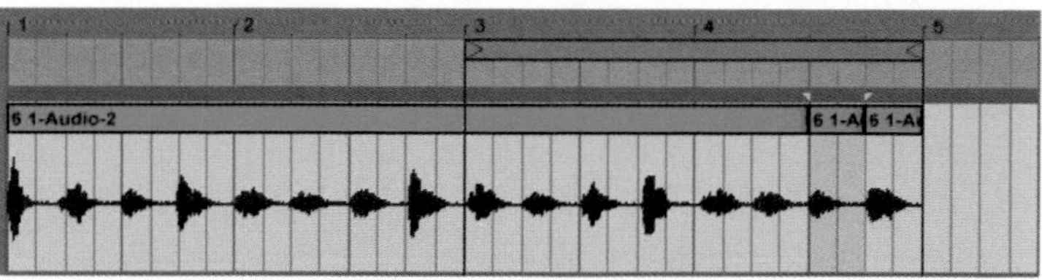

Figure 9.43 Highlight the selection that you wish to replace and paste the segment you previously copy selected from the other take.

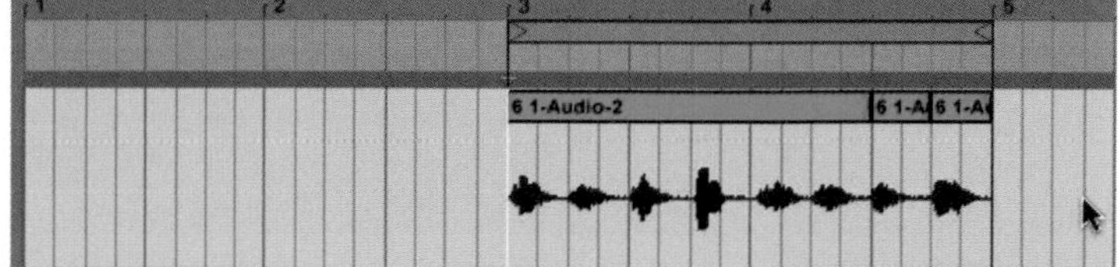

Figure 9.44 Resize your clip back to a two-bar take, the "comp."

4. *Drag* the left edge back into position so that only the second take is showing (the "comp").

There you have it, your first comp consisting of three audio clips.

9.4.4 Punch Record

Now let's say that you don't like any of your performances of the word "2" on beat 1 in the second bar of counting. Well then, you'll just have to rerecord it using Live's built-in auto Punch-In/-Out Switches or Punch Record it manually using Global Record or Arm button. Let's look at both ways in the Arrangement View.

9.4.4.1 Auto punch in/out

1. Set your punch in/out points using the Loop Brace. Set it to bar 4 and 4.2. Drag the Loop Brace with the mouse or use the Loop Start/Punch In fields on the Control Bar.

Figure 9.45 Loop Start/Punch-In fields.

2. Activate the Punch-In/-Out Switches located on both sides of the Loop Switch. Punch-In Switch prevents recording prior to the punch point, and Punch-Out prevents recording after the punch point. Make sure that the Loop Switch is deactivated.

Figure 9.46 Activate the Punch-In/-Out Switches.

3. Place your Insert Marker two bars before the punch-in point to serve as your preroll (count it). When punch recording, Live does not have a count-in.
4. *Record*: Record-enable your track, click Global Record, then press Play. Speak the number "2" at the punch-in to replace the original recording and Live will take care of the rest.

Figure 9.47 The resultant "Punch."

You may have noticed that if you we're not precise with your "2" OVR that the beginning of the transient may have been cut out of the recording. This is because we used the punch-in switch and set our punch-in exactly at bar 4. To ensure this doesn't happen, or that the end doesn't get cut off on the punch-out, set your punch points a little wider, then trim the clip back down using the bracket tool with your mouse after finishing the punch recording. Extra room for recording is okay, even if it overlaps the part you were happy with. Live is nondestructive, but you can never get back what you don't record.

As an alternative to using the "official" Live punch switches, you can simply do it manually using the Global Record button.

9.4.4.2 Manual punch in/out

1. Place the Insert Marker two or more bars before the point at which you want to record; in this case, the Insert Marker is placed around bar 2 for recording at bar 4.
2. Deactivate Loop Mode.
3. *Record*: Record-enable your track, then press Play. When you get to the punch, point click on Global Record to punch-in and click again to punch-out when finished.

Hot Tip *Try setting up a custom KEY Map to activate the Global Record button, or use the default keyboard shortcut F9. Doing so will give you quick access to select Global Record on the fly when punching in and out, all without having to grab your mouse when inspiration ignites.*

You could also reverse step 3 by using record-enable to punch-in/out while running in Global Record mode. Either way works.

9.5 Exporting and printing

When you are finished working on your song or production, it's time to start preparing and exporting your full mix, stems, or clips to disk. This includes determining how you are going to handle MIDI tracks and even effects. In general, the rendering/exporting process provides for many options, such as rendering an entire song, individual track, clip(s), or even video. So what are your options?

Most of the time you will render your tracks "offline" meaning that the process happens faster than the actual time it takes to play through the song or clip, and you won't hear playback during the rendering process. When rendering an External MIDI device, the process will always happen in real time, since it has to physically create sound to be recorded into Live. So as far as MIDI is concerned, you have a few options regarding bouncing/rendering or "printing" the source.

The ability to render your tracks offline as opposed to printing them to track brings up a good philosophical question. Which method should you choose to "bounce" your MIDI tracks down to audio and when do you use it? The answer is, you have options. Whether you render or print isn't really the question, that's more a matter of preference and convenience.

In ▶ 9.3 we talked about freezing and converting MIDI to audio, so that is one option. One of our favorites! More importantly is the "when…?" You could work through your entire project, then render your MIDI tracks at the end or not at all depending on the final delivery format of your song or production. If all you ever need is a stereo mixdown then you may never bounce down your MIDI to audio. If you're delivering audio stems to another producer or engineer, for example, then you could bounce them down to stems at the end.

Regardless, we always recommend rendering or printing your MIDI, especially external MIDI audio to track once you are done creating with the device. In this way you have lesser devices to worry about and can work on your song away from your studio. Otherwise you may not have that instrument, plug-ins, or settings available elsewhere. Remember, MIDI is just instructions for your instruments to interpret and convert into audio. Every composer/producer has a different approach and preference for rendering MIDI to audio.

The rendering process can be executed from both the Session and Arrangement View. The Arrangement View process is traditional by nature, designed more for Exporting your entire song/Set or large segments of clips over a time range, such as multiple clips on the same track as well as across multiple tracks. The Session View is more discrete in that you can only render a single row of Session clips at a time, those that are to be played back over a set length of bars and beats.

9.5.1 Rendering/exporting audio

Whether rendering your MIDI to audio, audio to disk, a single clip, or exporting an entire arrangement, the process of exporting audio is virtually the same across the board.

To render/export from the Arrangement View:

1. Select a clip or highlight a time range in the track display that you want to render in the Arrangement View track display. This can be a whole clip or just a selection. For exporting from the Arrangement View, make sure that the Back to Arrangement button is not red!

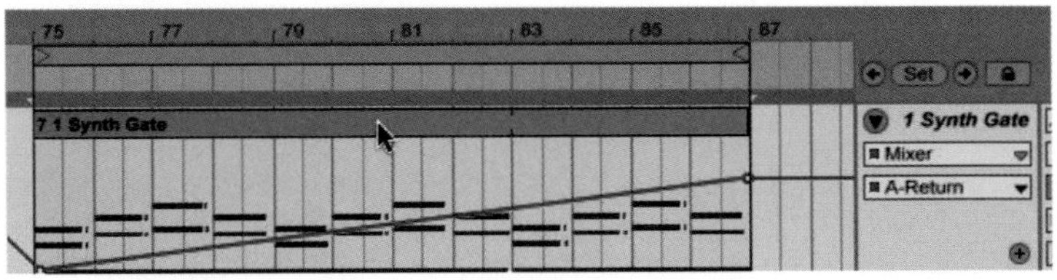

Figure 9.48 Select an Arrangement clip or time range you want to render.

In the Session View, launch your track clip or clips then pause playback with the spacebar or Control Bar Stop button. Your paused clip's Launch button should remain green (in standby).

Figure 9.49 Green Launch button indicates a clip is paused or in standby.

2. Select Export Audio/Video from the File Menu ([⌘+shift+R] Mac/[Ctrl+shift+R] PC).
3. If you are rendering from the Session View, you will need to set the render "Length." In the Arrangement View, length is not an option; your selection is your render length – "Length [Bars-Beats-Sixteenths]."
4. Now set the track source you want to be rendered. Choose an output source to render: Master, a single track, or "All tracks." The rendering of all tracks individually is also known as creating stems. If you are rendering your entire mix, then you will choose the Master Track as your Rendered Track and almost always use the Arrangement View, just be sure to select the entire time range or section of the song.

Use the Loop Brace as a quick selection tool for rendering in the Arrangement View. This is a good habit to get into. Set the Loop Brace to contain the front and end of your full arrangement or selection along the timeline. This assures that all your information along the timeline and within your arrangement will be rendered and exported completely.

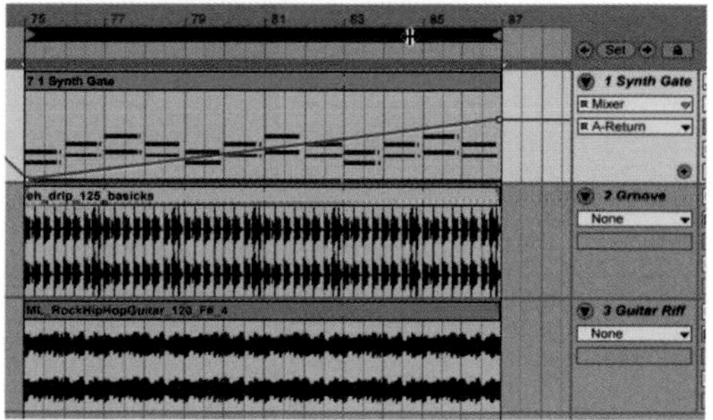

Figure 9.50 Use the Loop Brace when exporting from the Arrangement View.

5. There are a few destination formats that we should review here as well: Normalize, File Type, Sample Rate, Bit Depth, etc.
6. Press "Ok" then choose your save destination and select "save."

Once the rendering process is complete, you can import the audio file(s) back into an audio track in Live to continue working with it or ship it off as a stem for the production phase. For additional information on exporting and rendering MIDI, launch to 3.7.1.

9.5.2 Printing

Printing your MIDI tracks, Effects Returns, and other outputs to track is a valuable tool and often preferred when you don't want to export and then import those audio bounces (renders) back into your Set. It's also great if you want to print a *submix (internal mixdown of multiple tracks into one track)*. This can also be done through Group Tracks 11.3. Let's print our MIDI, Synth Track from the last exercise.

Figure 9.51 Arrangement View.

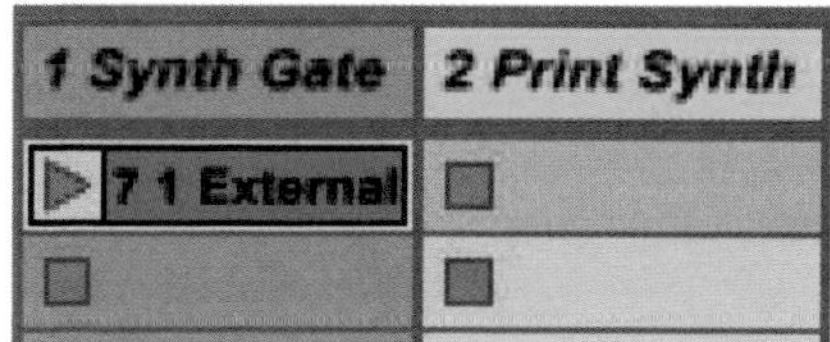

Figure 9.52 Session View.

1. Create a new audio track and name it "Print Synth Gate."
2. Set the Monitor Section on your new audio track to "IN." This will monitor audio input through the new track while in playback.
3. Route the Synth Gate MIDI Track's *Audio To* to your new "Print Track."

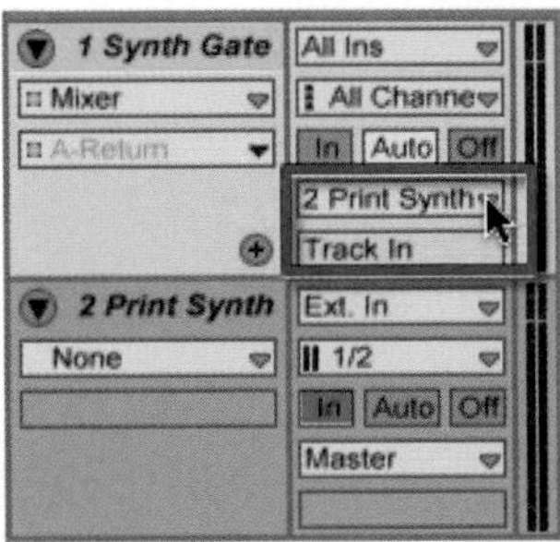

Figure 9.53 Audio To chooser routed to the "Print Track."

4. Place your Arrangement Insert Marker at the beginning where you want to print your clip. Make sure that "Back to Arrangement" button is not red! In the Session View, you will print to an empty Clip Slot.
5. Make sure that the Arrangement Loop Switch is "Off" or clip Loop Switch is "Off" for Session View Printing.
6. Record-enable the Print Track (Print Synth) and then for printing in the Arrangement View, activate Global Record and press play. In the Session View, launch your clip.

Figure 9.54 Synth Gate printing to the Print Synth track in the Arrangement View.

Of course you can also print MIDI to audio from the Session View to the Arrangement View. It follows the same exact process except that you'll be Global Recording the MIDI and printing the audio at the same time to their respective tracks.

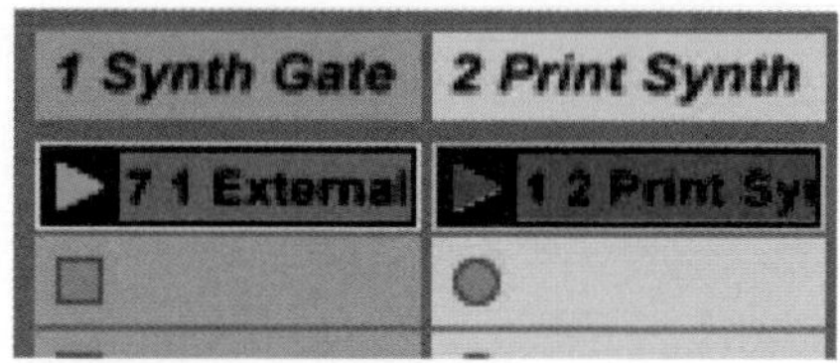

Figure 9.55 Synth Gate printing to the Print Synth track clip slot in the Session View.

Finally, you might be wondering about the Send Effects. If you want to print them with the track, you will need to route the Return to the same Print track, just make sure that all of your other Track Sends are not being sent to the Return at the Time of Printing. You could deactivate all of the other tracks while printing. You could also print an Effects stem, but that's not as common. There are a number of creative ways to go about this. Experiment and see what works for you.

9.6 Musical concepts

By now you should be thoroughly convinced of Live 8's ability to function as a DAW, more exactly, a multitrack recording and production software package. Let's put it to the test and show you how it can be even better than the traditional multitrack recording software.

9.6.1 Produce: real-time *stack tracks*

One very powerful technique when recording audio clips into the Session View is to use Live's real-time clip dragging capabilities while recording a performer, e.g., a guitar player or other session player to create accompaniments or layering takes for cool double-tracked type effects layers on the fly in real time. Unfortunately, dragging clips while you yourself are recording makes this process a bit difficult because your hand would need to be free to control the mouse to click + hold/drag a clip. So we'll apply this concept to working with a session player or additional musician and with you at the controls.

While recording clips to a new track, try dragging a previously recorded clip (take) from the same track you are recording on to a new audio track while the current track is still recording. Launch this clip to sync it up with your current recording clip on your main record-enabled track.

In this manner, you can instantly create "Stack Tracks" or "unison parts" on the fly in real time! This is a powerful tool in the production process because it keeps the workflow moving forward and the creativity flowing while never stopping to redirect a completed audio clip. As a matter of fact, you could predesignate two additional tracks in your Session View ahead of time and pan each track hard

Figure 9.56 Dragging clip takes in real time while recording.

left and hard right. Once you record two new parts as clips, simply drag them one by one to the panned track for an instant "stack" or "duplicated" track all the while keeping Live in play/record mode. This process really shows off Live's power in launching and dragging clips in real time.

In this chapter

10.1 Musical foundation and structure 215

10.2 Scene launch preferences 217

10.3 Tempo and time 219

10.4 Capture and Insert Scenes 220

10.5 Musical concepts 222

SCENE 10

Working with Scenes

The transparency of the workflow makes you feel that whatever you're hearing in your head is possible. Live 8 lets you make it happen quickly...without losing the initial inspiration to sub menus and bloated complicated software.

I would never use anything else...oh yeah...and it's fun!!

Shawn Pelton, Session musician, drummer for SNL

10.1 Musical foundation and structure

The communication of music and sound is most often dependent on some sort of structure or form that makes up its musical foundation. Even in the most abstract genres a unifying principle or nucleus of organization exists. John Cage himself outlined a specific structure for his piece 4′33″ – Four Minutes, thirty-three seconds, which is perceived by the listeners as silence. If we can agree that all music has some sort of musical foundation and structure, then we can agree that mainstream popular music is no exception. Most genres adhere to a specific and accepted song form. Such terms as *verse, chorus, hook, bridge, vamp, intro,* and *break* are household names. You could even go as far as to say that we have all even heard of sectional structures such as AABA, ABA, verse–chorus, or 12-bar blues. The point is that for centuries, musical foundation and structure have not only guided the composer, singer/songwriter, producer, or performer but have also served as musical vehicles for the listener. With this philosophy, Ableton has brought forth a method to creatively build a musical foundation in relation to structure to organize your music with clips, while also including a solid and unique way to deliver your musical ideas to fruition with *scenes*.

Figure 10.1 Launching a scene in Live 8.

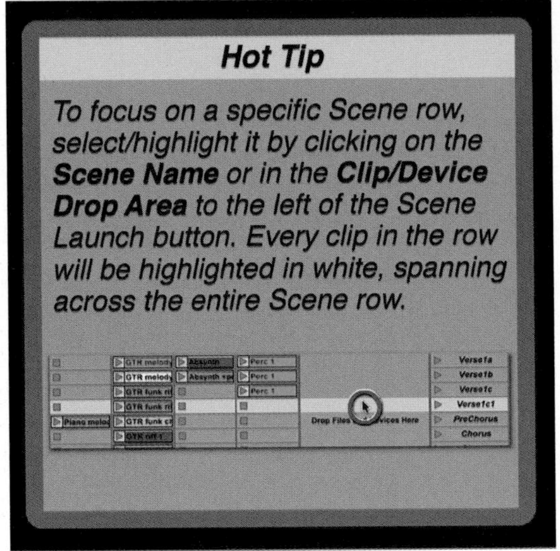

Hot Tip
To focus on a specific Scene row, select/highlight it by clicking on the **Scene Name** or in the **Clip/Device Drop Area** to the left of the Scene Launch button. Every clip in the row will be highlighted in white, spanning across the entire Scene row.

Represented as rows in the Session View, scenes are titled and launched from the vertical column in the Master Track. All clips existing in the same row make up a scene, and it can be launched altogether at once from the respective Scene Launch button. Only one scene can be played at the same time! Scenes are generally used to contain and launch song sections or launch a row of clips that form a part or section of a song, for example, intro, verse, vamp, or break, etc. They can also be used to launch an entire group of clips that make up a multitimbral (multipart) instrument or vocal, such as a drum kit beat/groove where the drum beat pattern is divided amongst each part of the drum kit across multiple tracks. Beyond organization, structure, and simultaneous clip launching, scenes will allow you to apply a few very unique, "on the fly" tricks while working in the Session View.

Using scenes is very simple. They follow the entire same rules that govern clip launching. To launch, click the Scene Launch button just like you would do for any single clip. To stop playback of a scene, use the Stop Clips button or the Control Bar Stop button/spacebar as we have already discussed. If you need to delete a scene, select it and press the delete/backspace key on your keyboard. To make launching scenes even more fun and manageable, color code them and keep them organized by theme, musical idea, instrumentation, or whatever your imagination and creative mind desires!

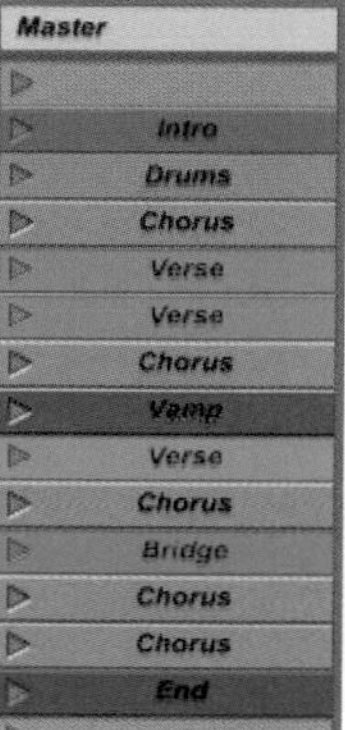

Figure 10.2 Color code your scene list.

10.2 Scene launch preferences

Making music on the fly is all about workflow, the ability to effortlessly make creative musical decisions without the interruption of a nonmusical task. By now, you know that Ableton believes in making this a priority. With that in mind, we'll take a quick look at a few scene launch preferences to streamline our creative workflow.

10.2.1 Select Next Scene on Launch

Under the Live>Preferences>Launch tab, you can customize how Live behaves after a scene is launched. This option is called "Select Next Scene on Launch." When this feature is set to "On," Live will automatically select the next scene below the one just launched to prepare for the next scene launch. This assumes that you are launching your scenes sequentially. This feature allows you to launch sequential scenes without having to manually select them before launch, thus eliminating a step. The next scene's activator button will flash until it's "downbeat" timing has arrived, then the new scene will activate and start playing accordingly. The Select Next Scene on Launch feature only works when using the Return Key on your computer keyboard to launch scenes. A common approach is to use the computer keyboard arrow keys to navigate between scenes and use the Return Key to launch. Alternatively, scene launching can be assigned to a MIDI controller. For details on Key/MIDI Mapping, launch to ▶Scene 12.

Hot Tip *Make your performances more intriguing by experimenting with various resolutions in the Global Quantization Menu. For example, change the Quantization Resolution between scene launches amidst playback, using higher resolutions like*

1/16 or 1/8. With several tracks and multiple clips, this increases the complexity of your music. For ambient and nonrhythmic Sets, try selecting "None" in the Global Quantize Menu. This allows a freeform and legato approach to launching scenes. Remember, the higher the resolution setting, the more precise you need to be when triggering your Scene Launches. It's a good idea to practice with the Metronome to get a feel for this approach. Also try using MIDI Map functionality to trigger clips and for selecting different Global Quantization Resolutions.

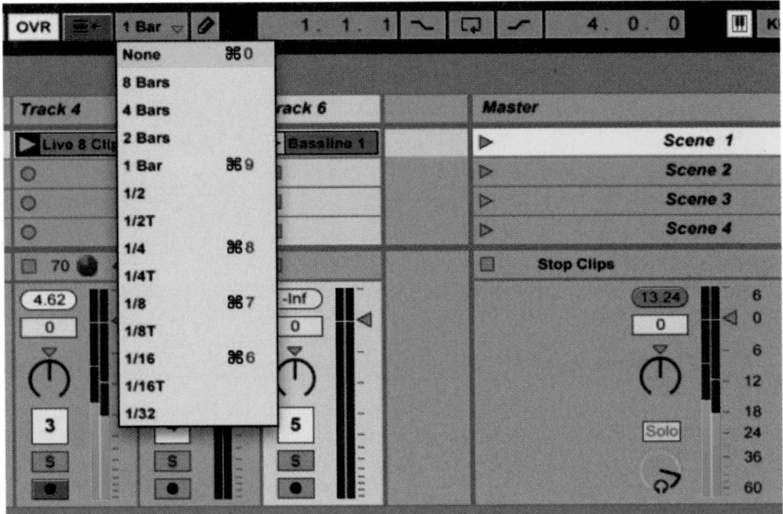

Figure 10.3 Global Quantization Menu and resolutions.

10.2.2 Record on launch

Using scenes for recording clips is essential to multitrack recording and for launching multiple track clips to play back while activating record on other tracks simultaneously; for example, when you want to record along with other clips that make up your backing tracks or a beat groove, let's say a keyboard part to accompany your rhythm section (drums and bass). This is common practice for recording because you'll usually want a count-in, and then hear the music you are playing/recording along with. This means that you'll most likely record with multiple clips lined up in a scene row. To make this process easy and efficient, you can set up Live so that when scenes are launched, they automatically trigger track/clip recording (auto activate Clip Record buttons) for any record-enabled track. From the Launch preferences tab, set "Start Recording on Scene Launch" to "On" to use this feature. When set to "Off," you will have to launch a scene first, then click a Clip Record button to record.

If you need to record on multiple tracks at the same time, hold [⌘] Mac/[Ctrl] PC + click on the Arm button when record-enabling each track. The *Record on Launch* automated feature streamlines the multitrack record/playback process into a one-click operation. Keep in mind that you can always step outside the "scene lines" by launch-recording additional independent clip(s) on other tracks at any time after launching a scene. The scene will continue to cycle (loop) while you manually select a new clip to record or overdub. This feature allows for great flexibility when recording parts on the fly. Not only does it give you the ability to multitrack record audio or MIDI in the Session View but also provides you the creative freedom to record all your clip ideas as a scene at a time and as your ideas change from scene to scene in your Set.

As an alternative to scene launch recording, you can use Group Tracks to launch and record on multiple tracks at once. Launch to ▶Scene 11. Take this feature one step further by MIDI Mapping a footswitch to navigate and activate the scene launch function. Launch to ▶Scene 12 and learn all about using mapping features.

10.3 Tempo and time

Working with tempo changes in your old linear digital audio workstation (DAW) involves programming a tempo map of sorts, which is something you could never imagine doing in real time. Sure, you could sequence out a song for a stage performance but you are still locked to a linear map. First, Live's Session View is by no means linear and on top of that, nothing you do has to be predetermined or remain permanent in a performance situation, or any situation for that matter. In Live, it's all about flying through tempo and time at the click of a Scene Launch button.

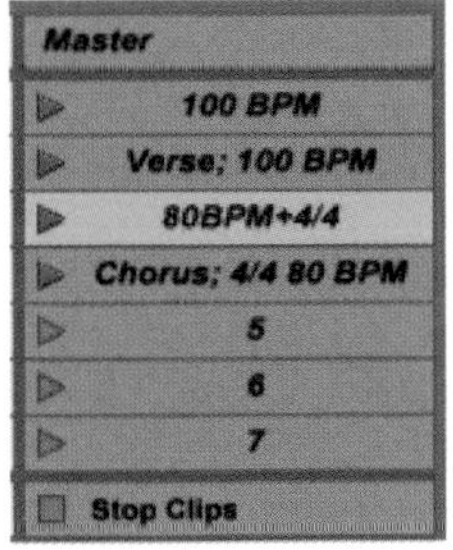

Figure 10.4 Scene names programmed with tempo and signature changes.

Scene names are used to label or describe a scene, but even more impressive is the unique ability of a scene name to store and impose a specific function upon launch. When a scene's name includes a user-defined tempo followed by the word BPM, the indicated tempo will be executed when the scene is launched. The same holds true for time signatures. Simply enter the desired signature, e.g., 3/4, 4/4, or 6/8, etc. Make sure to use the "/" (backslash symbol). To add a tempo or signature change into a scene's name, right click or ctrl + click and select "Edit Launch Tempo," or select "Rename" from the same contextual menu to rename a scene with a tempo or time signature ([⌘+R] Mac/[Ctrl+R] PC).

A scene can be named in a number of different ways in regards to the actual wording of the name label. It can be just the launch tempo or can include both the name and tempo as long as "BPM" is included in the name, that is, "Verse; 100 BPM" or just "100 BPM." For time signature changes, scenes should be labeled as a fraction, for example, 2/4, 4/4, etc. Both tempo and time signatures can be included in the same name in any order but must be separated by at least one character or space that is, "80 BPM+4/4" or "Chorus, 4/4 80 BPM." Once a tempo or time has been added, the scene's Launch button will turn orange to reflect the launch function. Once you've established a tempo or time into your scenes, you will see how easy life becomes when you need to change up the beat or timing of your music. As always, you can rename a scene on the fly to execute this feature in real time while your Live Set is playing or recording. If for some reason, you don't see the orange indicator, your nomenclature is incorrect.

Hot Tip *A quick way to enter names and symbols in multiple scenes is to use the Tab key to advance from one scene to next while the rename function is active. Enter a scene name by ([⌘+R] Mac/[Ctrl+R] PC). Don't press Enter/Return or navigate out of the scene naming area just yet. After you type the new name, select copy ([⌘+C] Mac/[Ctrl+C] PC) and then press the Tab key to advance to the next scene. The next scene name will be highlighted for text input. Type in a new name or paste ([⌘+V] Mac/[Ctrl+V] PC) to duplicate the name from the prior scene. This is useful when you want to change a digit or letter to differentiate each scene. Continue through the scene names as needed using the Tab key. If you make a mistake or change your mind while typing, press the ESC key to revert and exit renaming.*

10.4 Capture and Insert Scenes

Scenes work well for organizing, building, and launching song sections, multi-track loop segments, vocal double tracks, etc., not to mention the various ways for going about creating such elements. Because Live allows you to literally improvise your musical ideas, than just start firing off separate and randomly positioned clips to start building a nice hook or groove created as Scenes. Often, you will accidentally create a really cool song section because you launched a clip inadvertently against another clip, but you'll have no idea what is making the clips groove so well together. If you stop all the clips, there is a good possibility that you will never know or have the ability to recreate that "magic" vibe or moment. Maybe you're in a live performance, and you build up this incredible section on the fly and want to come back to it. The crowd yells, "do that again." Well, that's where Capture and Insert scenes comes in handy. So how does this work?

Just start firing off multiple clips until you have a groove you like. These can be spread across multiple tracks and over dozens of different scenes. Once you're satisfied with what you're hearing on playback, right click or ctrl click in any scene while your clips are still running and select "Capture and Insert Scene" ([⌘+shift+I] Mac/[Ctrl+shift+I] PC). This will autocreate a new scene below the scene row that you just clicked. Live will also number or copy the name of the scene you selected to capture into. For example, if you select "chorus" then "Capture and Insert Scene", it will be autonamed "Chorus1." In our case, we selected an empty scene, so Live named it "1." The new scene will be made up of all the clips that were playing at that particular moment in time when you commanded them to be captured.

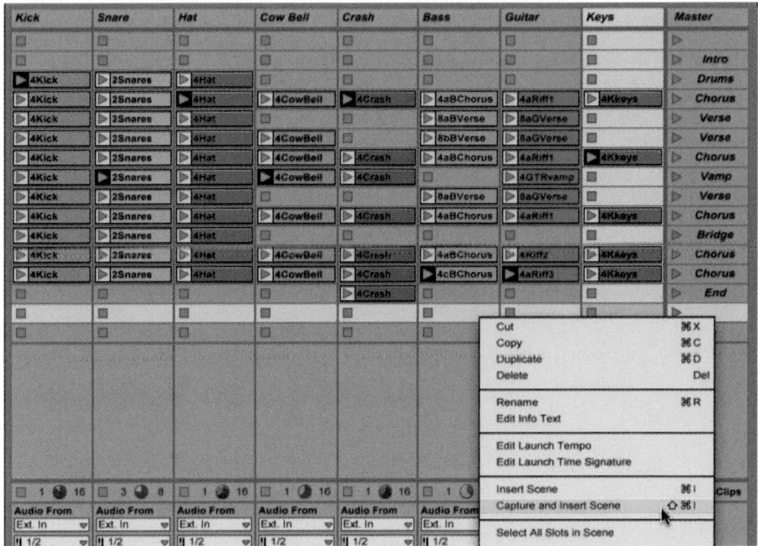

Figure 10.5 Launch any combinations of clips, then select "Capture and Insert Scene" from the contextual menu.

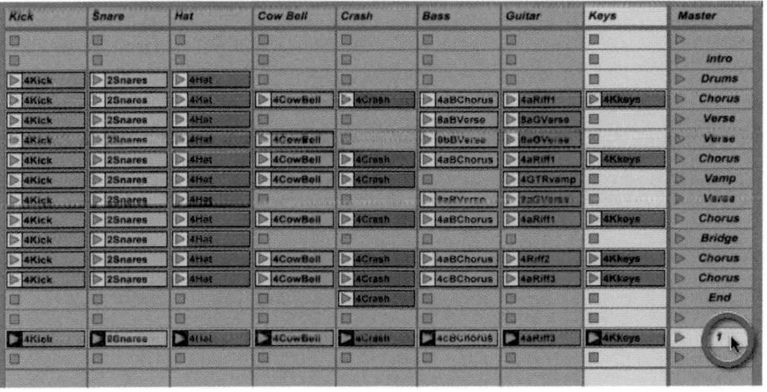

Figure 10.6 "Capture and Insert Scene" creates a new scene out of your entire active clips. Click on a scene (empty or filled) and Scene select the command.

A really cool trick is that you can capture scenes generated from the Arrangement View right into clips in the Session View in real time. Whatever is playing in the Arrangement View can be captured into the Session View without ever having to launch a clip in the Session. Simply start playback from the Arrangement – with clips of course – and then flip back to the Session View. Although the Arrangement is playing, choose Capture and Insert scene the same way as before. A new scene will then be created and playback will be taken over and generated from the Session View. The captured Arrangement clips are copied over just as if you dragged them in regards to the clip length and their contents. Any clips, session or Arrangement clips, can be captured as a scene! In the studio, scenes work great for auditioning groups of clips against one another, either while jamming alongside or rehearsing with a vocalist or session player. With the Capture and Insert scene function, the Live user can keep moving forward without stopping to audition other ideas allowing a producer to stay in the moment.

10.5 Musical concepts

Having the ability to call up a song section or group of clips intended to play together is brilliant. It all comes back to the nonlinear approach provided by Live's Session View. Scenes often represent the essence of a musical structure, unifying the culmination of a musical idea that provides direction for a composition or song, and necessary stability for the creator, producer, or performer. Let's take a look at scenes from these perspectives.

10.5.1 Create: multitrack instruments and recording

Scenes allow you to link your clips of musical ideas so they can be played simultaneously and recalled at any time. Scenes also serve as a neat and tidy way to launch your multichannel/multitrack instruments and multitrack recordings. In this way, a scene contains a row of clips that make up one instrument/recording such as a drum kit or grouped multivoice or layered vocals. In this scenario, the instrument or recording is spread across multiple tracks because it requires the sum of two or more tracks to complete its overall produced sound. For example, the kick on track 1, snare on track 2…or vocal layer 1 on track 1, vocal layer 2 with effects on track 2, and so on. You get the picture. The point to understand is that scenes don't have to contain entire song sections, rather they can be smaller parts of the whole, all the way down to a one-bar riff or an instrument/vocal if need be. Literally anything that needs to play simultaneously.

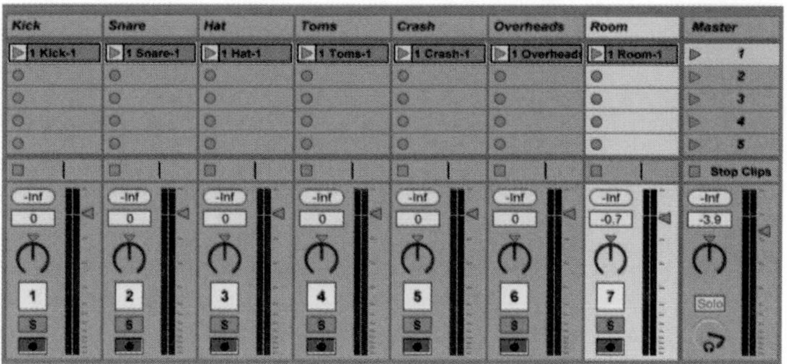

Figure 10.7 Here, the tracks are all part of a single multitrack (multichannel) instrument recording. This could be audio or MIDI. The scene row represents each track part that will launch simultaneously to make up the whole instrument.

Here is the concept at the early stages of creating a beat from a composing/ writing standpoint:

When you first begin working with a MIDI drum kit or live studio kit, its parts will often be spread across multiple tracks, that is, one track or channel per voice or mic input. We say often because you could sum the studio kit down to two-track stereo or route all the MIDI channels to Omni or the same channel. This is not a new concept, but just to reinforce it, remember that each track is serving as one part of the whole instrument. In this way, while creating or recording your drumbeats, you will be launching and recording them as a row of clips across multiple tracks forming a scene. The same concept applies to multivoice or double-tracked vocals and instruments that you're trying to either develop or record to track. Once again their collective clips would be isolated as a scene or as part of multiple scenes that make up an entire song section.

As you advance through the production process, these examples will be incorporated as Group Tracks, but during the creative phase it's quite possible that they are laid out as Scenes while being developed. Really, there is no right or wrong way to work with Scenes. What's more important is that you understand their potential. For more on Group Track, launch ▶ Scene 11.

Let's take a closer look at how scenes can be used to launch a multichannel/ track MIDI instrument as opposed to scene(s) launching entire song sections. Download Live Set **CPP_Impulse-MultitrackMIDI Project** from our Website to follow along or build your own similar instrument. We have also provided basic Impulse Kit Loop MIDI files for importing to play back and trigger your own original Impulse Drum Kit.

Whenever your are working with a MIDI drum kit, you have a choice to work with all the drums and MIDI notes triggering on one track/one clip or you can assign an individual track to each drum voice of the kit and then route each track to the same MIDI instrument (playback source). The same goes for importing

MIDI files, more specifically MIDI drum loops. Many MIDI files you'll work with will be MIDI Type 1, meaning they consist of multiple vertically synchronous tracks or simply multiple tracks intended to play in sync relative to the timeline (bars:beats). In the Session View, this will appear as multiple vertical tracks with clips laid out in a horizontal scene row. In the Arrangement View, this appears as multiple horizontal tracks and clips, which will play in sync vertically along the timeline – just as any other DAW does. Think of it as the kick on track 1, snare on track 2, hat on track 3, toms on track 4, etc. In many situations, using multiple tracks to trigger MIDI drum voices makes it easy to isolate and work with drum kits. As far as MIDI routing is concerned, this same concept applies to "Multi-Instruments" or "multis" (instrument devices that host multiple instruments within one interface that are triggered via multiple MIDI channels) often used with some popular third-party sample players, drum synth/samplers. The concept and routing scheme is basically the same as the drum kit concept. Launch to ▶Scene 15 for the routing of multi-instruments.

You will see in our example two different ways that the tracks can be assigned to handle a drum kit played and triggered by scenes. This is all about MIDI organization and dealing with multitrack MIDI file configurations. On the left, the kick track is located where Impulse has been inserted. The Snare and Hat tracks are simply routed to Impulse loaded on the Kick Track. In the example on the right, the Kick, Snare, and Hat are all assigned to a dedicated Impulse track. Either way it is up to you on how you want to work. The important point is that all clips on the first scene row are functioning together to make up the same groove as opposed to an entire song part, which also applies to audio tracks just as if the drums were a live drum kit with multiple mics.

If our MIDI instrument example was a keyboard instrument, you might program the left hand on one track and the right on another although it is less common. The point is that each one of these tracks is routed to the same instrument, making use of a scene to playback all its track parts at once. A scene can include as few or as many clips as you'd like. Just remember that the rows in the Session View are what constitutes a scene. In this case, the scene handles a drum kit spread across multiple tracks.

Before you start importing a dozen clips for your multitrack drum kits or audio tracks, note that you can import multiple clips at once from the File Browser or Finder/Explore window and lay them in as multiple tracks/scene rows as opposed to stacked into one track vertically (Session) or one after another horizontally (Arrangement). Simply hold [⌘] Mac/[Ctrl] PC while dragging them into the Session View or Arrangement View and they will automatically spread out to make a scene row, or place them across multiple tracks as well as autocreate new tracks if necessary.

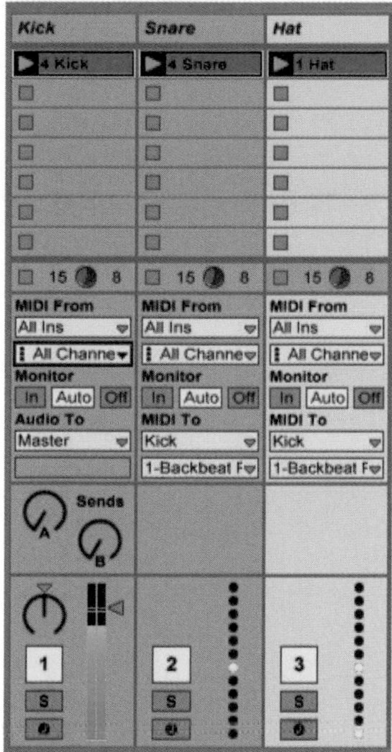

Figure 10.8 The Kick track hosts the instrument device, monitoring its output, and is arbitrarily used to trigger the kick sample.

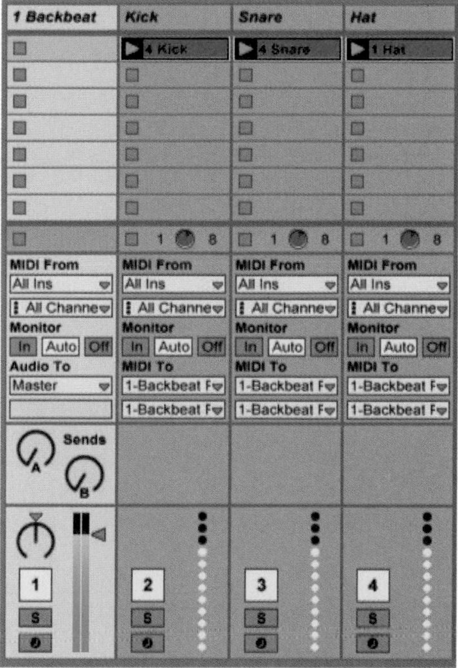

Figure 10.9 A separate MIDI track has been dedicated to host the instrument device.

10.5.2 Perform: mapping scenes

MIDI Mapping quickly comes to mind when taking your scene launching functionality to the performance stage and beyond. Each scene can be MIDI mapped to a controller key, control surface pad, knob, or slider, etc. This technique opens up a vast world of possibilities, especially, when your scene launches are tied in with the Global Quantization chooser. By mapping a rotary controller knob to the Global Quantize Menu and a controller Pad to your scene Launch function, you can creatively launch your scenes on the fly and in succession at every bar or down to the 1/32 value. Scenes keep your mind at rest and keep all your clips tidy and ready to go on a moment's notice. Some Live users use scenes to launch cues on stage, which their band mates can use as a reference for timing and structure. Other performers use scenes to color code and identify heavily structured and involved song arrangements via clips. You can even color your Group Channels in Live 8 for another level of fine tuning. Launch to ▶Scene 12 to learn more about mapping controls and launches for live performances.

In this chapter

11.1 Group Tracks 229

11.2 Launching Group clips 232

11.3 Mixing concepts 235

11.4 Musical concepts 237

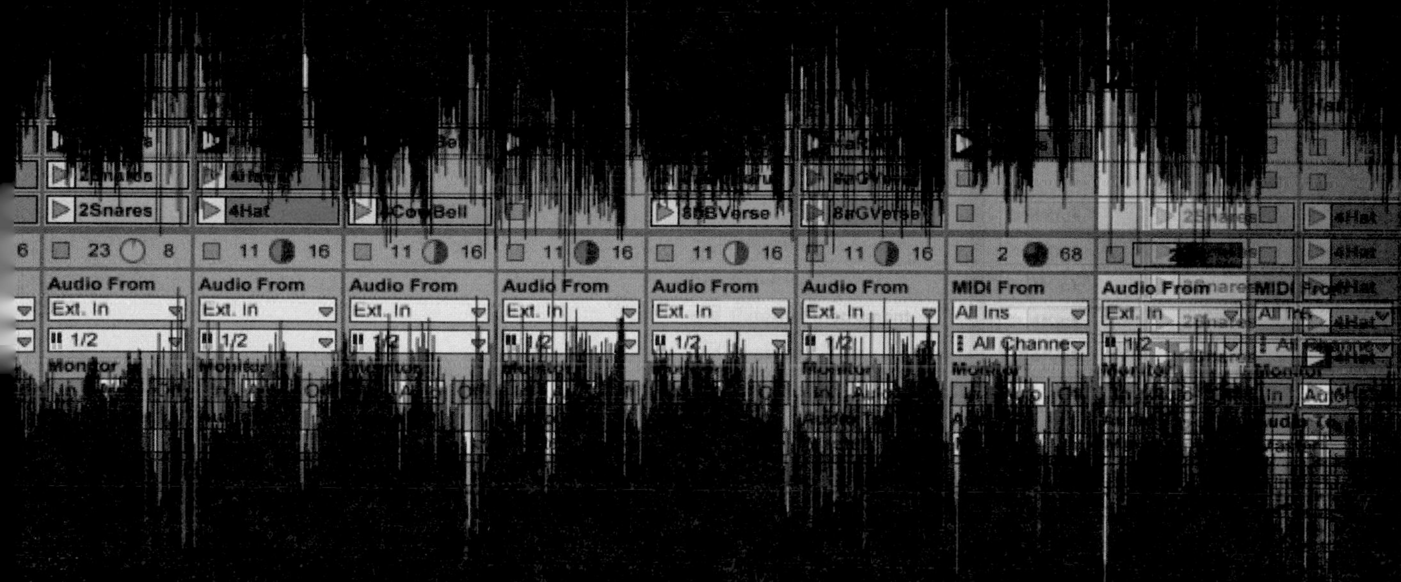

SCENE 11

Grouping Tracks

After exhaustive usage of "industry standard" DAWS, I found Ableton Live to be the holy grail of change when it comes to fusing and implementing impulsive ideas. Using Live 8's Warp function prompted my immediate conversion to Live.

Marco Garnett, Musician and Programmer

11.1 Group Tracks

The ability to launch a row of clips (multiple tracks) all at once without having to launch an entire scene or to use Follow Actions is a beautiful thing. That's what Ableton has provided with the new and celebrated Group Tracks feature in Live 8. Now it's possible to treat a multitrack MIDI or acoustic instrument, the drum kit analogy, or a multitrack vocal recording with layers or double tracking as a single track with a single launch button. Now, this may not be revolutionary and it's not like groups haven't been implemented in other music production software, but Live 8's Group Tracks are a fantastic addition to an already inventive program. To that end, Live's Group Tracks possess more power than most other digital audio workstations (DAWs). Although Group Tracks usually serve as an organizational tool and method for global mixing and edits, in Live 8, they also function as pseudo audio tracks complete with mixer controls and the ability to host audio effects. This makes them great for submixing and for global control over your groups. Sounds a bit like an Aux track, doesn't it? Let's not forget the organizational improvements to your workflow environment. Now you can hide all those tracks that get in your way when working with a large Live Set.

11.1.1 Group Tracks in Session View

Group Tracks are shared between both the Session View and Arrangement View. You can combine as many tracks, MIDI and audio, together into the same group. They can be created outside a group or within a group and deleted or moved out of the group. To create a Group, select the tracks that you wish to combine, then right click or ctrl+click and select "Group Tracks" from the contextual menu ([⌘+G] Mac/[Ctrl+G] PC). To ungroup, select "Ungroup Tracks" from the same contextual menu.

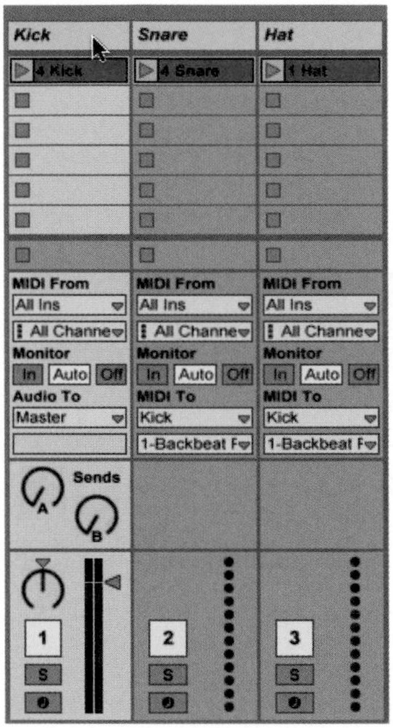

Figure 11.1 Multiselect the tracks you want to group and choose "Group Tracks" from the contextual menu.

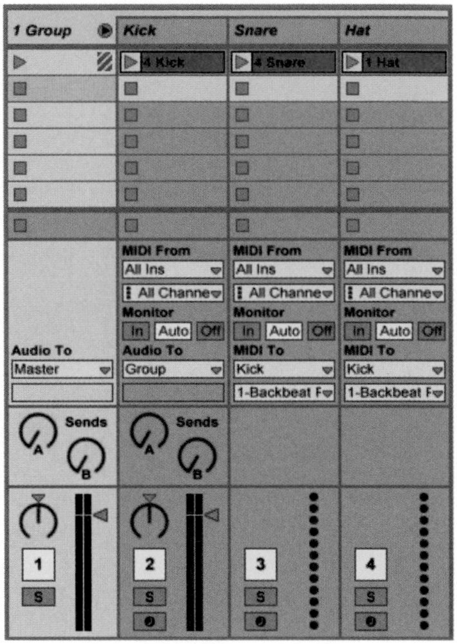

Figure 11.2 Group Tracks encompassing the selected tracks.

Once you group a set of tracks, you will have a new track column appear, the Group Track, and then to the right of the Group Track you will have all your grouped tracks. This is indicated by a few unique aesthetics. The Group Track is made up of Group Slots. These may look like Clip Slots, but they aren't. In fact, they cannot hold clips either. What you will notice is that Group Slots display a striped color box that indicates the existence of clips within its grouped tracks. The color is derived from the first grouped track on the left. You will now also see the Group Unfold button to the right of the Group Track's title, which now extends as a bracket over the group. Use the Group Unfold button to show or hide the group. When hidden, you are left with only the Group Track. Note, in our example, that the entire Group Slot is filled up now with the color of our grouped track clips. When multiple colors are involved, the Group Slot color will be that of the left most clip inside the group.

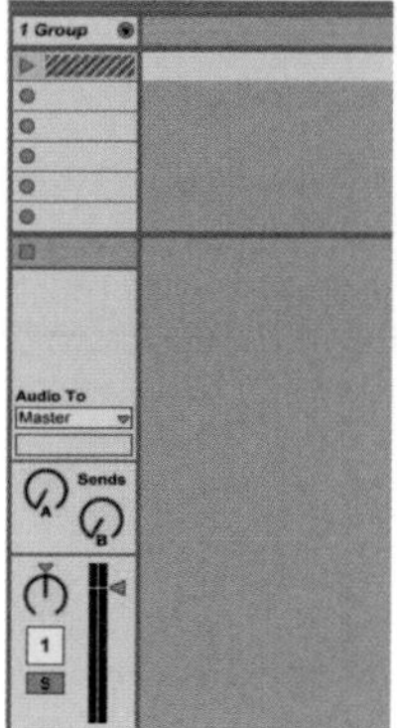

Figure 11.3 Colored Group Slot represents the presence of clips within its grouped tracks.

11.1.2 Audio Routing

When you group your audio tracks, their Audio To is rerouted to the Group Track unless they have been assigned custom audio routing before being grouped. As you can probably already guess, this is why Group Tracks are ideal for submixing and adding group-based audio effects. However, this routing configuration is not mandatory. You are free to route the audio outs of each grouped track to any available audio track or the Master track. One reason for doing this is to use the Group Track simply as a way to organize your tracks. From the Group Track, Audio To is routed automatically to the Master track, although it too can be custom-routed. Because it has a Mixer Section, it is afforded all the luxuries given to an audio track, except an Arm button.

11.1.3 Group Tracks in Arrangement View

Based on the parallel mixing principles shared by the Session and Arrangement Views, Group Tracks are carried over between both the views as mentioned.

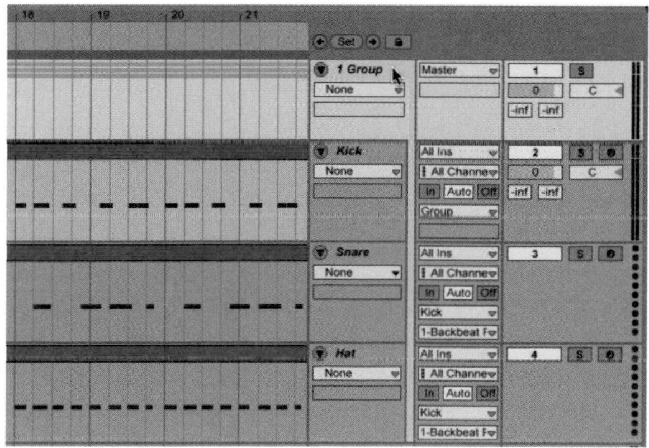

Figure 11.4 Group Track in the Arrangement View.

In the Arrangement View, you will see the same groups as just in a horizontal layout as expected.

For the most part, Group Tracks and grouped tracks function in a traditional sense, like a traditional DAW if you wish. In this way, they are ideal for submixes and organizing your Arrangement View real estate. You can see how much space is saved when using Group Tracks to fold up your clutter.

Figure 11.5 Fold Group Tracks to save space.

As far as the arrangement clips area is concerned, the Group Track will display a representation of the contained track clips. When folded, the clips are represented as an overview in horizontal lanes, whereas when unfolded, clips appear as gray transparent lines in the Group Track Display.

When a Session Group Track is unfolded, it will be unfolded in the Arrangement View and vice versa. Be aware that when you are Global Recording a Session into the Arrangement, Group clips do not visually unfold in real time. If you want to see your recording unfold, you have to unfold the group track.

11.2 Launching Group clips

An advantage of Group Tracks is the ability to launch a group of clips, which when played together make up a single instrument such as a multitrack drum kit, for example. Launch to ▶ 10.5.1 Multitrack instruments and Recording. We've talked about how at times a row of clips can collectively make up an instrument or multitrack audio recording. With Group Tracks, it's easy to launch all track clips with the click of a single launch button. To launch all the contained grouped tracks within a Group Track, activate the Group Launch button. These look and behave just like Clip Launch buttons. Stop your entire group by pressing the Group Stop button – an easier way than using

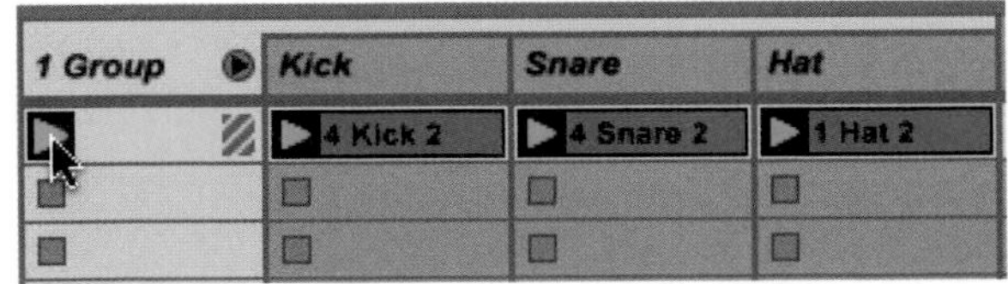

Figure 11.6 Group Launch button.

individual Clip Stop buttons or Stop Clips buttons! Out with the old, in with the new…bring it on!

Now for the fun part: Once you have launched a group, you can launch any other clip in your session as well as stop individual clips in your group while others play or advance to the next clip within one grouped track. Taking this a step further, you can now launch additional groups on the same scene row.

11.2.1 Groups versus Scenes

Nesting rows of clips in a scene…

One of the most exciting aspects of grouping tracks is the ability to simultaneously launch multiple clips on the same scene row without having to launch the scene, which is helpful when, for example, a row of clips doesn't make up an entire scene. This is great for both performance and recording. That being said, you can now rethink how you might produce and perform your Session clips. With the integration of Group Tracks, you may launch an entire scene, and then launch a Group Track from another scene without having to launch all the other clips on that row. Here is a comparison: before Group Tracks, scenes would be used to launch multiple track clips at once. In that way, each scene row contained specific clips that were generally designed to play at the same time. The difference is, now Group Tracks allows you to use a single scene row to launch groups of clips rather than giving them each a unique row to avoid launching other clips. You also have the ability to stop clips more efficiently with the grouped track clips, all stopping when the Stop Group button is pressed. This new approach benefits your multitrack instruments and your song parts. On top of all that, your workflow environment will become organized and clutter-free with all the folding and submixing you'll start doing.

Hot Tip *If you find that you really like how a few of your individual clips/tracks are grooving together, why not group them and use them as a modular idea that you can launch whenever you want?*

Here's a look at how Group Tracks work. We have three different Group Tracks in our Session View, all of which have clips in the same row. To start out, we launched the first group "Kit" that consists of an Impulse-based Instrument Rack.

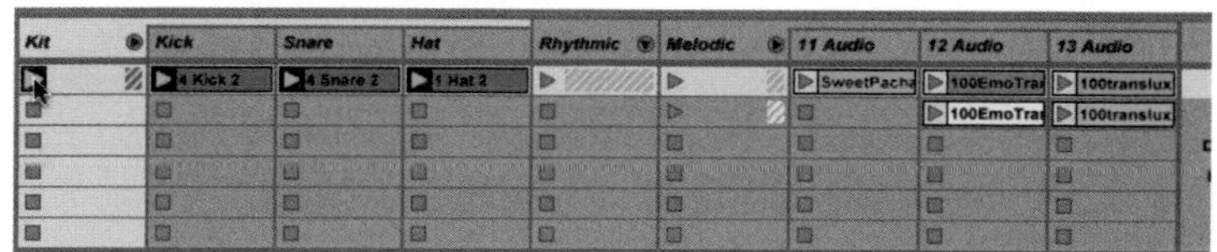

Figure 11.7 "Kit" Group Track launched.

Before the advent of groups, these three tracks within the Kit would have been a scene of their own.

After activating the Group Launch button, we'll go ahead and launch the next group that resides in the same row, "Rhythmic" Group Track.

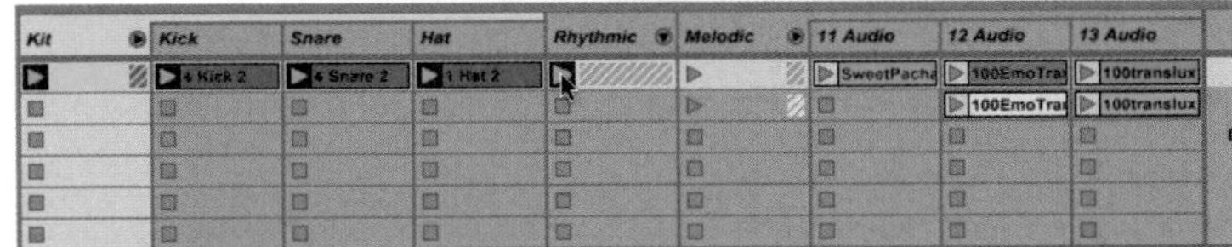

Figure 11.8 "Rhythmic" Group Track launched.

Still working from the same row, we launch the third and final Group Track "Melodic." Once again, we've done all this without leaving the row or launching a scene. Although all three groups are playing back, we'll fold up groups 1 and 2, then launch the next scene row for the Melodic, leaving Kit and Rhythmic Groups running.

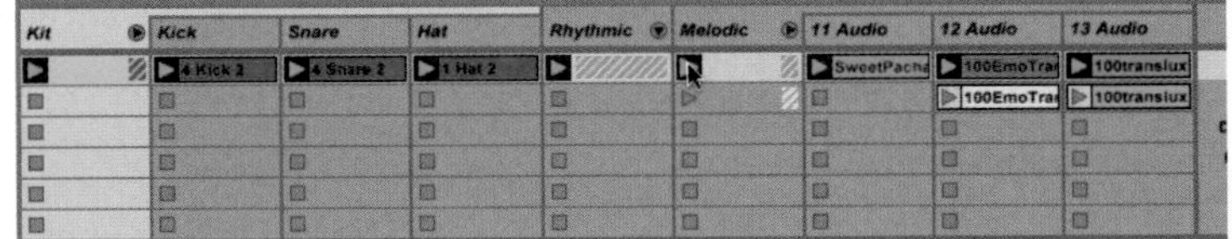

Figure 11.9 "Melodic" Group Track launched.

To do this, we will click the second Group Launch button. This activates all the clips on that row, leaving the other groups alone. If we had used a Scene Launch button, all clips playing from other rows and tracks would have been stopped. The only way to prevent this would be to remove the Stop buttons on those tracks below the clips so that the Scene Launch wouldn't trigger their Stop buttons. Of course, you will still want to remove Stop buttons for clips that you wish to continue playing within a Group Track so that scene or Group Slot launches don't stop those clips you wish to remain playing  6.4.1. Note you cannot remove Group Stop buttons.

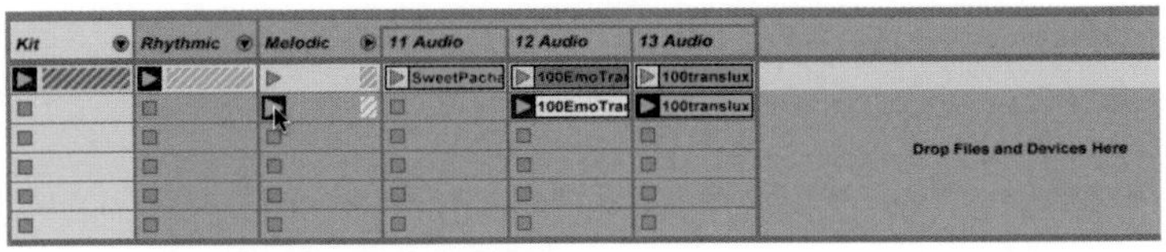

Figure 11.10 Launch a new row of clips without launching a scene.

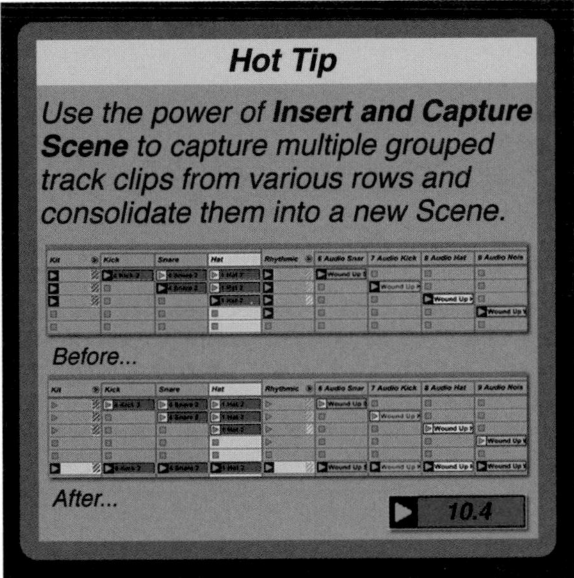

We would like to warn you at this point not to take for granted the notion that Group Slots are similar in function to traditional clips. They are to some degree, but there is an anomaly. At times you will see more than one Group Launch button lit up on the same track at once. This would appear to break Live's golden rule for clips on a track, but it doesn't. You still can't launch more than one Group Slot at once in the same track, but Live uses the Group Track/Group Slots to also indicate what rows have active clips as you will see in our example.

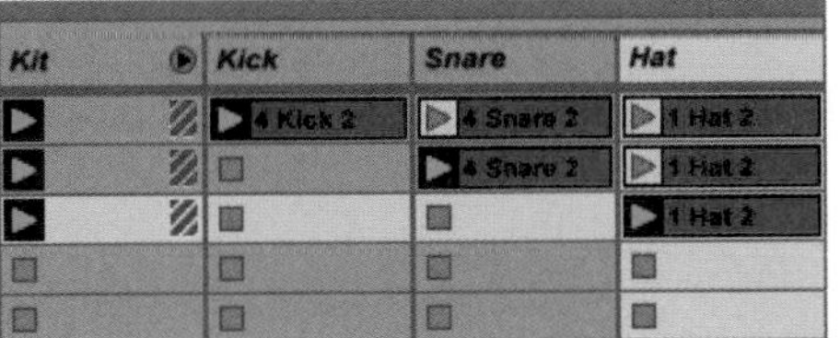

Figure 11.11 Multiple Group Slots activated indicating active clips on different rows within the Group Track.

11.3 Mixing concepts

Mixing with Group Tracks makes mixing, submixing, creating group-based effects, and automating tracks very easy and super efficient. As we mentioned, you can add audio effects to the Group Track, which affects all its contained tracks. You can also use Group Track sends as you would use nongrouped track sends. In this case, it would be a group-based Send. Mixing and automating your Set's overall mix is a piece of cake with Groups. All you have to worry about is the Group Track instead of all the contained tracks unless you want to create individual (internal) automation and tweak your mix. Dial in your internal mix, then use the Group Track to mix your entire Set.

11.3.1 Submixes

We talked a bit about submixing with the Return tracks in clip 6.2.1. With the institution of Group Tracks, you have little need to submix with Returns or any other audio track for that matter but feel free to do so if you wish. The general concept is to create a mix of all tracks to be grouped – the internal tracks. For example, try balancing the drum kit. Mix the kick, snare, cymbals, overhead mics, etc., then group them and control their overall mix for your Set with the Group Track fader (group them before creating the internal mix if you wish). This is a very common production technique and it is useful in lieu of mixing down drum kit multitracks to two tracks. This way you can submix and print them when needed either as a performance or for a traditional multitrack mixing session within the arrangement.

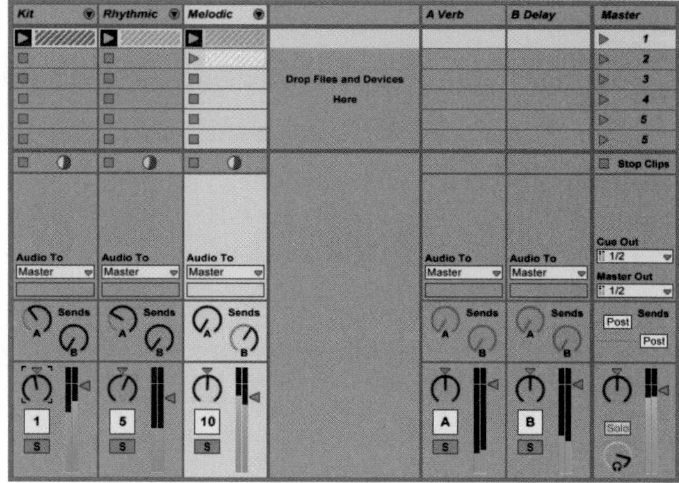

Figure 11.12 Use Group Track to submix grouped tracks. The Group Track/Submix itself can then be sent to Return tracks for Send effects if desired.

Submix your tracks in the Session View or Arrangement View and don't forget how simple it is to automate them in real time or draw a few breakpoint envelopes to submix your tracks.

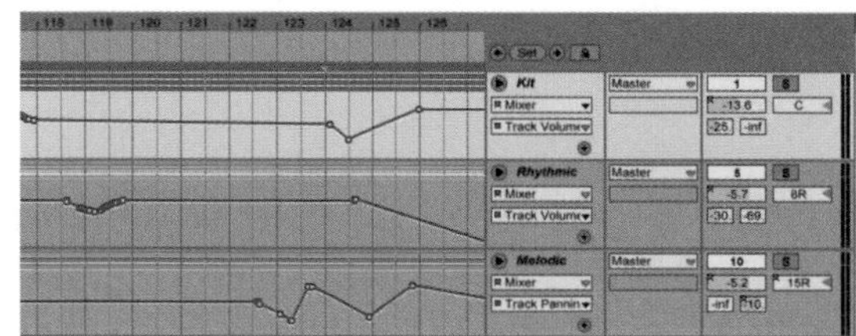

Figure 11.13 Group Track envelope automation.

Hot Tip *Sending multiple Group tracks' output into a new audio track with applied audio effects can really tighten up your project's mix in Live. We're talking about the* *power of multiple Group Tracks submixed into a single audio track – nesting them together. Try the following example:*

1. *Adjust your drum track(s) volume faders and other track parameters to a setting you like;*
2. *Create a new audio track and call it Submix Drums 1;*
3. *Set your input chooser on your new audio track to IN. This will monitor your audio through the new track while in playing mode;*
4. *Select all your drum-related tracks and group them ([⌘+G] Mac/[Ctrl+G] PC);*
5. *Then use your Group Track output chooser to route your audio on that Group Track to your new audio track;*
6. *Place a Limiter or Multiband Dynamics audio effect on the designated Return track and monitor all your drum audio this way;*
7. *Adjust your volume accordingly on your new "Submix Drums 1" track and listen to how your drums begin to blend with the mix in a much tighter way.*

11.4 Musical concepts

Group Tracks will play an important role in your future experiences with Live 8 and beyond. Ableton has truly found a way to merge the concept of an aux track, audio track, and a traditional group into one track type. With Group Tracks, the Session View becomes that much more unique because Group Tracks have both a use for mixing and performance.

11.4.1 Create: printing Group Tracks

Printing Group Tracks is a necessary part of the production process. Because they function as a submix of their contained tracks, they are perfect for building and creating audio stems to be delivered to a mixer or additional producer for adding new layers. This is also important for sending your stems for use in another DAW software application. One of the most common uses of stems is for a multitrack studio drum kit that must eventually be mixed down to two-track stereo. When that time comes, it's great to be able to print your entire group for use in your session or someone else's. The general process is very quick and easy. The hard part is deciding which method works best for you. To print a Group Track's output to track – Session or Arrangement View – route the Group Track's audio output to a new audio track and then record the Group's output into the new track or use the Export Audio/Video function to export the submix and then reimport the render back into Live if needed to continue working with it. For step-by-step printing and exporting, refer to clip 9.5.

In this chapter

12.1 Remote Control 241

12.2 MIDI Mapping 243

12.3 Key Mapping 247

12.4 The Relative Session Mapping Strip 248

12.5 Mapping Browser 249

12.6 Musical control 251

SCENE 12

Controlling Your Universe

Live 8 has become a main part of my production process. It allows me the luxury of being completely fluid and to produce records as if I'm DJ'ing to give my production that extra bite and feel.

Dean Coleman, Remixer and Producer

12.1 Remote Control

There are many ways to take control of your tasks in Live without the necessity of the mouse. Let's not forget that a mouse was not designed to be a musical instrument nor was it designed to emulate human control. As a musician, you may feel more in control when you're not tied down or limited by the mouse. In addition, many of you are using laptops or are sitting/standing behind a MIDI controller device. In either case, there is a wealth of power and creative flow waiting to be unlocked. Both MIDI and computer keys can be mapped for remote control of Live. This includes control functions such as Launch buttons, switches, buttons, input fields, and variable parameter values as well as the parameters and controls for the Control Bar, Clip View, Session View, Arrangement View, Mixer, tracks, clips, and scenes – virtually everything you would want to control!

Live uses a tactile approach for mapping actions and controls to the computer keyboard or a MIDI controller. It's all about "select and assign."

Figure 12.1 Key and MIDI Map Mode switches.

Remote control is governed by Key and MIDI Map Modes, depending on which control device you want to map control to – the computer keyboard or a MIDI controller. Now before you make your mouse obsolete, let's take a closer look at how control surfaces and mappings are set up and assigned in Live.

12.1.1 Setting up control

To set up Live 8 for remote control, go to Live's Preferences>MIDI/Sync tab. You may recall looking over this tab in clip 2.3.5, where we discussed setting up your MIDI controller.

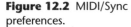

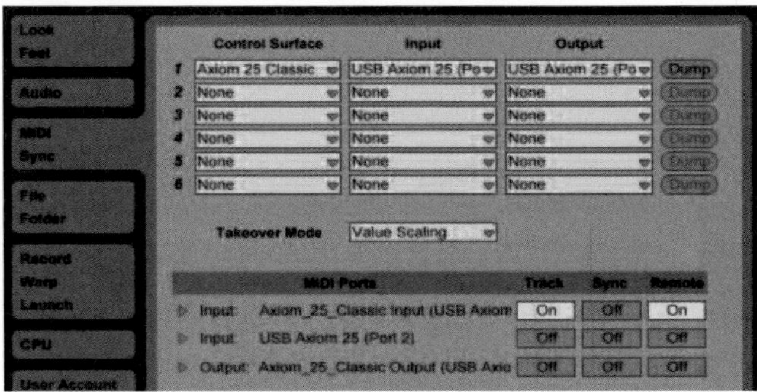

Figure 12.2 MIDI/Sync preferences.

As far as remote control is concerned, you can select up to six natively supported control surfaces. If you have a controller with knobs, a slider, wheels, pads, and faders, there is a good possibility it's listed here. Choose your control surface from the Control Surface chooser menu. If your control surface is in fact natively supported, select it, then set up the appropriate Control Surface Input/Output and MIDI Ports, and set "Remote" to "On." If your device uses motorized faders or real-time digital displays, turn on "Remote" for its output port. If your device is not supported, you can still use it, just be sure to turn "Remote" to "On" for the MIDI Input Port(s) you are going to use. As a reminder, the "Track" switch enables your device's MIDI input for the assigned MIDI port to communicate with Live's tracks.

Some control surfaces may require you to execute a "preset dump" (MIDI SysEx Dump) from Live if they aren't natively supported or set up to work as expected. This will allow you to install usable presets or programs that conform to Live's instruments and controls. Follow your controller's instructions, then click Dump to complete the setup of your device. Live will then automatically set up your device for use. These "Instant Mappings" will allow you to remote control Live's various functions and parameters from your control surface based on the physical controls and configuration of your device. See your MIDI controller's documentation on how to receive MIDI dumps. Of course, you can customize and remap assignments as you see fit. With our Axiom 25 controller, Dump installed three programs into the controller so that we could easily control Live's transport, launch scenes, track volumes, or trigger Impulse's sample slots, for example. Using these presets eliminates the need to customize MIDI mappings for those specific controls and parameters.

When a Control Surface has been set up in Live's preferences, a blue hand will appear next to a device's name. This indicates that a MIDI control surface is currently controlling it. When you select a different device, the hand will appear with the new device, leaving the old. If you want to maintain control over a device even when it is not in focus (selected), you must select "Lock to Control Surface…" in its contextual menu.

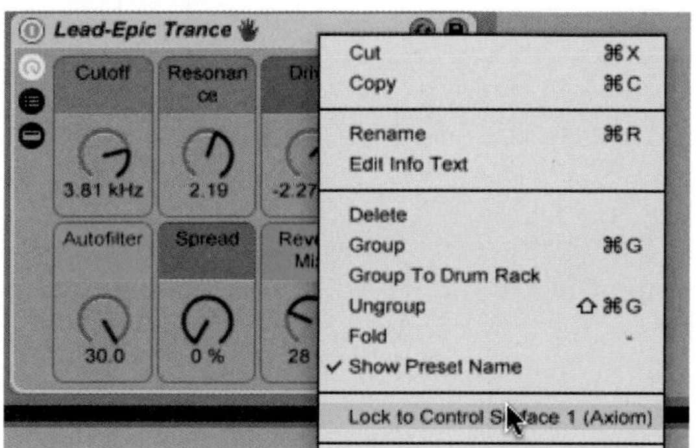

Figure 12.3 The blue hand icon signifies that a device is being controlled by a control surface. Lock a control surface to the device in the contextual menu.

12.1.1.1 Takeover modes

Now before we move on, let's take a moment to choose a Takeover Mode. This determines how your control device responds to incoming data values. There are three modes: None, Pickup, and Values Scaling. Each option determines how Live behaves and reacts to sudden jumps in values sent from your controller. For example, if you assign your keyboard controller's mod wheel to control a send knob in Live, the moment you move the mod wheel, it will send a new value to the knob, thus moving it. The purpose of Takeover is to determine when and how that data is acted upon by Live. Should Live react suddenly or gradually? Take the time to determine which setting is best for your performance demands. For our particular device and style, we'll use Pickup.

12.2 MIDI Mapping

As you already know, MIDI Mapping is the process of assigning MIDI notes and controllers – hardware faders, slider, knobs, wheel, buttons, pads, etc. – to Clip Slots, scenes, transport controls, and many other controls and parameters within Live 8's user interface. These assignments allow you to remote control their functions and actions by your external MIDI device in place of the mouse. Common MIDI controllers are, but not limited to, any MIDI keyboards, trigger pads, control surfaces, as well as the Ableton Performance Controller (APC40) by Akai.

The Akai/Ableton APC40 Performance Controller is designed to control Live's Session View in an intuitive and tactile way. The APC40's custom layout of buttons is based around an 8 × 5 grid that emulates Live's Session View track, Scene, and Clip Slot interface. Even though all the buttons, sliders, and knobs on the APC40 are already assigned and mapped to specific parameters in Live's Session View, they can be reassigned to initiate other control via the MIDI Map Mode function in Live. Therefore, you can assign various knobs to take over control of your favorite key functions. Launch to ▶ Website for more on the APC40 Performance Controller and visit www.ableton.com or www.akaipro.com.

Whether using a MIDI keyboard or a dedicated control surface, all these devices will have some sort of physical controller(s) such as faders, knobs, keys, wheels, and pads that can be mapped to a control, instrument, or function in Live. Being that a MIDI controller can also function as a traditional input device/instrument and a remote controller at the same time, MIDI mapping assignments take priority over the default note input settings. For example, if your MIDI keyboard note C3 is mapped to launch a scene, then it will no longer be able to input or play C3 as a note for an instrument.

To make a MIDI remote control assignment (MIDI Mapping), click on the MIDI Map Mode switch located in the upper right corner of the Main Live Screen.

Figure 12.4 MIDI Map Mode off.

Figure 12.5 MIDI Map Mode on.

It will turn blue when activated ("On"). Subsequently, all "mappable" parameters and controls will also glow lavender/blue, seen as a highlight scheme across Live's user interface. This signifies all the parameters that can be mapped via your MIDI controller, control surface, or other MIDI addressable hardware. Keep in mind that you can "multi-map" multiple parameters to the same key or MIDI control. For example, you could map a single knob on your controller to both the Pan on one track and the Send on another track or any configuration of tracks or the same track. You'll find that such multi-map assignments allows for a MIDI control to easily control multiple parameters, which is just one way to creatively and efficiently take control of your Live universe.

Once you have entered MIDI Map Mode, click on any highlighted parameter and then press or move a MIDI knob, slider, or note on your controller (your choice) to assign that control to the selected Live parameter. After making your mapping assignments, click the MIDI Map button once more to exit Map Mode. You can assign the same physical controller to multiple parameters if needed except when it causes a direct conflict. This would be the case if you try to assign

Figure 12.6 Session View – MIDI Map Mode.

Figure 12.7 Arrangement View – MIDI Map Mode.

the same key, let's say C3, to clips on the same track. In this case, you must give each clip a unique assignment. Next to each assignment in the user interface, a label will appear indicating the specific note or control that has been assigned to remote control it. This is only visible while in MIDI Map Mode.

Figure 12.8 Mapping label.

To expedite mapping, use the key command [⌘+M] Mac/[Ctrl+M] PC to enter MIDI Map Mode. After you've made your assignments, use the same key command to exit MIDI Map Mode or click on the MIDI Map switch again. Feel free to add more or edit the mappings as you like. If you find a prior mapping control number/message is already affixed to a parameter you want to assign, press the Delete/Backspace key or simply assign a new one to overwrite it. This can also be done in the Mapping Browser as discussed later.

Hot Tip *Drag and drop audio effects in real time without exiting Key Map or MIDI Map Mode. This proves useful when you want to quickly experiment* *with automating specific effect and mapping controls while remaining in Map Mode.*

Once assignments have been made, the *Key/MIDI In Indicator* will light up yellow upon remote control input activity.

12.2.1 Control behaviors

MIDI hardware controllers/control surfaces are often designed with a multitude of controllers, such as buttons, continuous controllers (knobs, wheels), keys/notes, and pads. Each type of controller sends MIDI information differently. Notes, for example, are sent as simple *on* and *off* messages. Continuous controllers, on the other hand, send a range of MIDI data in values of 0–127. Some MIDI controllers also send incremental data such as ± increments, where a button adds/subtracts a single value at a time as opposed to a free range of continuous data values. Live refers to incremental controllers as relative controllers and the continuous controllers as absolute controllers, but this varies among different controllers.

The type of controller and message data being sent determines how a control or parameter in Live will respond. With the numerous assignable controls and parameters in Live, certain MIDI controller types are more suitable for Live's parameters and controls than others, which will have a different effect on Clip Slots, switches, radio buttons (button with multiple options), and variable parameters. For example, MIDI notes/keys can launch clips (based on their launch mode settings), activate switches, navigate Radio button options, and toggle between a Variable Parameters' Min and Max values (track volumes, sends, etc.). For variable parameters, you will notice that a MIDI note message will toggle between the Min and the Max values, but nothing in between. To better understand how your MIDI controller integrates into Live, we suggest experimenting with different mappings and controller types to determine which controllers are best suited for the task you want to remote control. Don't spend too much time trying to locate the various parameter types. Who cares what a Radio button is, just dive in and slap a MIDI slider, button, key, or pad onto things and see what they do.

By now, you've read about the great way to control your Set in both the Arrangement and Session Views by using your MIDI controller/control surface. This, with the addition of your computer keyboard commands, will streamline your entire workflow to "untethering" you from your computer for "handsfree" manipulation of Live, well, mouse-free to be exact. One way to accomplish this is to map Scene Launches and Track Activators (track On/Off/mute buttons) in the Session View. With a keyboard for instance, you have a huge amount of flexibility due to the sheer number of individual keys that exist across the note range. Assigning each one of these keys – white keys to the Scene Launches and black keys to the track mutes (Activators) – can open up a new way to work. From experience, we can tell you that those Live users who have adapted this workflow are absolutely hooked on the results. Firing off a scene with the press of a keyboard key is seemingly effortless and can bring forth a free form way of working to your projects. You must try this for yourself! Engaging the Global Record button during this process guarantees that all your key actions and MIDI mapped parameters will be recorded.

12.3 Key Mapping

If you don't have a MIDI controller, never fear. Your computer keyboard can act as your controller, minus a continuous controller of course. Key Map Mode is activated by clicking on the *Key Map Mode switch* on the Control Bar located next to the MIDI Map Mode switch or [⌘+K] Mac/[Ctrl+K] PC.

Figure 12.9 Key Map Mode off.

Figure 12.10 Key Map Mode on.

When activated, the Key switch will turn orange, as will all the "mappable" parameters and controls across Live's user interface.

Figure 12.11 Session View – Key Map Mode.

Figure 12.12 Arrangement View – Key Map Mode.

Key mappings function just like MIDI mappings. To make assignments, enter Key Map Mode then click on a clip, control, or parameter. After you select one, type the key on the computer keyboard that you wish to make the assignment to. For key mapping to work, you must make sure that you have deactivated the Computer MIDI Keyboard. This is the keyboard-like button located next to the Key button. When this is activated (yellow), your computer keyboard will function as a MIDI note input device rather than a remote control. In this situation, you would not be able to execute your key map assignments. Although activated, any keys that share both mappings and MIDI input functions will cause the Computer MIDI Keyboard button and Key to light up orange when the conflicting keys are pressed.

12.4 The Relative Session Mapping Strip

Figure 12.13 Session View – Relative Mapping Strip.

In the Session View, when you're in MIDI/KEY Map Mode, you will see on the Master track just below the Stop Clips button an area called The Relative Session Mapping Strip. This contains four assignable buttons that can speed up navigation and launching of scenes. These buttons are Scene Launch, Scene Down, Scene Up, and Scene Select, and they will improve your workflow when working in a large Live Set.

Scene Launch: When a controller/key is assigned to this button, you can launch any selected scene with the controller. Set the "Next Scene on Launch" preference to "On" in the launch preferences to further enhance and speed up navigation with this feature.

Scene Up/Down: This assigns a controller/key to navigate up and down through the scenes column.

Scene Select: This displays the number of the currently selected scenes. Assign a continuous controller to Scene Select and scroll through scenes.

In the same row as the Relative Session Mapping Strip, you will see that each track has an assignable Track Launch button located above the track's Sends. Assign this to a Key or MIDI controller to launch a clip from that track located in the highlighted (selected) scene row. For example, whatever clip lies in the row that is highlighted will be launched when the mapped MIDI/Key controller is used.

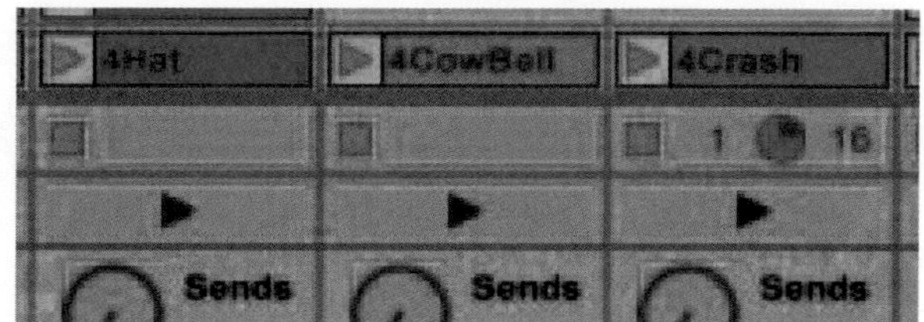

Figure 12.14 Track Launch buttons.

Let's put this feature to some good use by enabling MIDI Mapping and assigning a control to navigate and activate the Scene Launch function. Here is a walk-through:

1. Enable MIDI Mapping: You'll notice that just above the Stop Clips on the Master track that the Relative Mapping Strip appears – Scene Launch, Scene Up/Down arrows, and Scene Select.

Figure 12.15 MIDI Map Mode, Relative Mapping Strip.

2. Click on and assign a separate control of your choice to Scene Up, Scene Down (Up/Down arrows), and Scene Launch, so you can access and trigger them remotely.
3. It takes some experimenting to find the method and mapping style you are most comfortable with, but once you find it you'll enjoy "mouse-free" scene navigation and launching to its fullest potential in Live.

12.5 Mapping Browser

With all the possible mapping scenarios, Ableton has provided an organized browsing system: Mapping Browser. All mapping assignments are managed from this browser, which will appear on the left side of the window when in mapping mode.

MIDI Mappings

C...	Note/Control	Path	Name	Min	Max
1	Note C0	Kick	Slot 3		
1	Note D0	Kick	Slot 4		
1	Note E0	Kick	Slot 5		
1	Note F0	Kick	Slot 6		
1	Note C1	Snare	Slot 3		
1	Note D1	Snare	Slot 4		
1	Note E1	Snare	Slot 5		
1	Note F1	Snare	Slot 6		
10	Note C3	Crash	Track Launch		
10	Note D3	Bass	Track Launch		
10	Note E3	Guitar	Track Launch		
10	Note F3	Keys	Track Launch		
10	Note G3	Kick	Track Launch		
10	Note A3	Snare	Track Launch		
10	Note B3	Hat	Track Launch		
10	Note C4	Cow Bell	Track Launch		
1	CC 102	Kick | Mixer	B-DELAY	-inf dB	0.0 dB
1	CC 103	Snare | Mixer	B-DELAY	-inf dB	0.0 dB
1	CC 106	Kick | Mixer	A-REVERB	-inf dB	0.0 dB
1	CC 107	Snare | Mixer	A-REVERB	-inf dB	0.0 dB

Figure 12.16 Mapping Browser.

The Mapping Browser displays all active note/control assignments and to what parameter they have been assigned. These are labeled by "path" (where it is located) and "name" (what it is). Min and Max values are set from the browser for variable parameters. As a default, an assignment addresses the entire range of values possible for a Variable Parameter.

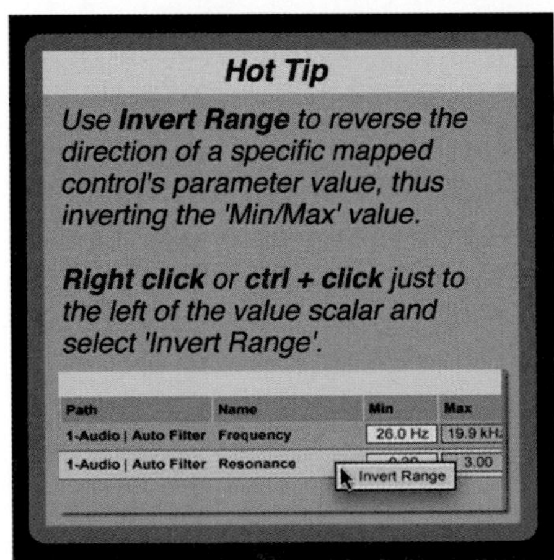

Hot Tip
Use Invert Range to reverse the direction of a specific mapped control's parameter value, thus inverting the 'Min/Max' value.
Right click or ctrl + click just to the left of the value scalar and select 'Invert Range'.
Path
Name
Min
Max
1-Audio | Auto Filter
Frequency
26.0 Hz
1-Audio | Auto Filter
Resonance
3.00
Invert Range

Adjusting the range limits the selectable range of values. Use the "Invert Range" feature to reverse the direction of a specific mapped control's parameter values, inverting the "Min/Max" values. Right click or ctrl + click just to the left of the value slider (scalar) in the Mapping Browser. This is a great way to manipulate parameters that need to oppose one another when automating, like EQ Bandwidth and Gain, or an LPF (low pass filter) Frequency and Resonance. Try assigning one knob control to the Low Pass Filter in Auto Filter and another to Resonance in Auto Filter, and then select "Invert" in the Mapping Browser for one of them. Now you can turn both knobs inward (counterclockwise) while the value actually increases for both instead of moving in opposite directions.

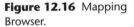

12.6 Musical control

No matter what your end goal for using Live 8 and Suite 8 may be, having control is paramount. Taking musical control of your songs should be an effortless intuitive process that allows you to spend more time in the music than in user menus and setup. That's what Live does. It makes your experience enjoyable so you can focus on creating, producing, and performing your music.

12.6.1 Perform: Mapping Locators

When working in the Arrangement View, you'll find that a Locator is a great way to pinpoint and start a new section of your music or a particular part of a song along the linear timeline. With MIDI/Key Mapping functionality, you can remote control designated Locators to be triggered from your computer keyboard, MIDI controller, or control surface to jump to and start playback from a Locator. You can also navigate to and from Locators when playback is stopped to establish a new start location for playback. Look at this as a way of quantizing a start position while playing.

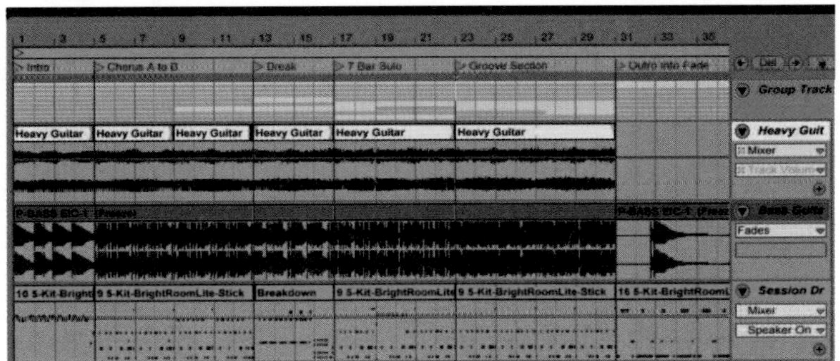

Figure 12.17 Locators along the Beat Time Ruler

Select and create Locator positions along your arrangement by placing your Insert Marker at the position you wish to place your Locator. Click the Set button to add a Locator and continue to add as many as you need for your Set.

Now comes the interesting part. Select the Key Map button ([⌘+K] Mac/[Ctrl+K] PC), then click on any one of the new Locators to map it to a computer key.

Assign any key you wish on your computer keyboard. One tip is to select a key that will be the first in a row of keys for launching all your Locators in succession. Continue to assign keys to all your Locators and when you're done, you will be able to quickly select specific locations where you want to begin playback. If you want to MIDI Map these Locators instead, just assign them to a note on your keyboard controller or drum pads on your surface controller – same functionality, just MIDI Mapping instead of Key Mapping. Open the Mapping Browser to get

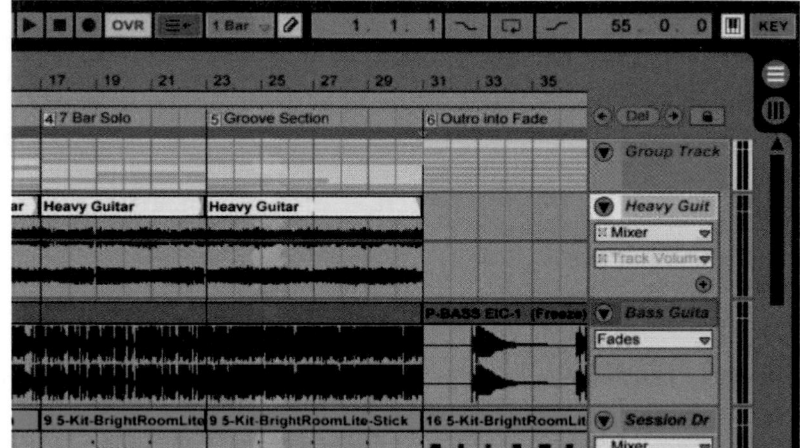

Figure 12.18 Mapping Locators.

a view of specific Mapping parameters. Here, you'll see all your Locators and their names if you named them…hint, hint.

Figure 12.19 Locators in the Mapping Browser.

When using mapping with Locators, try out various resolutions in the Global Quantize Menu for launching with Locators.

2Snares
4Hat
8bGVerse
8aGVerse
Audio From
MIDI From
Ext. In
All Ins
1/2
All Channe
Monitor

SCENE 13

Warping Your Mind!

For me, it's all about the intuitive new warping engine in Live 8. Now Live has the ability to detect and quantize transients and extract grooves or vice versa. I can now do in seconds what used to take hours; having that ultra-fast workflow onstage is essential.

John Jacobus, Composer, Guitarist & Remixer

13.1 Elastic time

It is now time to change and upgrade your perception of audio. Let's begin with a little background on controlling the tempo of audio. There was a time when the only way to change the tempo of an audio recording was to alter the sample rate, or if you were an audio ninja, you could conform a performance with hours of editing magic. Altering the sample rate unfortunately changes the pitch of the recording and can degrade the quality of the playback. As an alternative to this, software companies have developed time-stretching/pitch shifting algorithms implemented into their software. This allows you to lengthen or shorten an audio file to fit the tempo of a song by altering the sample rate and pitch, shifting it back to the correct pitch.

In addition to all these advanced algorithms, developments have led us to the acclaimed Rex/ReCycle technology by Propellerhead Software that allows you to slice up an audio file into little itty-bitty slivers (chunks or pieces of sound) at each transient point so that the tempo can be changed without affecting the quality of playback or the pitch. There is no shifting or stretching of the sample slices, rather the space between the slices is stretched when the tempo is changed thus preserving the quality and pitch of the audio. Audio files can then be quantized and locked to a grid so they can be effortlessly looped in a digital audio workstation (DAW). When the file is sped up, the slivers trigger faster, truncating and fading the slivers at their end points. When it is slowed, it spaces out the transients leaving little tiny gaps, which can be covered up with fades and reverb-like tails. With this technology, you can alter the tempo of your loops without degrading or pitch shifting the clips. Although it's a powerful and useful technology, unfortunately it's a third-party application and void of its own full-fledged DAW, well for now at least. Although Rex files can be imported into other DAWs, the technology is really designed around loops and doesn't quite cut it for whole performances, songs, complex harmonic elements, or nonrhythmic material.

Not only does Live 8 support the use of Rex files, (this is where Ableton steps in and revolutionizes the concept of tempo control) it also liberates audio from its historical confines of time and tempo. Where once the tempo of an audio recording or loop was married to its original tempo, Live gives you the ability to change the tempo and pitch independently and allows you to extract the feel from one performance and apply it to another. Ableton is committed to advancing the technology of time stretching and pitch shifting, consequently making audio in Live truly elastic.

13.1.1 Warping

Working with audio in Live 8 allows for flexibility in that you can quickly sync sample content to your song tempo or let it playback at its original tempo. This is very important, considering that not all samples are intended to be time stretched as is the case with ambient textures, effects, or one shot percussive impacts, etc. When rhythmic synchronization is important, activate Warp for desired audio clips and let Live 8 time stretch your audio to match the tempo of your Set. Stretching and "elasticizing" your audio has never been more powerful! Activate the Warp Switch located in the center section of the Sample Box to engage warping.

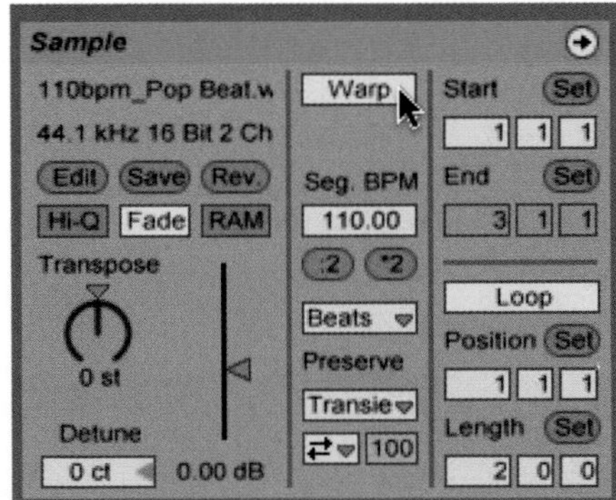

Figure 13.1 Activate the Warp Switch in the Sample Box.

Once activated, audio clips will follow the Set's tempo by employing one of Live's time-stretching algorithms to the clip's sample. Warping allows the sample to be manipulated and/or quantized to alter its playback, timing, feel, and of course loop clips. When deactivated, the audio clip will play back at its original tempo. Even when working with rhythmic or tempo specific content recorded or created at your current tempo, it's a good idea to switch Warp on. This will give you tons of flexibility, especially the ability to loop your clips. Of course, each sample may require some tweaks to get the most out of the various warp algorithms. Warp settings are configured from the Sample Box. To streamline this process, tell Live when to automatically assign Warp Markers to samples and to what mode they are set. You will find these settings under Live's Warp/Record/Launch Preferences tab.

Hot Tip With Auto-Warp Long Samples turned "On" in the preferences, Live will automatically warp your audio in files – with a total time frame of approximately 45 seconds and longer – to the best of its calculated ability. By turning this preference "Off" you will adjust this warping process yourself. Having this "Off" can be a time saver when you already know exactly where you want to place your Start Marker to Warp from and so on, or when you want to work with a specific region along the timeline of your "long" sample. By adjusting the Warp Markers yourself – manually stretching your audio – you can work in a mode that best suits you. Try both methods. As an additional note, you will find that turning off Auto-Warp Long Samples helps with extended remix material like stereo files that play longer than 10–12 min.

13.1.2 Master versus slave

Warping is ideal for tempo/beat matching, looping, and pitch/time-stretching audio material without a noticeable drop or loss in fidelity. By default, all warped audio clips play in sync with your Set's tempo. When working with Arrangement clips, activating the Tempo Master/Slave Switch located in the selected Arrangement clip's Sample Box can change this. Initially, all Arrangement clips are set to "slave" meaning that they are warped so as to conform to the Set's tempo. When set to "Master" the Set's tempo adjusts to the clip's original tempo. For example, if the Set is at 100 BPM and the clip's original tempo is 120 BPM, the Set will automatically switch to play back at 120 BPM, the clip's original tempo. You can set as many warped clips to master as you wish, but whichever clip is at the bottom of the Arrangement has priority.

Figure 13.2 Tempo Master/Slave switch for Arrangement clips only.

If for some reason you decide to delete a tempo master clip but want the Set's tempo to remain unchanged, you must Unslave Tempo Automation. This is located from the Control Bar's Tempo contextual box (right click or ctrl + click on the tempo field).

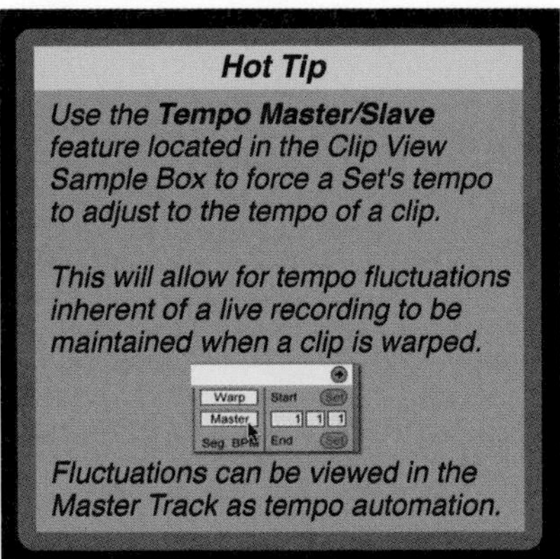

Hot Tip
Use the Tempo Master/Slave feature located in the Clip View Sample Box to force a Set's tempo to adjust to the tempo of a clip.
This will allow for tempo fluctuations inherent of a live recording to be maintained when a clip is warped.
Warp
Master
Start
End
Fluctuations can be viewed in the Master Track as tempo automation.

To get a better understanding of how flexible the Tempo Master/Slave feature is within the Sample Box, try working with an audio clip containing sample material of live drums loops that are a bit loose and out of time. Switch the "Slave" button to "Master," then unfold the Master track View at the bottom of the Arrangement View. Select *Song Tempo* in the Control chooser and examine the fluctuations in the Song's Tempo automation lines and breakpoints noting the radical changes in tempo along the timeline.

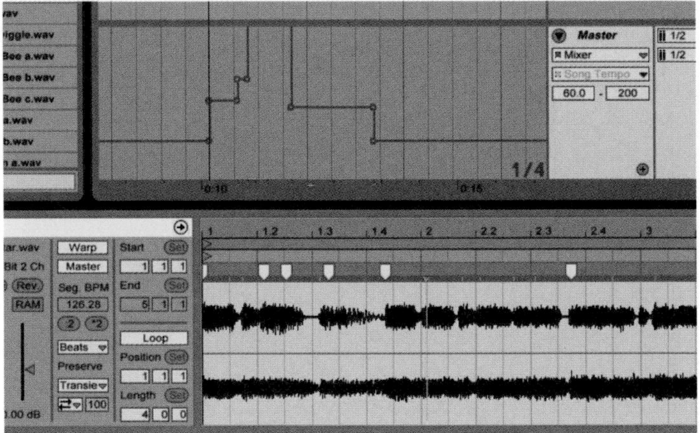

Figure 13.3 Song Tempo derived from a clip's Warp Markers that was set as Tempo Master.

13.1.3 Transients

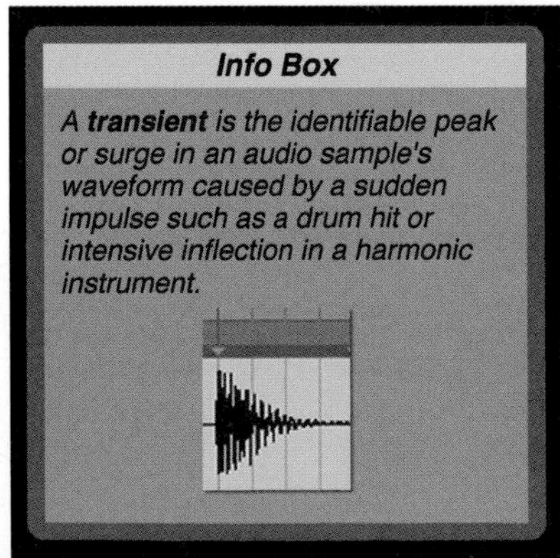

Info Box
A transient is the identifiable peak or surge in an audio sample's waveform caused by a sudden impulse such as a drum hit or intensive inflection in a harmonic instrument.

An audio sample's waveform is made up of transients or sonic events. These transients can be thought of as the "peaks and the valleys" in the waveform. A transient is defined as the identifiable peak or surge caused by a sudden impulse in the acoustic energy such as a drum hit or intensive harmonic inflection.

The attack transient is the onset or attack phase of these peaks within a sample, most easily identified from the very first sound created from an acoustic source.

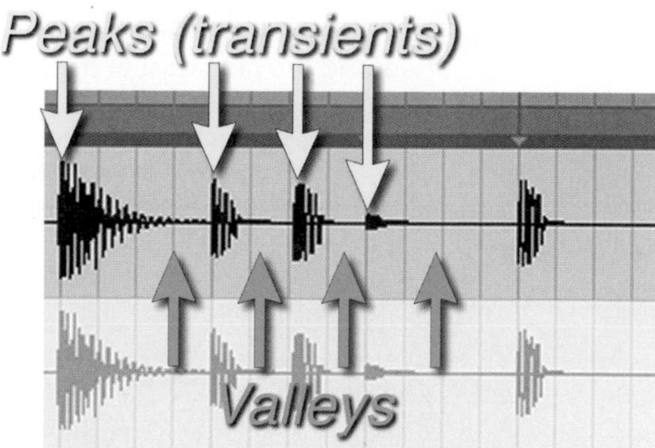

Figure 13.4 Transients as seen in a waveform.

In its simplest form, a transient could be where a note or percussive hit begins, e.g., a drumbeat or a keyboard chord. On a grand scale, it could be the dynamic accents of an entire song.

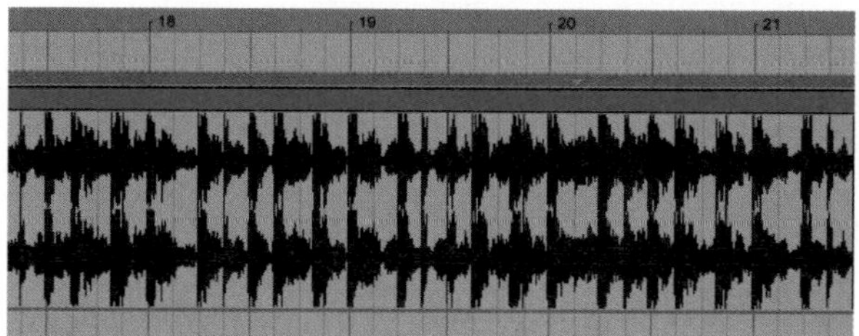

Figure 13.5 Transients of an entire song waveform.

Live 8's warping technology is centered on Transient Analysis. Transients (Transient Marks) are what Live uses to warp samples. When you first import an audio sample, Live analyzes the audio waveform to determine where all of the attack transients are to be found. On the basis of this analysis, Live identifies the original tempo of the sample and references it against the timeline based on the calculated attack transients.

Figure 13.6 Transient Mark in the Sample Editor.

These transients are then marked out as little gray vertical lines or triangle handles along the time ruler at the top of the sample display. Most of the time, Live will place the Transient Marks accurately. If not, they can be moved, deleted, or added. All of these functions are accessed under the Sample Editor's contextual menu, except the move function. To move a Transient Mark, click + shift + drag the Transient Mark to the desired location. This will move the Transient Mark without adding a new one or moving an existing one (described in the next section).

13.1.4 Warp Markers

When you mouse over a Transient Mark a *Pseudo Warp Marker* will appear. Keeping in mind that the Transient Markers represent how Live has derived the tempo of an audio clip, Pseudo Warp Markers act as visual handles for moving transients.

Figure 13.7 A Pseudo Warp Maker appears when the mouse pointer is placed over a Transient Mark.

You can slide or move them forward or backward from grid line to grid line along the time ruler thus converting them into Warp Markers. Double-clicking on a Pseudo Warp Marker will also create a Warp Marker. Once moved slightly from left to right the Pseudo "gray" Marker will automatically turn yellow. No need for double-clicking when designating a Warp Marker. Yellow square shaped Warp Markers – usually at the front of a warped sample – indicate they were added automatically by Live 8's warp engine.

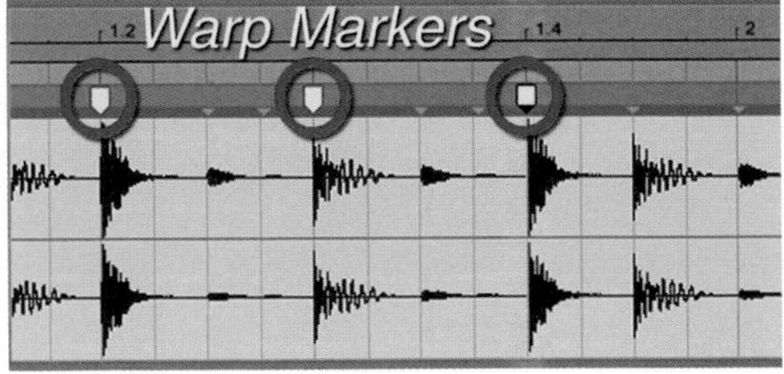

Figure 13.8 Drag or double-click a Pseudo Warp Marker to convert it into a Warp Marker.

To delete them, double-click on a Warp Marker. Think of a Warp Marker as a physical anchor holding a transient or any part of a sample in place against the timeline/grid. They are most commonly used to bind a transient to a beat or rhythmic value within a bar. Moving a Warp Marker will move its anchored point. Moving one forward/earlier in time (to the left) that is bookended by other Warp Markers will compress the audio preceding it, while stretching/expanding the audio that follows it. This is because the sample will remain anchored to the adjacent Warp Markers. In the same way, if a Warp Marker is moved backward, the audio preceding it will stretch and the audio behind will be compressed. This is the case for all Warp Markers that are between two adjacent Warp Markers. To anchor surrounding Transient Marks, hover over a Transient Mark to bring up a Pseudo Warp Marker, then when you press + hold [⌘] Mac/[ctrl] PC you will see the adjacent Pseudo Warp Markers appear.

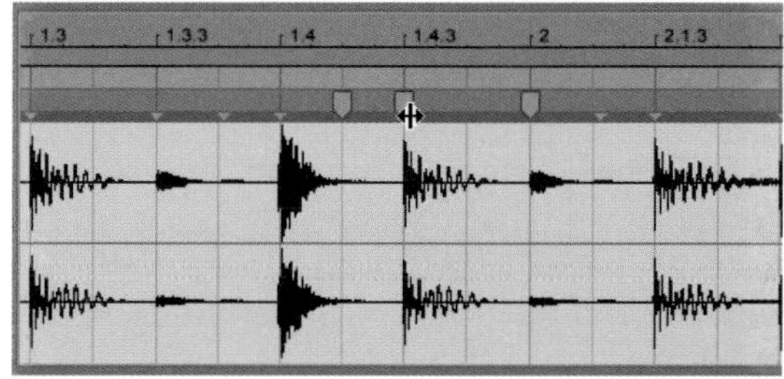

Figure 13.9 Press + hold [⌘] Mac/[ctrl] PC to visually bookend a Transient Mark with Pseudo Warp Markers.

Press + hold [⌘]Mac/[Ctrl] PC + double-click to autocreate the appropriate adjacent Warp Markers in place of the adjacent Transient Marks. You will now have three Warp Markers.

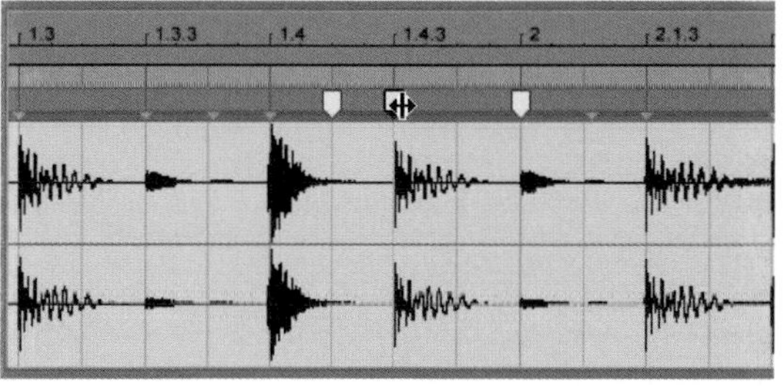

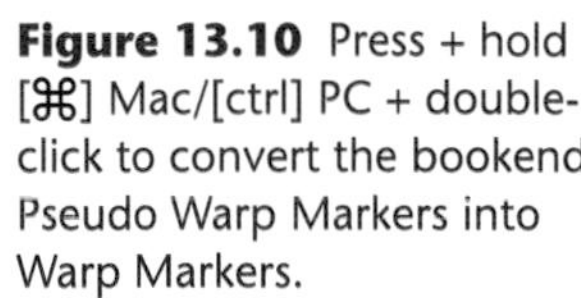

Figure 13.10 Press + hold [⌘] Mac/[ctrl] PC + double-click to convert the bookend Pseudo Warp Markers into Warp Markers.

The beauty of Live is that this combined effort to create and customize Warp Markers can be automatically ported into another Set. All you must do is save the Warp Markers to the clip. This is done in the Sample Box. Simply *click* the Save button and Live will save the Warp Markers in the analysis file (.asd) of the sample so that they are automatically recalled when you bring it into another Set.

13.2 Warp Modes

Warp Modes are vital to how Live algorithmically divides and stretches an audio file for warping and tempo synchronization. Provided are six Warp Modes, each designed specifically for different categories of audio content such as rhythmic, monophonic, polyphonic, harmonic, or melodic. These modes have been provided to achieve the highest quality of time-stretching possible without noticeable stretching artifacts or degradation. Choosing the best mode and setting for your sample will help Live with transient analysis and detection.

To achieve such a feat, Ableton has employed the power of what they refer to as "granular resynthesis techniques," loosely based on the writings and works of Curtis Roads, Barry Truax, and Iannis Xenakis into some of their Warp Modes. To get into the specifics of granular synthesis is a bit much for this book, but "granular" means grains, like pieces of sand. Sand is made of refined earth granules. In audio, a grain is a very small duration, or segment of an audio sample, often the smallest sliver possible. When an audio file is broken down into these snippets of grains (slivers), it can be expanded or compressed at a "microscopic" level. This means that segments of grains can be omitted for compressing time, or looped in rapid succession for the expansion of time. Granular-based Live 8 Warp Modes use a unique algorithm to determine the method and quantity of selection, how they overlap, and how grains fade or crossfade.

Warp Modes and their associated parameters are selected from the Sample Box Warp Mode chooser. Here you can assign the algorithm (method or process) that is best suited for the content of your audio clip. Each is clearly labeled in the chooser menu based on the goal of the warping algorithm's analysis process.

There are six modes to choose from in Live 8: Beats, Tones, Texture, Re-Pitch, Complex, and Complex Pro. Each mode enlists its own set of parameters and settings located below the chooser. These Modes have been explained time and time again in several publications and books. Here, we'll try to explain them in a new way.

13.2.1 Beats Mode

Beats Mode has been designed for predominantly rhythmic and percussive content, as you might have guessed. It works especially well for beat loops and other beat driven content. Beats Mode comes with multiple parameters: Preserve: Granulation Resolution, Transient Loop Mode, and Transient Envelope.

Figure 13.11
Beats Mode.

The Preserve: Granulation Resolution chooser offers seven different options that can determine how a sample is virtually divided into metronomic chunks that align to the tempo grid. In general, the algorithm slices up and sectionalizes each transient into segments, or the "peaks and the valleys" of the audio waveform display. Note that this is the concept behind Warping in general. The Preserve control determines how a sample is virtually divided and bound to the grid. When using Transients as the Granulation Resolution, your sample will be divided by its transients to determine how it should be warped. Resolution can also be set to a fixed note value, which forces a division upon the sample regardless of its transient content. The default setting is "Transients," meaning that the time-stretching algorithm will divide the sample by its transients to warp it. This can also be set to fixed note values as seen in the example.

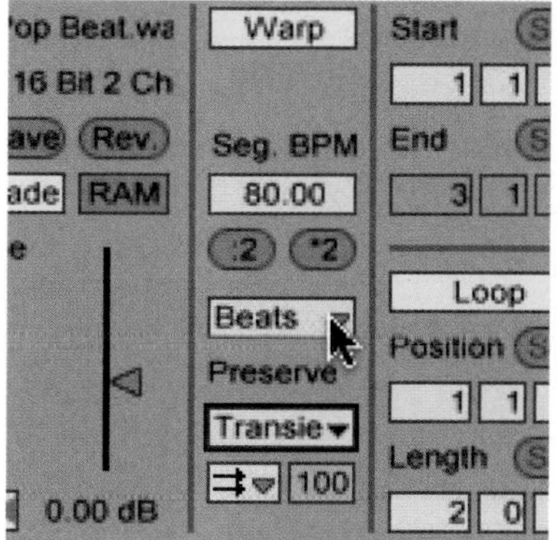

Figure 13.12 Set to Beats Mode.

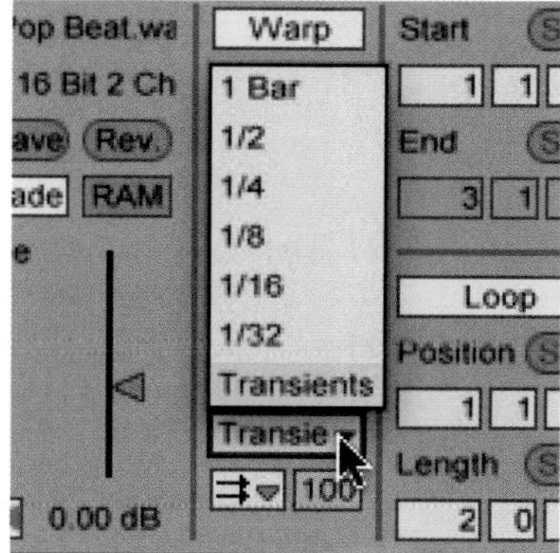

Figure 13.13 Preserve setting: Granulation Resolution – Transients.

The Transient Loop Mode chooser, located to the lower left below the Granulation chooser, helps to manage how the stretching algorithm handles the endpoint of each "sliced" division in the sample as chosen by the Preserve setting. This sets the way in which the audio content between transients (valleys) is handled when each segment reaches its end point. As a warped sample has been divided into transient segments, the warp mode algorithm must fill in any missing information or gaps between transient segments, which were created when the sample was expanded (stretched). The Transient Loop Modes determines how theses micro gaps are handled by the warp engine, whether they are looped forward, back and forth, or left as is.

There are three modes that can be assigned to handle this process: Loop Off, Loop Forward, and Loop Back-and-Forth. When set to "Off" the audio between each divided segment will play to the end of its content, then stop. Any remaining time before the next segment will result in silence. This can sound like stutters or a gating effect during playback when the playback tempo is set to be slower than the tempo of the clip. With Loop Forward chosen, the audio will begin to loop, rather than just stop, once it reaches the end of the decay, until it is time to play the next transient. With Loop Back-and-Forth the audio will start looping in the reverse direction when it reaches the end point of the transient decay. It will exclude the attack transient, rather than play back to the very beginning of the loop segment each time it gets to its endpoint.

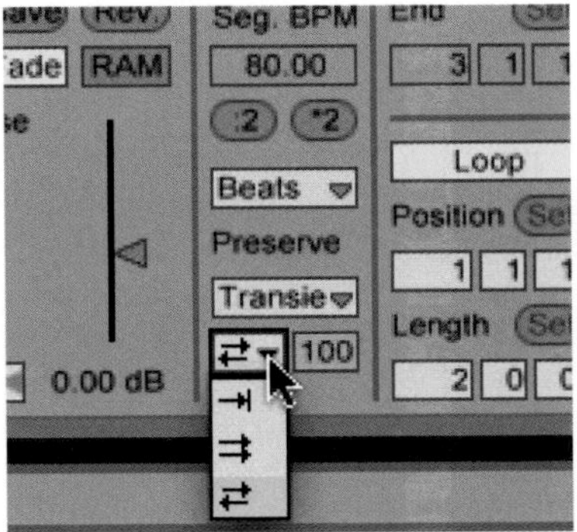

Figure 13.14 Transient Loop Mode.

On the basis of the stretching percentage you apply, these audible results are usually undetectable most of the time but the options in the new Transient Loop Modes chooser allow discrete and exacting control over these gaps, giving you more control and flexibility than ever before.

To the right of this chooser is the Transient Envelope Box, which allows you to control the amount of fadeout at the end of each divided segment. This is theoretically designed to smooth out and minimize any potential pops, clicks or other audible artifacts in the audio playback while applying Warping to an audio sample.

13.2.2 Tones

Tones Mode has been designed for melodic or monophonic audio content. This includes instruments such as leads – vocals, guitars – and basses that are characterized by distinct notes/pitches. Tones Mode gives you control over the average selected grain size based on the content of the waveform. A larger grain size can help the time-stretching algorithm to smooth out samples whose pitches are somewhat blurred, but it will be at the expense of audible granule loops.

Figure 13.15 Tones Mode.

13.2.3 Texture

Texture Mode has been designed for warping audio content that is less focused around single pitches and more around "pan tonal" or polyphonic/harmonic content. This includes dense harmonic content, ambient soundscapes, sound design, and noise-based nonharmonic frequency content. There are two parameters associated with this mode: grain size to determine the size of each grain (slice) used, and flux (fluctuation) to generate randomness in the grain selection process.

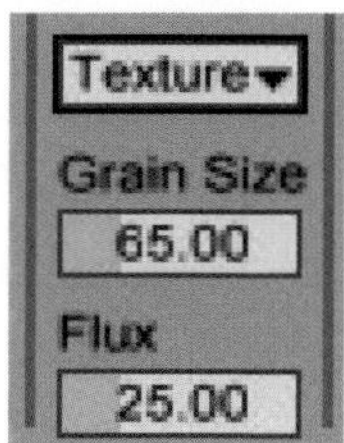

Figure 13.16
Texture Mode.

13.2.4 Re-Pitch

Re-Pitch Mode has been designed to alter the playback speed of a sample to sync directly with a song's tempo like a tape machine. Therefore, this is a frequency-based pitch shifting effect rather than a true time-stretching effect. The slower the tempo, the lower the sample's pitch becomes as it remains relative to the tempo and vice versa. Examine this in real time by adjusting Live's Set Tempo BPM manually while Re-Pitch is selected within a clip and hear the difference.

13.2.5 Complex

Complex Mode has been designed for warping entire songs and tracks, which inherently consists of a multitude of musical elements and frequencies. It uses its own unique algorithm for identifying a broad spectrum of content such as melody, bass, polyphony, harmony, rhythm, ambience, etc. When using Complex Mode, keep in mind that all of its goodness comes at the expense of CPU power. While this tax on horsepower may not be an incredible amount, it must be noted as such. If artifacts or other audible discrepancies are noticed during playback, adjust your Buffer Size in the Preferences to a larger numerical amount for the specific project at hand. Review these settings in Clip 2.3.

13.2.6 Complex Pro

Complex Pro is a new enhancement and warping feature in Live 8, which is a more advanced CPU intensive version of Complex Mode. It has been designed for warping the same types of audio content as Complex, but it is intended to be superior to it using "élastique Pro," a time-stretching engine by zplane (zplane.development). This provides sample accurate pitch shifting designed to preserve formants. In theory, this is better for complex polyphonic material because of its ability to minimize artifacts. With Pro you have additional parameters not available in the regular mode. The Formants Slider helps to maintain the sample's original tonal quality, the formants being the characteristics of a sound that are not naturally transposed. For example,

Figure 13.17
Complex Pro.

you'll prevent the effects that cause "munchkinization" or that chipmunk effect in the human voice. When set to 100% the formants should remain unchanged from their original state. This should allow a sample to undergo drastic transposition (pitch) while maintaining the tonal quality and characteristics of the sound. The Envelope Slider should also help shape the spectral content of the warped sample, theoretically improving the effect warping has on the sample. You'll also find that Complex Pro can be a friendly tool for sound design situations and other audio altering projects you may find beneficial. Complex Pro is a great warp mode enhancement for complete experimentation in addition to just warping your heavily concentrated stereo tracks.

13.3 Warping samples

Hot Tip
Batch Analyze entire folders of audio samples in the File Browser using the "Analyze Audio" feature. This is a great way to continue developing your Project while Live prepares your audio content in the background.
Right click or ctrl + click on a folder and select "Analyze Audio".

Audio files come in all shapes and sizes and are usually broken down into various samples or full-length songs/tracks. Some are short and some are long, while others are perfectly cut loops and others are not. No matter what type of audio files you are importing, they can all be warped in Live 8. With Warping, nonlooping audio files can become loops and songs can be matched to sync with other beats, loops, or songs. To top it off, any audio file can be quantized, adjusted, and regrooved – all thanks to Live 8's fantastic warping engine. When it comes to tempos and warping, Live 8 has been designed to do a lot of the guesswork for you. Live 8 makes educated guesses and presumptions that help guide and dictate how a sample's tempo is determined and handled within a Set. This is all a part of the analysis process, which is stored in the program's .asd file. If you prefer not to have Live create .asd files, then turn off this function from within the Live Preferences>File/Folder tab. The first time you import an audio file into the Session or Arrangement View, it will undergo analysis. This process will temporarily disable the file from playback. If you want to drop in audio files on the fly, you should choose to batch analyze your audio content before you perform, etc. To do this, open up a File Browser and navigate to a folder containing your audio files and right click or ctrl + click in the File Browser areas (not on the file) to bring up a contextual menu. From there, choose "Analyze Audio." Analysis will begin, as you will see in the status bar. Once all of the files in that folder are analyzed, they will be ready for immediate import, playback, and editing.

13.3.1 Loops

Dropping perfectly cut audio loops into Live 8 is the fastest way to get your Set grooving. This is because loops are easy to work with and Live 8 does a good job of analyzing shorter samples and loops to determine their original tempo. Once a loop is imported to your Set, Live will place Warp Markers to an audio file (often at the start and end) based on the Auto Warp settings and the files length, anchoring it to the grid. The loop should now play back in sync with the Set's tempo – in theory at least. If the loop has been imported into a Session Clip Slot, it will automatically loop when launched. Occasionally Live will miscalculate an audio loop's original tempo by a factor of two, being either twice or half the correct tempo. For this reason, you can quickly double or halve the audio files original tempo by pressing the Double or Halve Original Tempo buttons in the Sample Box below the "Seg. BPM" section as seen in our example. If you know what the tempo should be, then you can type it into the input field. Click on the first Warp Marker in the Sample Editor, and then type the correct tempo into the BPM field.

When an audio loop is not cropped to a perfect loop length, then you'll have to do some adjustments in the Sample Editor, moving Sample Start and Warp Markers around to make it work perfectly as a loop. The most common edits are removing silence from the front or end of the audio file, but often you will need to isolate a loop region (selection) within an audio file. This will be the case when the sample is longer than necessary or if you want to choose a more desirable section to loop. A common case is a lead-in/fill to a drum groove or when a sample file is not cropped. Now, before we actually start warping and fixing these minor issues, let's talk about the Warping Commands.

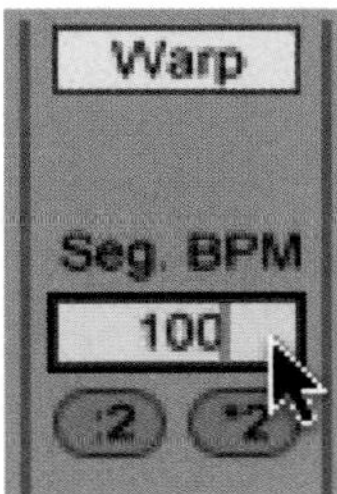

Figure 13.18 Type in the correct tempo.

13.3.2 Contextual Warp Commands

What can Warp Commands do for you? Let's say that you import a sample and it plays in sync with the tempo of your Set, but the downbeats are off. A quick remedy is to place a Warp Marker at the first down beat of your song (audio file). Then open the contextual menu from the sample editor and select Set 1.1.1 Here. Now your song should align correctly on downbeat with the metronome. If the song has a pick up or lead-in to the downbeat (incomplete measure/bar at the beginning of the song leading into the first downbeat), move the clips' Start Marker backward to include that portion of audio.

Warp From Here Probably your most frequently used Contextual Warp Command will be Warp From Here. This warps all the audio from the selected Warp Marker to the end of the audio file just as you did in the previous example. To

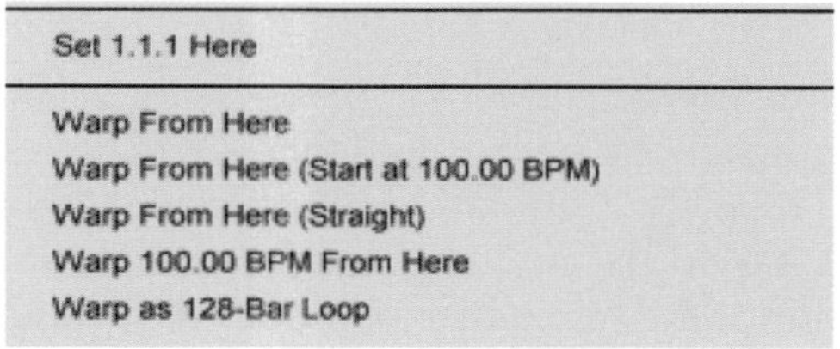

Set 1.1.1 Here

Warp From Here
Warp From Here (Start at 100.00 BPM)
Warp From Here (Straight)
Warp 100.00 BPM From Here
Warp as 128-Bar Loop

Figure 13.19 Select Set 1.1.1 Here to align your warped sample with the grid-based downbeats.

reiterate its function, let's say that you imported a song and it's just not in sync. So, you will have to take matters into your own hands. It's best to listen closely and look for the downbeats of the song and locate the Transient Marks for them.

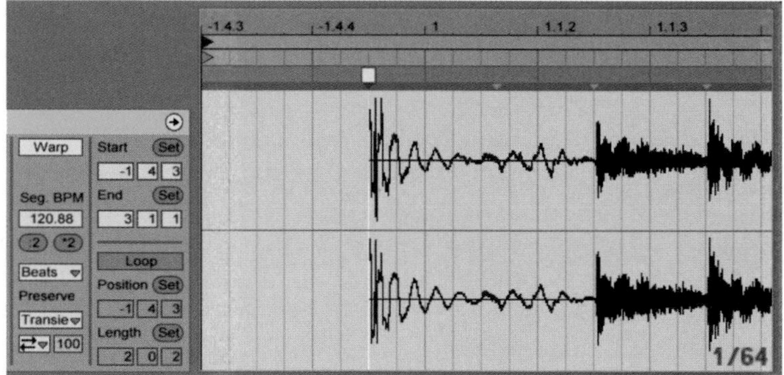

Figure 13.20 Downbeat out of sync.

Starting from the beginning, you will want to manually add or move Warp Markers to the nearest downbeat on the grid. Once you have set the first one you can try the auto warp command "Warp From Here."

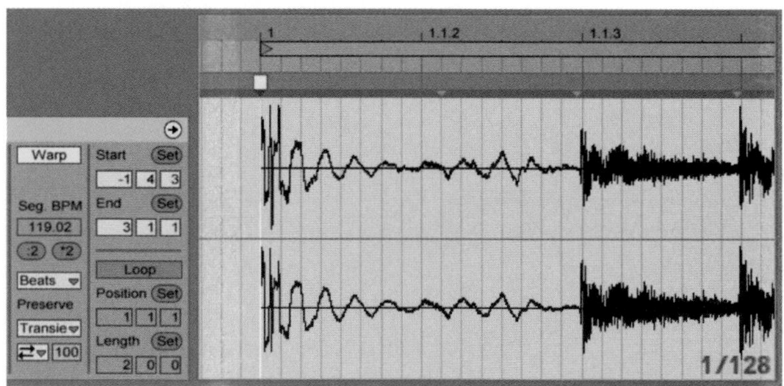

Figure 13.21 Downbeat synced at 1 (1.1.1).

Once you have executed that command, listen to the song and see if it is in sync. If not, you may have to add and/or adjust more Warp Markers. Don't be surprised if the song drifts out of sync over time. To fix this, just find the area where it starts to drift and add/adjust the Warp Markers to the correct grid lines then Warp From Here again. Keep repeating this as necessary. The good thing is that when you are all done with this, you can save the Warp Markers so you will not have to do this again.

Warp From Here (Straight) – It is very similar to the previous command, except that "Straight" tells Live that the song has one consistent tempo throughout the song. You might find this works better for those genres or songs that adhere to these types of tempo maps – Dance, Techno, Electro, and House, for example.

Warp From Here (Start at … BPM) – Probably your most frequently used Contextual Warp Command will be Warp From Here. This warps all the audio from the selected Warp Marker to the end of the audio file just as was done in the previous example. To reiterate its function, let's say that you imported a song and it's just not in sync. So, you will have to take matters into your own hands. It's best to listen closely, look for the downbeats of the song and locate the Transient Marks. Starting from the beginning, you will want to manually add or move Warp Markers to the nearest downbeat on the grid. Once you have set the first one you can try the auto warp command Warp From Here. Once you have executed that command, listen to the song and see if it is in sync. If not, you may have to add and/or adjust more Warp Markers. Don't be surprised if the song drifts out of sync over time. To fix this, just find the area where it starts to drift and add/adjust the Warp Markers to the correct grid lines then Warp From Here again. Keep repeating this as necessary. The good thing is that when you are all done with this, you can save the Warp Markers so you will not have to do this again.

Warp … BPM From Here – If you already know the actual tempo of your audio file then you may choose Warp … BPM From Here. This will make Live warp the file precisely to the tempo that you have entered, in this case the exact tempo of the original recording.

Warp as …-bar Loop – The last and probably most straightforward command is Warp as …-bar Loop. Live will automatically recommend a loop length based on the analysis of the file's tempo. On the basis of this information you can tell Live to warp the clip to the loop length it has suggested. This works well for warping perfectly cropped audio loops, especially those that extend beyond the traditional loop lengths (8 bars, for example).

13.3.3 How to fix "out of sync" audio in Live 8

Ok, now to fix our audio loop. To make the necessary changes you will need to identify the first downbeat and the beat where you want the sample to start.

Hot Tip *Keep the key command short cuts in your back pocket while warping. They will make manipulating and warping samples more expeditious. Here are two to keep in mind: (1) press + hold shift then click + drag to move Transient Marks without moving or creating Warp Markers; (2) press + hold opt/alt then click + drag to move Warp Markers free from the grid snap.*

1. Double-click to delete the first Warp Marker that was assigned by Live automatically when you imported the clip.

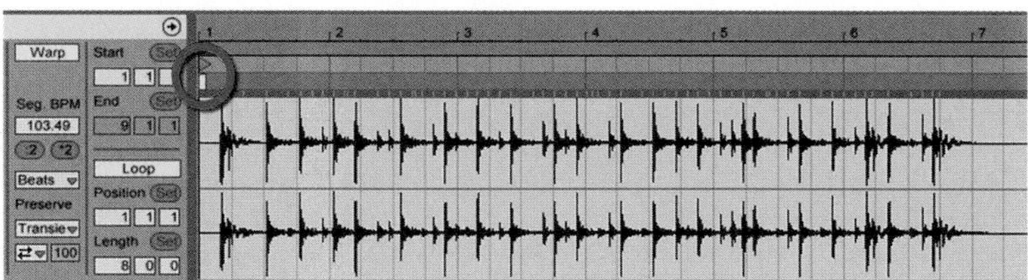

Figure 13.22 Locate and delete the first Warp Marker assigned by Live.

2. Insert a new Warp Marker at the Transient Mark located on the first beat where you wish to start sample playback from.

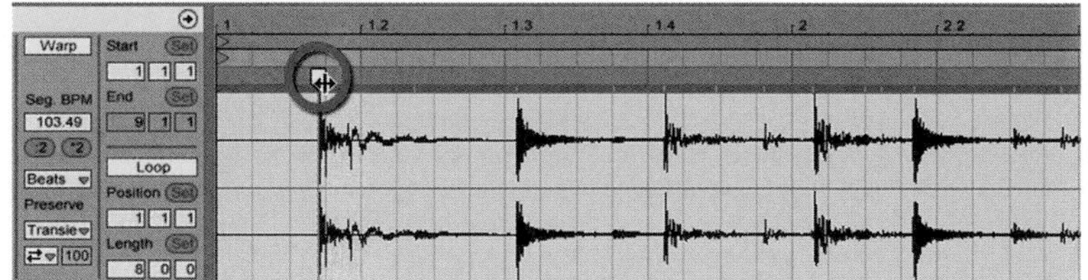

Figure 13.23 Insert new Warp Marker.

3. Right click or ctrl + click on the Warp Marker you just created and select "Set 1.1.1 Here."

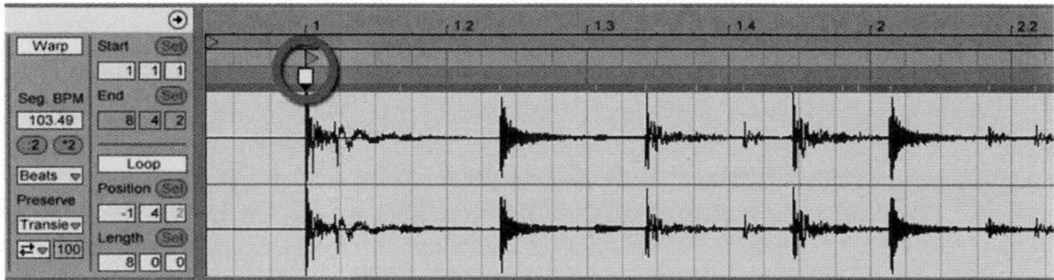

Figure 13.24 Set 1.1.1 Here.

4. Right click or ctrl + click again on the same Warp Marker select Warp from Here. Live will then rewarp the sample based on the new Sample Start point exactly where you inserted the Warp Marker. This will adjust the Sample Start and Bar 1 to be in alignment at your new Warp Marker position.

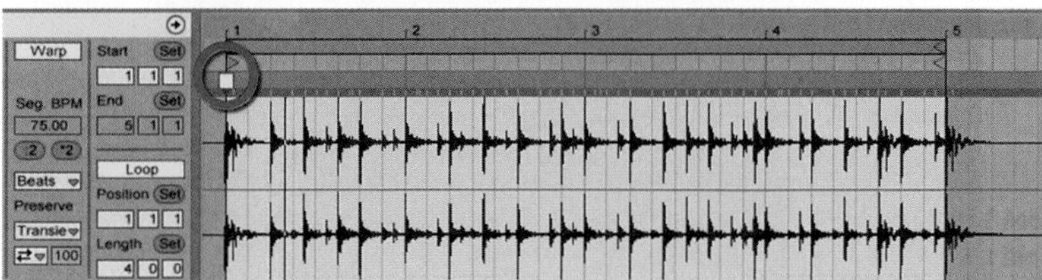

Figure 13.25 Warp From Here.

5. Adjust the Loop Brace so that it is set to cycle playback of the sample waveform as a perfect loop, choosing a specific selection that you desire. Bars 1–4 would be a full loop as seen in our example. When you move the Loop End point it will move the Sample End Marker too.
6. Make sure that the Loop is activated and then launch your clip. It should now play back in sync with your Set's tempo. Double check this by turning on the Metronome. If playback sounds too slow or too fast, try doubling or halving the loop's original tempo as this is often the cause. If Live has been precise with your sample it will be a perfect loop and will show a rounded number in the Original Tempo Box. If not, then you may need to fine-tune the last Transient Mark at the End Marker because it's not exactly on the beat.

13.3.4 Songs/tracks

Before you start warping every audio file you have in your iTunes library, understand that there is no "perfect fix" for warping full songs or tracks. A long sample or song can often breathe and fluctuate rhythmically over time. That's what makes music human. In any case, some files will be quick and easy, while others will require an audio ninja to warp the file all around. To that end, do not be misled by an algorithm and what it is supposedly designed for, because you never know which one will work the best, regardless of what Live might suggest. At this point, Beats Mode is always a great warping algorithm to start with. Don't hesitate to try it.

So far we have primarily dealt with importing short samples or loops. So let's start talking about long samples such as full songs and tracks. When a long audio file is imported into a Set, Live will autowarp it under the presumption that it is a song/track or long nonloop-based sample. Of course this setting can be changed in Live's preference, but this is the default setting and will work fine for now. With your warping chops ready, go ahead and import one of your own

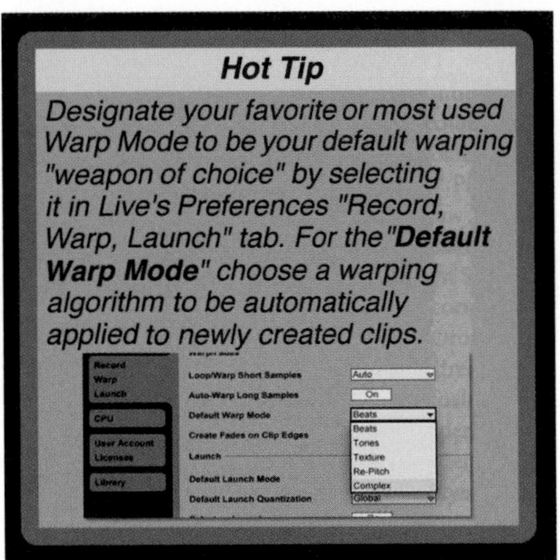

songs/tracks (mp3 is fine too). Play it back with the Metronome turned on. If it is warped correctly, then you're ready to keep on grooving away with your music. Unfortunately, full songs are complex and Live cannot be right all of the time. Sometimes a mismatch is as simple as the downbeat being off, while the tempo is correct. Other times mismatches are due to unwanted silence at the beginning of the file or any number of other things. In this case we'll have to make some tweaks to get the timing right. For our full song track, Live created only one Warp Marker, although our second example track was not so lucky.

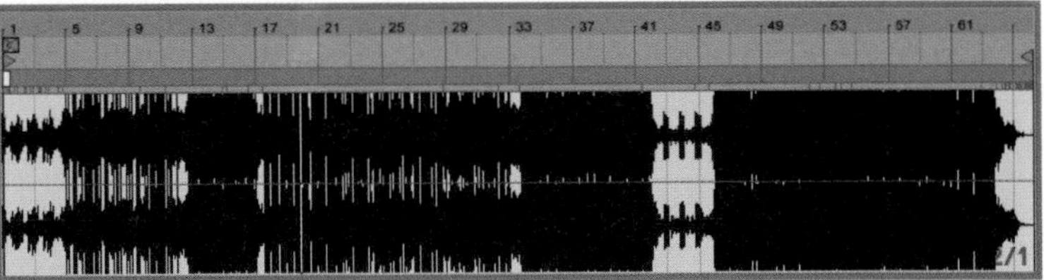

Figure 13.26 Full audio track (song) in Live's Sample Display with one Warp Marker.

The second track has tons of Warp Markers splattered all over the place like abstract art (that doesn't mean anything). The real issue is that unfortunately Live didn't detect the Start Marker right and started late in the song, skipping a few bars at the beginning of the song.

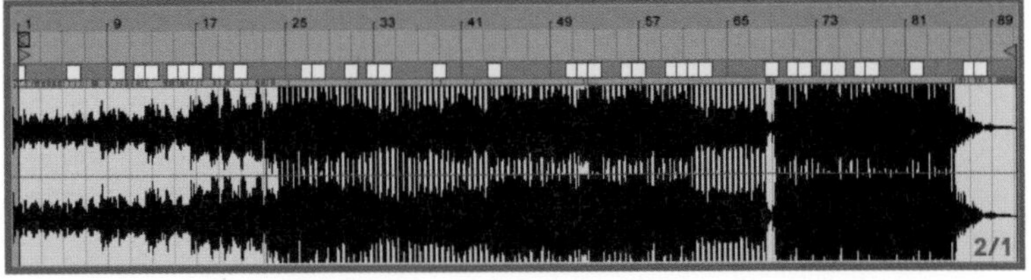

Figure 13.27 Full audio track with tons of Warp Markers and the wrong Start Marker.

No need to worry, we can fix all this. Remember, all of the necessary commands are available from the Sample Editor's contextual menu. Bring this up by right clicking or ctrl + clicking in the Sample Editor to command all of your warp manipulations.

Now our track is in tempo, but it doesn't start in the beginning of the song. This is an easy fix!

1. Zoom into the very beginning of the track and see what's going on with the Start Marker and Warp Markers.

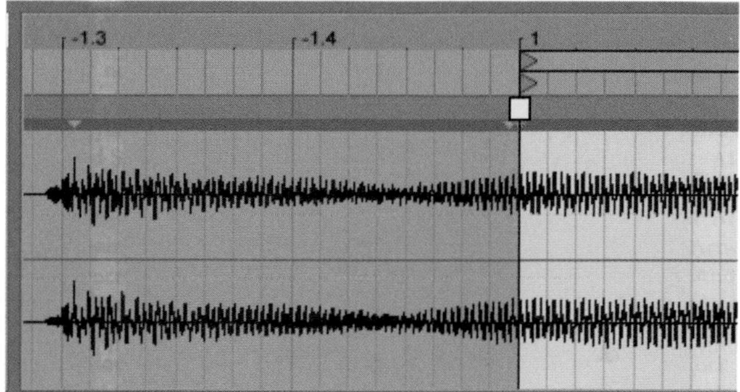

Figure 13.28 Zoom in on the Start Marker to see the correct starting position.

2. Because the song is in tempo, we'll just add a new Warp Marker where the start of the song should be.
3. Insert the Playback Marker at the exact point then double-click to insert a new Warp Marker. If the grid is preventing you from selecting your exact position, then turn it off from the contextual menu Fixed Grid "Off" ([⌘+4] Mac/[Ctrl+4] PC). Now you will be able to click and move your selected Warp Marker freely left or right along the timeline. You can also override grid snapping using the *alt* key modifier.

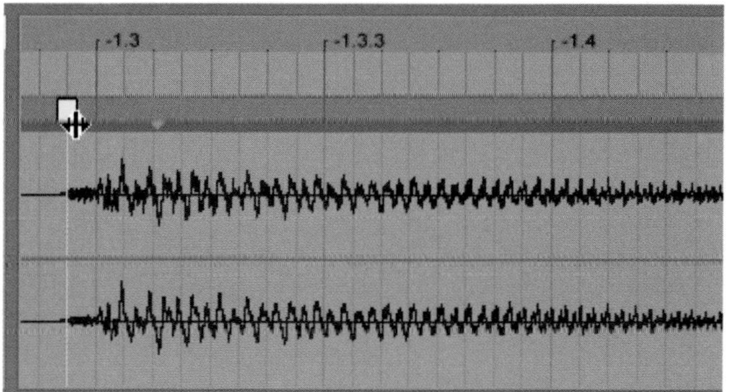

Figure 13.29 Insert and place a new Warp Marker at the start of the song.

4. Once the new Warp Marker is inserted at the beginning, select Set 1.1.1 Here followed by Warp From Here. Now, everything is lined up again and we can play back our track perfectly in sync!

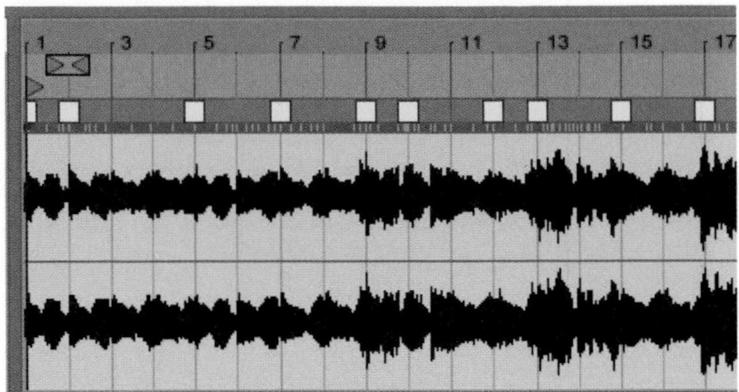

Figure 13.30 Select Set 1.1.1 Here then choose Warp From Here. The correct Warp Markers should appear.

13.3.5 Adjusting timing and quantizing

Once upon a time there was not much you could do once a "take" had been recorded to a track. The same problem existed for premade audio loop libraries, or any recording for that matter. Sure you could cut up an audio file and tweak the timing to some extent, but the fact remained that if you wanted total flexibility with timing or correcting timing errors you had to work with MIDI. As you should know, that is a thing of the past. Live 8 makes it possible to edit audio recordings with the same flexibility as MIDI. Now that's a revolution!

There are two ways to manipulate the timing of your audio files. You can manually move the Warp Markers like we have already done, or you can utilize a new Live 8 feature and Quantize your audio to the grid for you. Moving the Warp Markers is a great way to fix timing errors and is as easy as grabbing a Transient Mark handle or Warp Marker and moving it to the correct location. Quantizing an audio sample will adjust and warp every Transient Mark of the sample into a Warp Marker, placing them on the respective grid lines based on the resolution of the Quantize Settings. To do this, select an audio loop for this demonstration. We are using an acoustic drum groove that is not perfectly played. Select your clip to bring up Clip View and make sure that the Sample Editor is in view. From the Sample Editor you should see your sample's waveform and its timing/tempo drift problems.

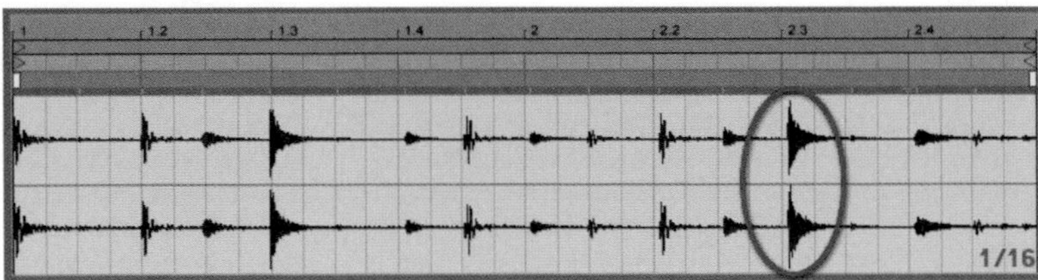

Figure 13.31 Downbeat out of sync due to "drift."

Bring up the Sample Editor's contextual menu (right click or ctrl + click) and select "Quantize Settings..." ([Shift+⌘+U] Mac/[Shift+Ctrl+U] PC).

This will bring up a new menu where you can choose the desired quantization settings and the amount (quantization strength) for quantizing your sample. You could also choose "Quantize" to skip over the settings options. Keep in mind the smallest subdivision you'll need for your sample and how close to perfect you want it to be aligned to the grid, 100% being max strength. Just because you have this power of control over timing, doesn't mean you should always use it. Quantization takes the human feel out of your recordings or samples, etc. It is best to use this feature when absolutely necessary or when the genre calls for it, such as dance music, techno, or other electronic music where a precise "robotic-like" timing is desirable.

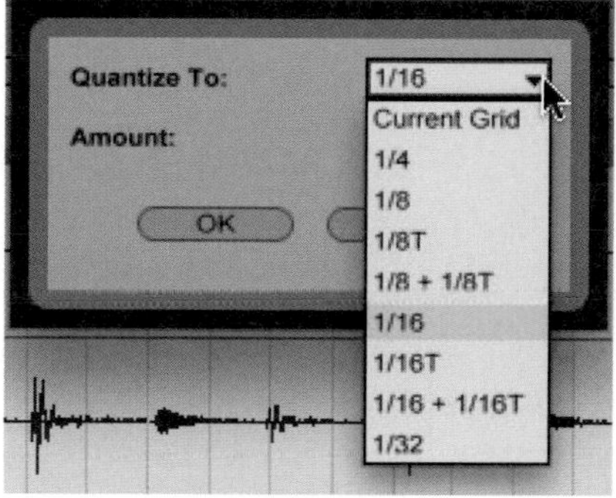

Figure 13.32 Choose a resolution based on your audio's smallest beat subdivision.

After you select your quantize resolution, press OK and you will see in the Sample Editor, a perfectly quantized sample.

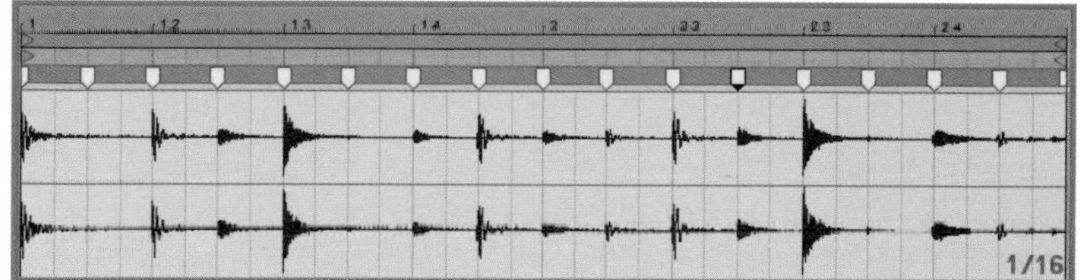

Figure 13.33 Quantized audio with Warp Markers.

Hot Tip *Quantizing your audio may take a little getting used to at first. To help you wrap your head around this whole audio quantizing concept keep in mind that the Grid alignment in the Sample View plays an important part in the quantization process. If you take the time to experiment with Grid resolutions versus your original audio's timing and feel, some pleasing results should occur more times than not.*

13.4 Musical concepts

The continual development and improvement of time-stretching algorithms and techniques to manipulate and quantize audio seems to be a never-ending quest for software developers. As you know, Live 8 just received a major overhaul and it appears that a similar transformation has taken place in many other of Live's contemporary DAWs. With every new development in this area, we get closer and closer to effortless and undetected pitch and time manipulation over audio, just as if it were MIDI. With that in mind we now have the ability to warp our audio for any number of reasons. Whatever your reason, think about how much easier it can be when you want to clean up the timing of a not so great recording session, or to put that drummer or bass right in the "pocket," with the rest of the band's multitracks to follow suite. Well, due to the elastic properties of warping, the philosophies of quantizing audio must be reevaluated, with you ultimately deciding how you want to make your tracks groove. Now that you have all that figured out, try quantizing your audio in real time. Say what?

13.4.1 Produce: quantize audio in real time

You can adjust your Quantize settings easily in real time using [shift+⌘+U] Mac/[shift+Ctrl+U] PC. This will bring up the Quantize window allowing small to large adjustments in your quantizing resolutions and strength. Aside from changing the percentage of strength, one trick is to make these adjustments based on altering and changing the Grid resolution within the Sample Display. To do this, set Quantize To to Current Grid. A tighter Sample Display Grid resolution will allow the Quantize Setting strength to latch to the closer grid lines along the timeline. A broader grid will allow for a more general snapping of your Warp Markers once engaged. Keep this in mind when using the Quantize Warp function in the contextual menu, because it can really make a difference in how this function works with your Sets. Speed up changing grid resolutions by utilizing key commands ([⌘+1, +2, +3, +4] Mac/[Ctrl+1, +2, +3, +4] PC). This can make for a more intuitive workflow.

In this chapter

14.1 Loops demystified 279

14.2 REX loops 280

14.3 Slice to new MIDI track 282

14.4 Working with loops 288

14.5 Looping in the Arrangement View 289

14.6 Loops with Unlinked Clip Envelopes 290

14.7 Looping concepts 292

SCENE 14

Loops, Slicing, and More Looping

Live's ability to drag and drop REX loops directly into a project is great. But it doesn't stop there: the ability to quickly and intuitively manipulate each slice of the loop without breaking a sweat keeps the creative flow going.

Kyle Z., Creative Director and Owner of Nine Volt Audio

14.1 Loops demystified

Much of computer-based music production is centered on loops and various loop technologies; therefore, it is important that we shed some light on the topic as well as dispel some misunderstandings about them. Loops serve as a compositional foundation for many composers, providing rhythmic or melodic material that they otherwise would not have access to or the ability to create. Well, at least the time to create anyway. Don't be fooled by the title computer-based music. Virtually all produced music spends time "in the box," meaning that a digital audio workstation (DAW) is the backbone of the studio recording regardless of the genre – yes, even for you analog purists (are we still having this debate?). That being said, acoustic singer/songwriters and performers will often use loops, whether or not they end up in the final mix or not. Many producers create their hooks/choruses in the studio by recording a live band or instrument(s), building loops out of the recorded material to use in the song or for studio musicians to record along with – even improve too!

Regardless of how or why you use loops, they are a fantastic resource for composers, producers, and performers alike. The most important thing is that you understand the concept of loops, what types of looping technology are available to you, and how to use them effectively. The word "Loops" doesn't mean "repetitive music," "cheating," or "unoriginal," rather quite the contrary. With the current loop technologies and programs like Live 8 and Suite 8, loops can be as original and creative as any other recorded instrument or performance. In all actuality, they are malleable, flexible, and completely customizable, not to mention performed by a real musician in the first place (oh, the irony). Think of loops as a means to an end, an instrument to create, produce, and perform with.

14.2 REX loops

We briefly mentioned REX formatted (.rx2) loops in Clip 13.1. They are created with Propellerhead's ReCycle, but REX loops can also be purchased from third-party developers as loop bundles of sorts. As a quick review, REX files are created from traditional audio files (.wav/.aiff) and are intended to be looped. They can be perfectly cropped to a loop/bar length or not, it doesn't really matter. Recycle will handle both types; it'll just be your job to cut and create the perfect REX loop out of the material you feed ReCycle. The process is fairly simple, especially considering that we've been talking a lot about Transients and Warp Markers, which is very similar to how ReCycle interprets samples/loops.

REX loops are created by establishing a start and end point within an audio file/sample to create a loop region, then by slicing the audio file/region at each transient. This cuts the audio loop into little chunks that become anchored by the slices (slivers), similar to the way Warp Markers anchor Transients in Live 8. The slices are locked to a time-based grid, allowing the audio to be stretched or compressed to any tempo without changing the pitch or reasonable loss of sound quality. This will be apparent in our example showing a loop sliced up in ReCycle.

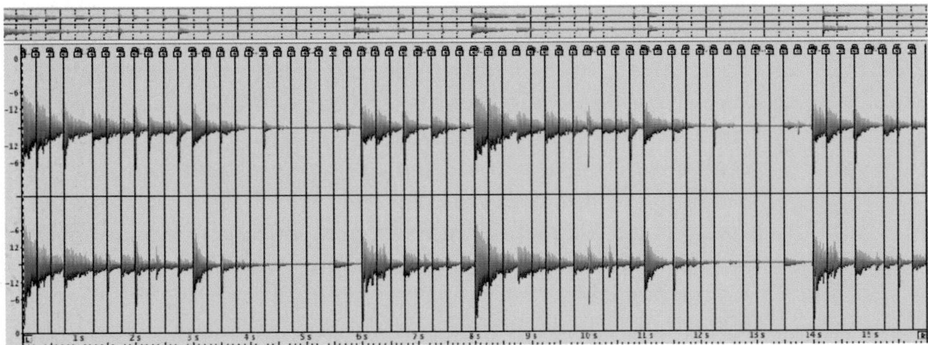

Figure 14.1 Sliced up ("rexed") audio loop in ReCycle.

That's not where the ReCycle technology stops. REX files also carry a MIDI component that allows the slices to be mapped individually to the MIDI keyboard triggered in a chromatic note sequence. This allows for some really creative, flexible ways to manipulate audio loops similar to how Live slices to MIDI ▶ 14.3, but we'll save that for later. Now that you know what REX files are, let's look at how they translate and can be used in Live. You will find REX files on our Website to use throughout this scene. Download "CPP_REXLoops" and use them to follow along and maybe get a bit creative along the way.

14.2.1 REX mode

Upon import of an .rx2 file (REX), Live 8 will automatically run the clip in REX Mode. This is unique to REX files and cannot be activated manually. Live handles REX files with ease, because all the calculations, tempo, and length information have been predetermined and stored in the .rx2 format. Looking at the Clip View Sample Display, you will see that Live has quickly translated and interpreted the loop perfectly with Transient Markers. If you were to open the loop in ReCycle, you would see that the Transient Markers in Live are the same as the slices in ReCycle. Just like Live's Warp feature, these loops sync perfectly to your Set's tempo and can also be speeded up, slowed, and transposed.

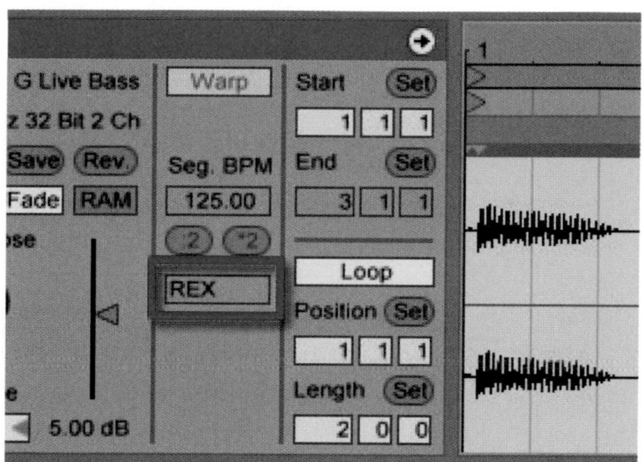

Figure 14.2 REX mode.

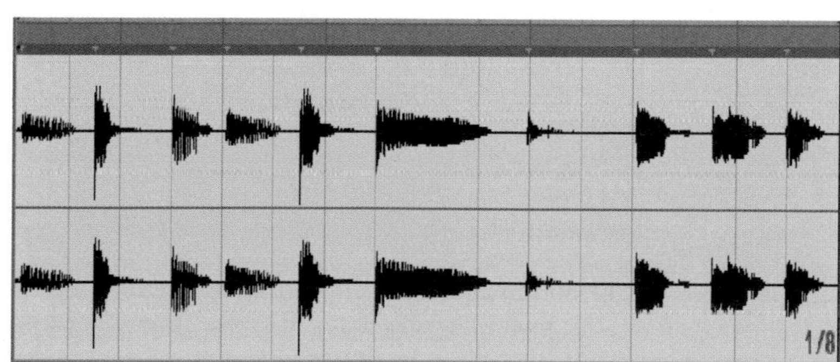

Figure 14.3 Transient Markers in Live.

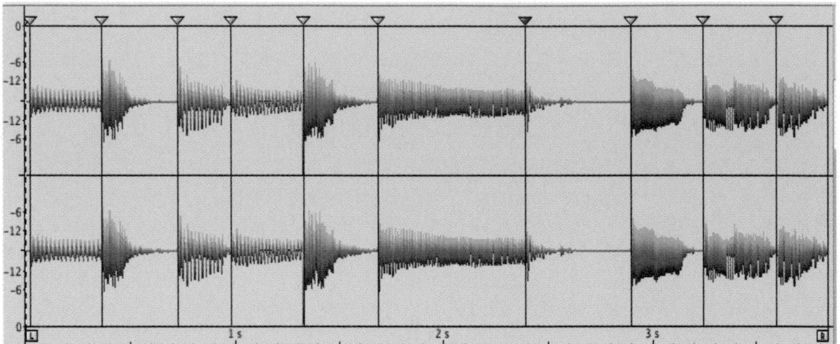

Figure 14.4 Slices placed identical to Live's Transient Markers.

It should be pointed out that Warp Makers and any warp-associated parameters are unavailable in REX mode, so you're stuck with the REX slice programming as is. At the end of the day, this is only one dimension of working with REX files in Live 8. For more flexibility and power, you can map the individual slices, or in our case, Transient Markers, to a chromatic MIDI note scale! Let's go over this concept next.

14.3 Slice to new MIDI track

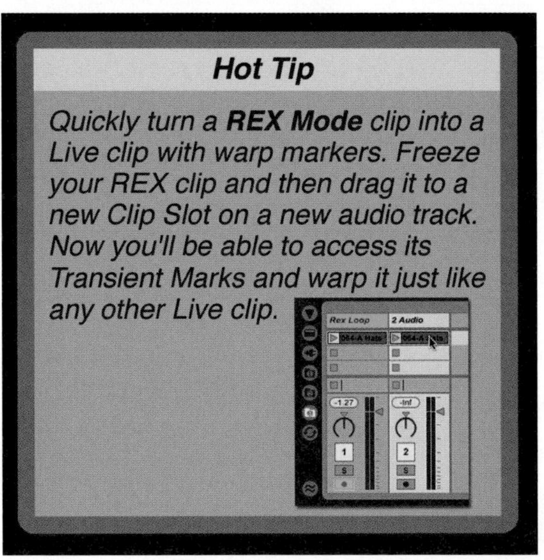

Hot Tip
Quickly turn a REX Mode clip into a Live clip with warp markers. Freeze your REX clip and then drag it to a new Clip Slot on a new audio track. Now you'll be able to access its Transient Marks and warp it just like any other Live clip.
Rex Loop
2 Audio

Live provides you with the ability to use and manipulate your own warped audio clips. This is made possible with the "Slice to New MIDI Track" feature. Any audio file that can be read and managed by Live is eligible to be sliced, including REX files themselves. Once slices are assigned to MIDI notes, you can then resequence them, trigger only specific slices, and/or customize them to any rhythm that's relatively logical and within reason. With this powerful feature, you have the ability unleash the true creative powers of MIDI and audio loops integrated right into your Live Set. In our example, you will see how flexible MIDI slices can be. The concept of slicing in Live is conceptually derived from ReCycle's slicing/REX technology.

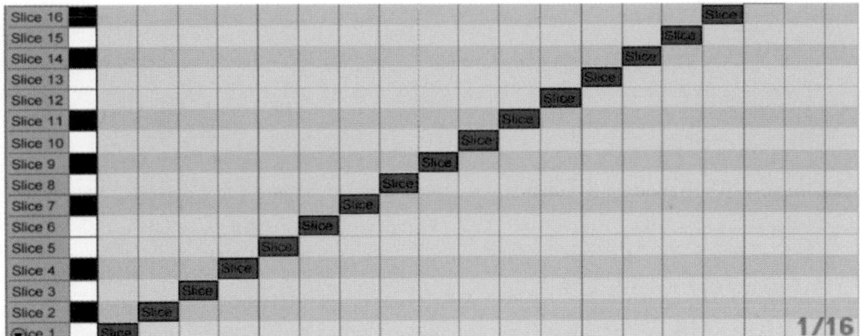

Figure 14.5 Original sliced MIDI clip.

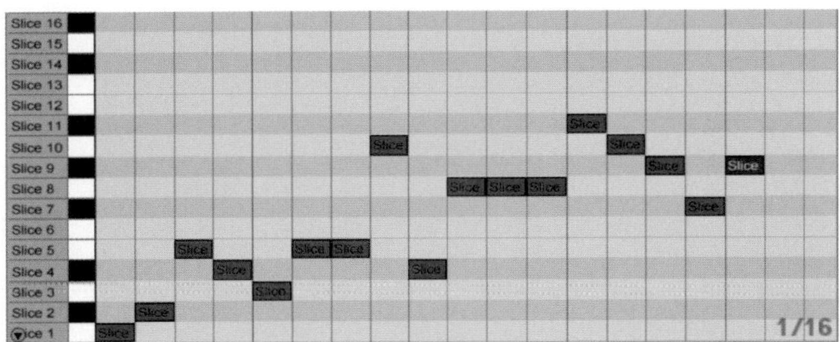

Figure 14.6 Edited MIDI slices.

14.3.1 REX loop slicing

Slicing a REX loop to MIDI is a very simple task with a ton of benefits to you. All you need is a REX formatted file (loop), which can be obtained from our Website in .rx2 format. Once you have located your REX loop from the File Browser, let's begin to slice and dice. Once again, this feature can also work with regular .wav, .aiff, and .mp3 files, as you will see later in 14.3.2.

1. Import your .rx2 loop into the Session View. You can do this by double-clicking on it in the File Browser or drag it to the Drop Area.
2. Bring up the Sample Editor's contextual menu (right click/control + click) and select "Slice to New MIDI Track."
3. This will bring up a new menu telling you how Live will slice the REX file. Live will slice your REX file based on the slice information that was translated from the file to REX mode. The slice will be a derivative of the loop's length in beats.
4. You will now have the option to choose a Slicing Preset. Live has provided Slicing Presets with very cool effects. You can also create and load your own,

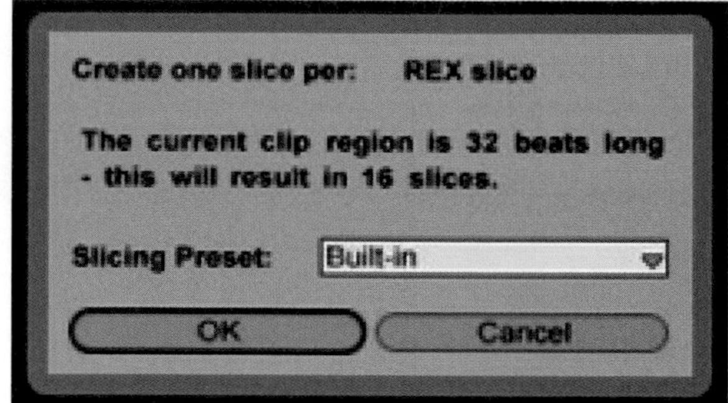

Figure 14.7 Menu shows how many slices provides a Slicing Preset chooser.

but for now just choose Built-in. This will create a new MIDI track and Drum Rack containing all your slices. Launch to ▶Scene 16 for more on Device Racks.

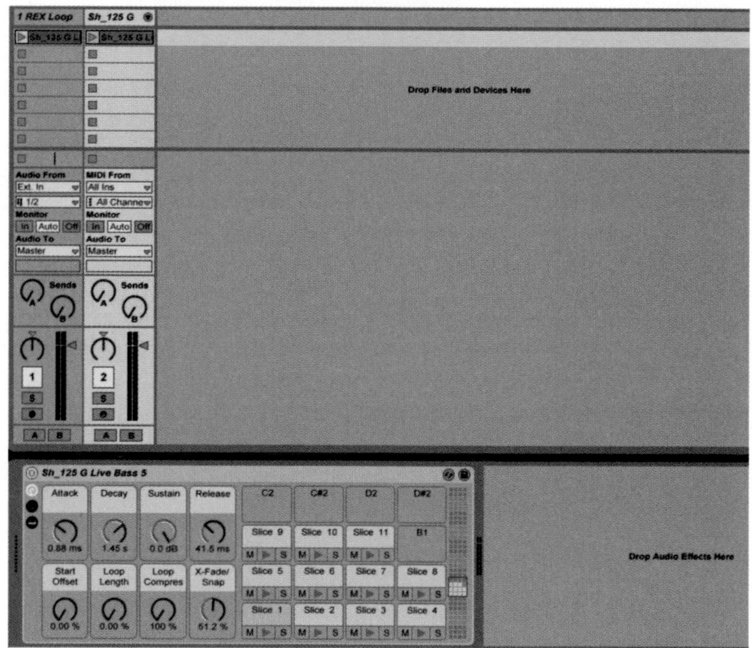

Figure 14.8 New MIDI track and Drum Rack containing the new MIDI slices assigned to its drum pads.

5. Launch your new MIDI track clip. It should sound virtually the same as the REX loop clip.
6. Unfold the new MIDI track by clicking the "Chain Mixer Fold button" from the track's title bar. With the I/O section and Mixer in view, you will be able to see how Live has sliced and routed the slices to each MIDI note using chains/tracks (Rack Chains) along with associated mixing parameters.

Figure 14.9 MIDI slice (note) rack chains.

7. Each one has been mapped starting at MIDI note C1 and ascends chromatically until the last slice. Select your new MIDI clip generated by the slice command, and you will see how the MIDI notes have been sequenced chromatically. Also notice that the note positions and lengths are relative to the slice or transient positions and lengths.

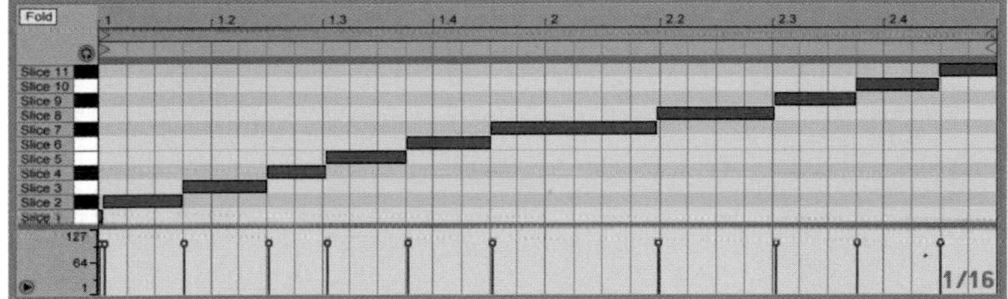

Figure 14.10 Chromatic note slices in the MIDI Note Editor.

That's it. You have successfully sliced your REX file to a new MIDI track. Of course, there is much more to learn about tweaking your new Drum Rack Device and all of its Macros. Launch to 16.3 for more on Drum Racks. For now, understand that Live creates a Drum Rack regardless of the loop content, i.e., a melodic instrument loop versus drum loop by design. Therefore, your REX file MIDI slices will be triggered via Drum Rack's virtual drum Pads.

14.3.2 Audio loop slicing

Thanks to Ableton, we don't have to rely on ReCycle or any other third-party developers to slice up our loops. Live 8 makes it possible to turn any loop into the equivalent of a REX type loop and with even more flexibility. Using a traditional audio loop (non-REX), slice it to a new MIDI track just like we did in the previous example. To begin, locate an audio loop via the File Browser.

1. Import your audio loop into the Session View to an empty Clip Slot or as a new audio track.
2. Launch your new clip to ensure your audio loop is warped correctly. This means that the best Warp Mode has been chosen and that the Transient/Warp Markers are positioned correctly.
3. Bring up the clip's or Sample Editor's contextual menu (right click/control + click) and select Slice to New MIDI Track. This will open up the slicing options for the audio loop.

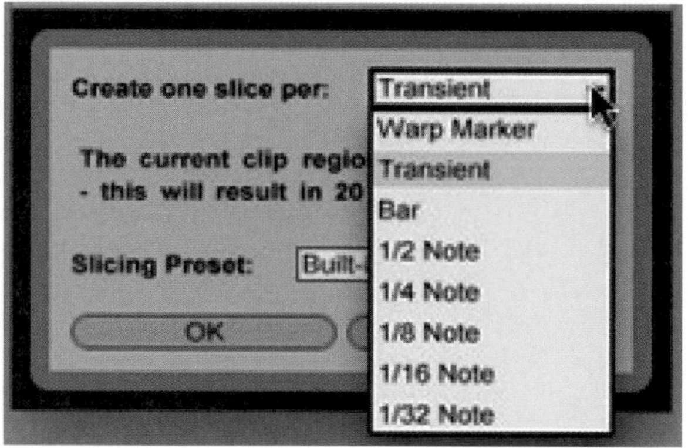

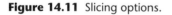

Figure 14.11 Slicing options.

4. Chose the Slicing Preset – "Built-in" for now – and how you want each beat (transient) to be sliced then click "OK." (Create one slice per the Warp Mode for which the clip is set to). This will create a new MIDI track and Drum Rack containing all the audio file's slices.
5. Launch your new MIDI track clip. It should sound virtually the same as the warped audio loop you imported at the start. Remember to mute your original track source you sliced (to the immediate left of your newly Sliced track). It's easy to forget sometimes, and you'll hear both tracks playing at once if you're not careful.

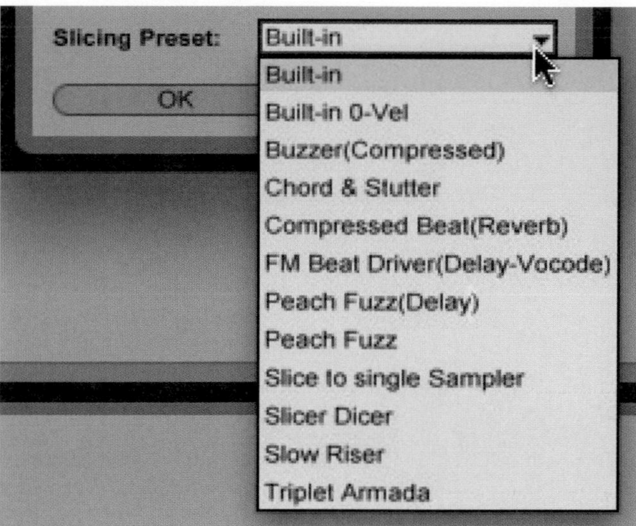

Figure 14.12 Slicing Presets.

That's it. You have successfully sliced your audio loop to MIDI and can now manipulate it just like a REX file and beyond. If you look at the MIDI clip, you will see that Live sliced it based on the preset you chose. For example, if you choose to use a note value as your slice resolution, then you will see that slices converted to MIDI notes in the Note Editor are placed at that resolution, such as a note on every 16th note for example. Don't forget that any audio file can be sliced and resequenced. For these types of files and depending on the nature of the loop and your plans for manipulation, you might decide to use Warp Markers, quantization, or grooves to adjust the timing before you slice to MIDI. This may make it easier to resequence some loops.

Before you move on, try slicing out on your own and experiment with each different Slicing Preset. It can be a lot of fun!

14.3.3 Resequencing

Now it is time to begin customizing your loops, which is why we sliced them to MIDI tracks in the first place. It is all about the ability to liberate yourself from the original patterns of your loops. If you haven't noticed already, editing the loop's MIDI note durations, positions, and velocities affect the once cemented loops. To resequence your loops, go to the MIDI clip's Note Editor and alter the MIDI notes as you wish, as shown in the three graphic example in Clip 14.3. You might change the beats around, altering the feel. You could also delete notes or repeat notes to add fills and variation.

Hot Tip

*Create that cool "**buzzing/stutter" effect**! Use various Grid resolutions when editing newly sliced Audio-to-MIDI notes in the MIDI Note Editor. With a tighter grid (higher resolution), try duplicating several notes consecutively and vary velocities for groups of notes. You'll quickly see that mixing up the grid resolutions for adding and subtracting notes in the Note Editor opens up some very creative possibilities.*

Remember that Live allows you to do all this in real time, while the program is playing – in this case looping. Customizing the sequence breathes freshness, eliminates repetition, and allows you to manipulate the audio as MIDI. Think about what you can now do if you start adding MIDI effects or slice the loops up with a creative preset! Or better yet, how about slicing a solo vocal or guitar phrase? You can see the potential here for complete control over your loops and audio.

Hot Tip *Try changing the Grid to Triplet Grid [⌘+3] Mac/[Ctrl+3] PC. This gives you the advantage of adjusting the notes to a new start point on the Grid and syncopating your notes for a completely different feel.*

14.4 Working with loops

You've just seen what is possible with audio loops so let's look at how to work with loops in Live 8. Because the loops can be launched in the Session View and Arrangement View, we'll go over how they are used in both and in what way they differ in their use.

Session clips essentially function as loops. Of course, they don't have to, but for our sake we'll stick to that assumption. They repeat infinitely unless you do something to stop them. Well, we wouldn't want that, especially to loop without variation over time. For that reason, it is necessary to find ways to vary them. So how do you do that? On a fundamental level, you will copy your clips (loops) and paste them into consecutive slots, one after the other to create multiple scenes. Each one can be a variation of the previous loop so that you can alternate between them, as you wish. This is obviously a no-brainer fundamental that you will always do as you build your song, but you will also use this approach along with other tweaks to add variation to your clips. One especially interesting way to bring variety to your duplicated Session clips is to create Follow Action groups. This allows you to predesign how many times a loop will cycle before automatically moving to the next clip as directed. This automated process resembles somewhat of a timeline approach that you might create in the Arrangement. The point here is to create variations over time, keeping your loops fresh. Each clip in the Follow Action group can be a segment of the previous clip, an intro, break, fill, or simply an effectual variation of the original. Launch to Clip 7.4.5 for more on Follow Action.

Another consideration for audio/REX loops is to slice them to MIDI, as we talked about earlier. Try resequencing and adding MIDI effects to create musical variety to potentially static loops.

14.5 Looping in the Arrangement View

The beauty of clips is that each one is independent of the other and for that reason loops can be changed and customized so easily that neither your workflow nor your music is interrupted, regardless of the view you are working in. Therefore, this process of loop manipulation translates seamlessly into the Arrangement View, because it too works with clips. It's only the actual launching and sequencing of clips that changes because the Arrangement View is linear, whereas the Session View is not.

Figure 14.13 Session clip launched and playing back.

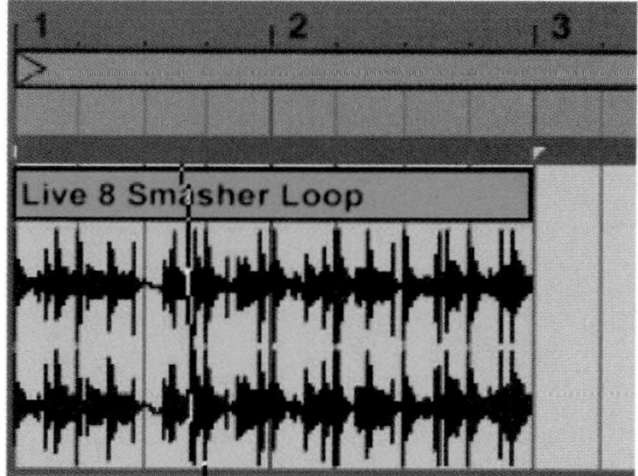

Figure 14.14 Arrangement clip playing back. (Linear)

As for the Arrangement, you can tweak and customize your clips just as we suggested earlier, but now you have the ability to edit them as whole or as partial clips, varying them along the timeline. This then brings us to some tactile operations that must be familiarized to create and vary our loops in the Arrangement. Most of this information you are somewhat familiar with, that is, the actual process of looping and editing clips in the Arrangement.

Working in the Session View is incredibly creative and a ton of fun, but you'll eventually find yourself working in the Arrangement View as well, to edit, enhance, mix, and remix your performances. Part of this process involves looping and manipulating your clips. There are a few different approaches to looping clips in the Arrangement. First of all, you can copy and paste clips one after another or use the Duplicate command resulting in the same outcome.

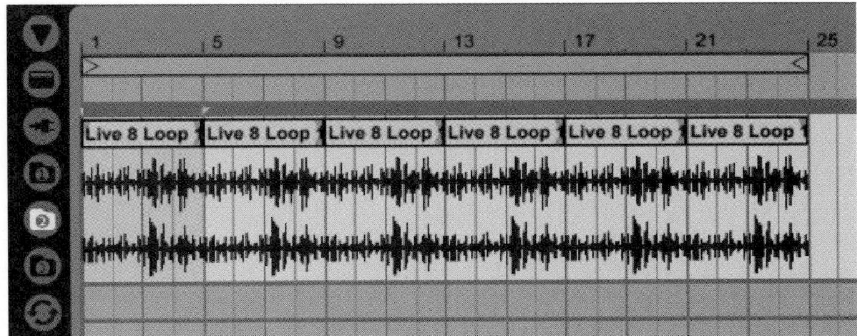

Figure 14.15 Duplicate clips along the timeline.

Once you have a number of loops laid out along the timeline, you can make subtle changes to each clip to help evolve your music over time. When, where, and how much evolution will be up to you and the structure of your song. Use the traditional keyboard shortcuts to copy and paste or [⌘+D] Mac/[Ctrl+D] PC to duplicate. Another way to loop your clips in the Arrangement View is to extend the clip from either edge by dragging its length out for as far as you want it to loop. Live will automatically draw out the clip as a single "looped" clip without breaks. What is unique about this is that the clip is still interpreted as a single, let's say four-bar loop, even though it has been extended beyond four bars. Essentially, Live references the original clip, as seen in Clip View. This provides a quick, efficient way to extend a loop over a selected range while minimizing added strain on your hard drive and system speed as Live is not actually building, consolidating, or recording new physical regions of audio to disk. There are a few innate limitations to this approach in that any adjustments or changes to the clip will affect each loop cycle of the looping clip. For this reason, event variations should be made by duplicating, copying, or by splitting a clip(s). In addition, consider taking advantage of Live's Clip Envelope automation feature for resequencing events and/or altering each duplicate, copied, or "split-up" clip.

14.6 Loops with Unlinked Clip Envelopes

Creative variations can be achieved with even a single looping clip using Unlinked Clip Envelopes. This creates progressive variation over each loop cycle of a clip, which can be customized to taste. The concept is that a clip can loop indefinitely while changing over time. This can be done with any clip in both the Session and Arrangement Views. The difference between this method and duplicating/copying clips is that envelope-based variations are made within a single looping clip as opposed to multiple variations of a clip.

Take a moment to walk through our exercise and we'll show you how looping with Unlinked Clip Envelopes is done. You can use one of the audio loops from our Website or use your own for this exercise.

1. Using a clip that is two bars long, open its Envelope Box/Editor from Clip View.
2. Click on the Volume button (Quick Chooser) to make sure you can see the pink shaded area in the Editor you'll be working with.
3. *Unlink Clip Envelope* – Click on the yellow "Linked" button directly under the "Region/Loop" section on the bottom left of the Envelopes Box to switch to "Unlinked." It should turn orange.

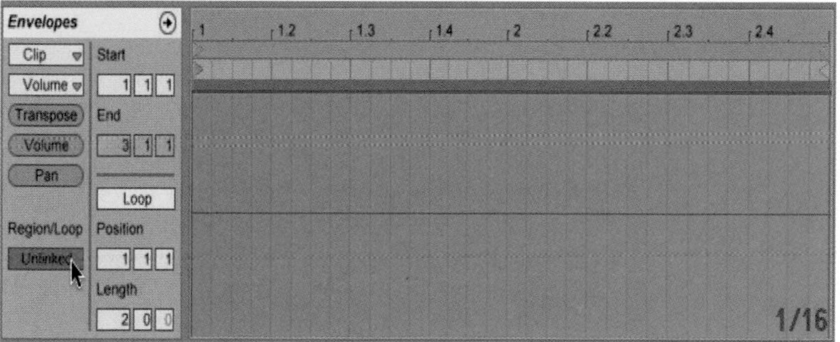

Figure 14.16 Unlink Clip Envelope.

4. With "Volume" listed in the Control chooser, extend the Loop Brace in the Envelope Editor to four bars (1–4) or type in "4" in the Loop Length input field at the bottom of the Envelope Box. You will want to zoom back out to see all four bars after you have extended the Loop Brace.

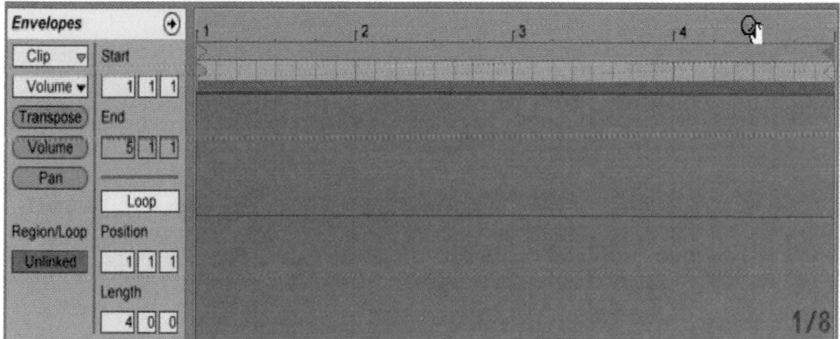

Figure 14.17 Unlinked Clip Envelope.

5. Double-click on the envelope itself (dark pink line) at bar 3 to insert a breakpoint. This will anchor the volume automation value at the end of the second bar.
6. Create another breakpoint at the end of bar 4 (start of bar 5), then click + hold + drag this breakpoint downward to create a gradual fade out with the envelope. You can also make this a sharp cut in volume by highlighting the entire range from the end of bar 2 to the end of bar 4 (3–4).

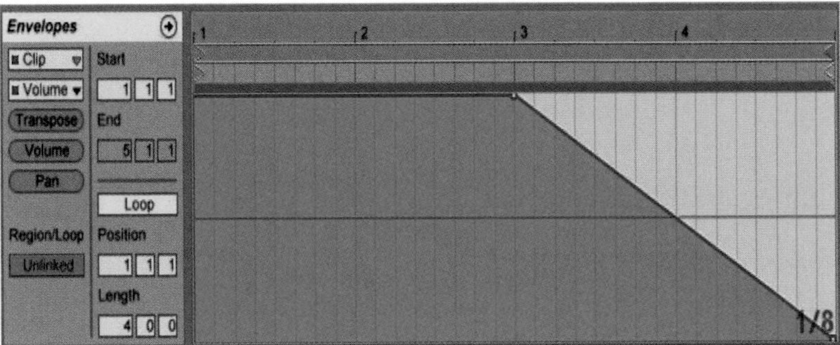

Figure 14.18 Unlinked Envelope with fade out starting at bar 3.

7. Playback your clip and listen to the results.

Now that you have completed this exercise, check your Set against our Set. Download "CPP_14-6_Unlinked-ClipEnvelopes Project" and see if they sound the same.

There are many uses for this exclusive Live feature. On a basic level, it allows you to use one looping clip and have it driven by "unlinked" envelope loop automation to customize its audio or MIDI events. Use it with any of your clips. Another interesting way to take advantage of unlinked envelope automation is to split a longer MIDI clip sequence in the Arrangement View into multiple clip segments for affecting each clip segment differently. Just split a clip as many times as you like at various points where you want to alter the automation effect or event change. You can then apply a different envelope effect to each new clip that is created. Make an insertion with the mouse pointer where you want to split the clip into a new clip, then *press* [⌘+E] Mac/[Ctrl+E] PC or choose "Split" from the Edit Menu.

14.7 Looping concepts

By now it is obvious that loops are very powerful tools for a musician. Love them or hate them, they are here to stay and there is no reason why a digital composer, producer, or performer should go without. Regardless of whether they end up in a final mix or not, they're an invaluable resource. Especially because they are so malleable in Live. At the least, a digital composer will at some time make their own if they don't use a third-party library. Look at loops as a challenge. Some of the greatest digital composers are those who not only have the resources, but also can create unique sequences from the simplest of loops.

14.7.1 Create: *Custom Loop Library*

Creating your own custom library of loops is easy and satisfying in Live 8. Once you begin to fine-tune your skills of slicing your loops with the Slice To New MIDI Track feature, you'll probably want to start archiving or saving your work. You can keep these as sliced MIDI clips or resave them back as audio clips. They both have their advantages, but for a moment let's consider converting your newly sliced and edited MIDI clips back to audio to save in the Library for future use. Here's how:

Create a new audio track next to your sliced MIDI track. Select the Freeze Track function for the sliced MIDI track then drag the frozen sliced MIDI clip to your new empty audio track's Clip Slots directly to the right. Presto, there you have it, a new custom audio clip (preset) that is looped and contains all of your hard work. Now it's ready to be saved and archived in the Live Browser. Select one of the File Browsers and right click or ctrl + click on a current folder or in the area just outside of a folder to open the contextual menu listing several options including "Create New Folder." This will keep things organized, although you don't have to create a folder.

In doing so, you'll be able to immediately rename the new folder. Enter a name then press the return or enter key on your computer keyboard. Now you can drag your new audio clip directly into your custom folder. Follow this same procedure for saving your sliced MIDI clip as a preset instead of converting it back to audio. Use this functionality to build up your arsenal of custom Sliced loops. Launch to ▶ 15.2.4 for more on custom presets.

If you think about what you have followed and read in this Scene, you'll see that your audio has gone from audio to MIDI and back to audio. Pretty cool!

In this chapter

15.1 Introduction to Live Devices 297
15.2 Working with Live Devices 298
15.3 Live 8 Instrument basics 303
15.4 MIDI effects 311
15.5 Audio effects 313
15.6 Device chains 325
15.7 Plug-in devices 332
15.8 External (MIDI) Instruments 336
15.9 Working with devices 338

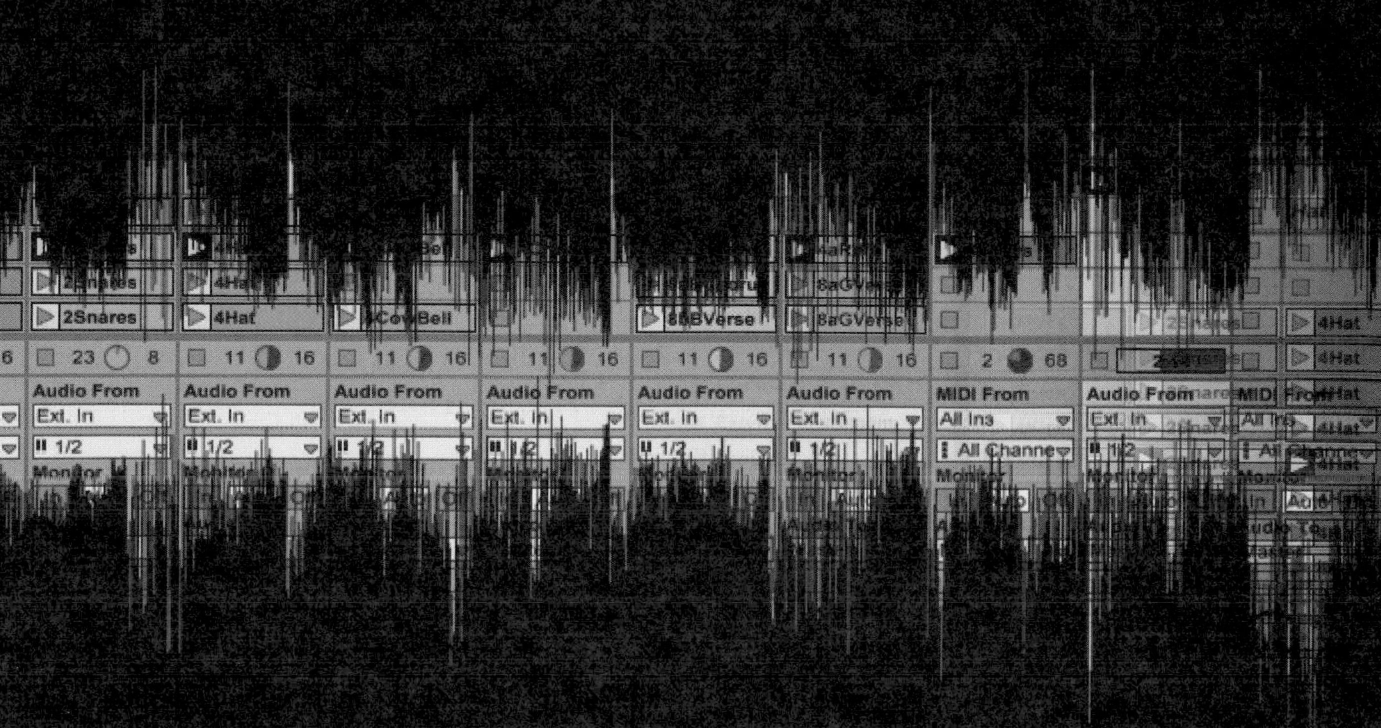

SCENE 15
Instruments and Effects

My favorite feature(s) of Live are the software instruments. I find myself using them all the time. When I'm mixing and I've been given a weak bass, I'll pull up Analog to give it some extra heft.

Phil Tan, Mixer and Producer

15.1 Introduction to Live Devices

Software instruments and effects are the backbone of the digital audio workstation. Whereas the project studio was once crowded with racks and racks of outboard gear, hardware samplers, synth keyboards, sound modules, reverb units, and drum machines, it's more likely now that your whole production studio fits in your backpack or at the most takes up a small corner in your bedroom; just you, your computer, an audio interface, and your MIDI controller. By no means are we suggesting that you abandon all analog devices or reduce your project studio down to a shoulder bag, but current trends have led us to the point where there are some amazing software-based sample players, virtual synths, and every imaginable DSP-powered (digital signal processing) effect you can imagine. So, some of you may have never even handled outboard gear except at the used and discounted section of your local pro audio store. That being said, we don't suggest running out and trading in all your hardware, not that anyone would take it, but you should consider giving up the latest release of your favorite video game and trading in some old titles to buy some RAM and a hard drive! While you're at it, don't forget to pick up or upgrade to Suite 8.

All right, enough of the philosophy. Live 8 and Suite 8 are all about making your music flow with ease. That's why Ableton has put in a lot of hard work over the years to develop a quality user interface with built-in instruments and effects integration. You have seen throughout our many examples and exercises that Live Devices (instruments and effects) are always accessible and never require your music and workflow to stop. Now all you have to do is master them, which shouldn't take too much time.

15.2 Working with Live Devices

Through the Device Browser (on the far left of your screen), you have access to all of your Ableton Instruments, Audio Effects, and MIDI Effects, which include factory and user-defined presets ▶ 2.4.

Keep in mind that Suite 8 is Ableton's premier package of instruments and effects bundled with Live 8 in one "box." Potentially – between Live 8 and Suite 8's bundle – all the essential instruments and effects, as well as the included presets you'll need, are right in front of you. You will eventually want to upgrade if you haven't purchased Suite 8 already. For more about Suite 8, launch to ▶ Scene 17. In the meantime, check out Ableton's Website and grab any free downloadable content/Live Packs you can find or need such as "Impulse," "Live7Legacy," "Basics version 8," and "Solid Sounds." It never hurts to have more free samples and presets. (www.ableton.com/livepacks) Review installing Live Packs ▶ 3.4.

15.2.1 Overview of Devices

Here are the basic guidelines for working with instruments and effects:

Instruments: They receive MIDI note information through MIDI tracks and convert that data to audio, which in turn is output by the MIDI track it's placed on. For this reason, they cannot be inserted on audio tracks, Returns, or the Master track. Instruments can only be inserted on MIDI tracks ▶ 15.3.

MIDI Effects: They work exclusively with MIDI notes and can only be inserted on a MIDI track. They generate some amazing performance-based manipulations, affecting how MIDI notes are interpreted and distributed prior to being converted to audio signals. When applied, they are placed first (before an instrument) in a MIDI signal chain. MIDI Effects are inspiring devices that literally manipulate and transform your music into dynamically complex and rich arrangements – a great way to start your day! ▶ 15.4

Audio Effects: They can be hosted by any track type, even MIDI tracks so long as they are inserted after the instrument ▶ 15.6. Located in the Device Browser are tons of unique audio effects, from the "classic" mixing and tracking processors to wildly manipulative devices. Combine these with the power of Device Racks and the possibilities are quite endless ▶ 15.5.

There are three basic ways to insert a device into a track as long as the track type supports the type of device being inserted. (1) Double-click on the device name in the Browser, which creates a new track. (2) Preselect a track and then double-click on a device or press the return key on your computer keyboard. The device will be inserted on that track. (3) Drag and drop the device to a track, to a Drop Area, or to the Track View. Use any one of these methods for both the Session and the Arrangement Views.

15.2.2 Track View

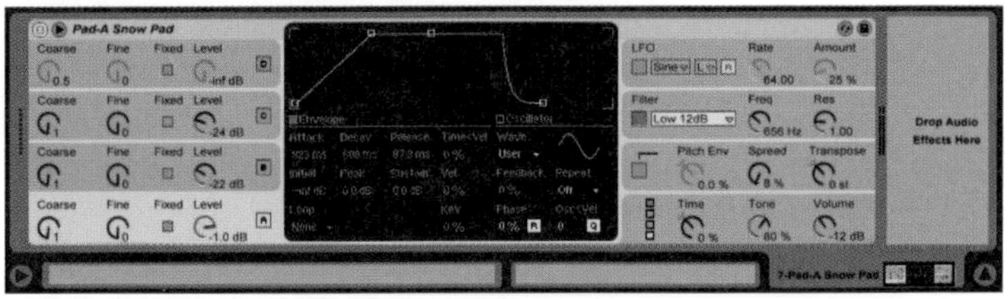

Figure 15.1 Shown here is the Operator instrument in Live 8.

Track View is where you'll interact with your selected instrument's and audio/MIDI effect's user interface. As you've seen already, Track View shares the same window space as Clip View, at the bottom of the main Live screen. Toggle back and forth with the Track View Selector or double-click on a track's Title Bar to bring it up. Feel free to insert devices directly to Track View by dragging and dropping. There are many ways to insert devices, so don't feel obligated to do it any one way. To remove a device completely, click on its Title Bar, and then press the delete/backspace key on your computer keyboard.

Hot Tip *Use the keyboard shortcut [⌘+Tab] Mac/[Ctrl+Tab] PC to quickly alternate between Track View and the Clip View.*

Because Live handles clutter and space exceptionally well, you can even make more rooms on your screen by hiding the entire Track View. Just click on the Show/Hide Detail View at the bottom right of the Main Live Screen and it will be tucked away.

Figure 15.2 Show/Hide Detail View and Track View Selector.

Every device has a Device Activator, Hot-Swap, and Save Preset button on its Title Bar. To temporarily bypass a device, click on its Device Activator button in the upper left-hand corner of the interface (yellow = "On"). On the opposite side of the Title Bar to the far right is the Hot-Swap button. This is used to efficiently audition effects in the same way you would for all Hot-Swap buttons.

Next to Hot-Swap is Save Preset. You can save your current device settings as a preset in the Device Browser. Some devices will have an additional *Parameter Fold/Unfold* button adjacent to the Device Activator. This will be on a per device basis.

Figure 15.3 General device Layout.

On the left and right edges of each device is a Level Meter (input and output). This will display the active signal flow (left to right), lighting up as MIDI or audio signals pass through a device or between devices in a chain (from device to device). It also lights up when a device is receiving an input signal (MIDI or audio). Level Meters are very helpful for identifying if and when a signal is reaching a device(s) in the chain. Note that MIDI and audio signals are each represented through a different type of Level Meter indicating the difference in signal.

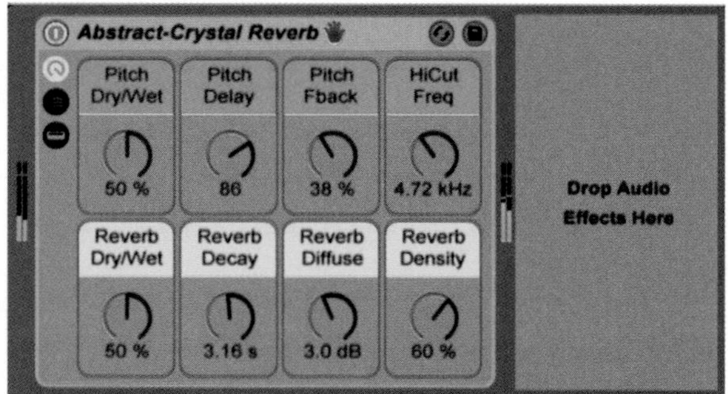

Figure 15.4 Audio effect device Level Meter indicating input and output.

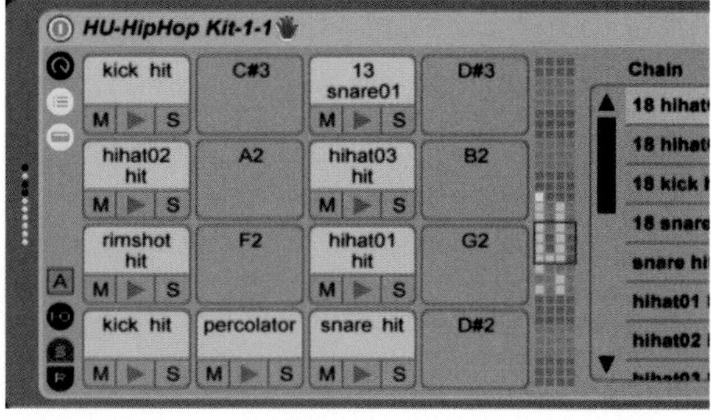

Figure 15.5 Instrument device Level Meter indicating input.

Depending on the device, there may also be a Sidechain Toggle located next to the Device Activator button, i.e., Compressor. Not all devices have built-in sidechain controls. Click the toggle to unfold a device's sidechain parameters/controls when applicable. From there, you will be able to activate a sidechain, choose an input source, and set specific parameters. Launch to ▶ 15.6.3 for more on sidechaining.

15.2.3 Device Presets

Presets store device parameters and settings. They are valuable tools for everyone, even if you think otherwise. Someone spent a lot of time building and refining them, so you might as well see what they have to offer. You'll find that they can be a great starting point and time saver. Presets are located in the device's respective folders in the Device Browser. To load a preset, double-click or drag and drop, it's all the same. One of the best ways to navigate presets is to load up any one to begin, and then click on the Hot-Swap button to start trying them all out. This is even a great (maybe lazy) way to add any instrument. At times, you may not feel like scrolling through the browsers or trying to remember where your device came from. All you have to do is click the Hot-Swap button and you're instantly taken to the exact folder structure of the device you are currently using.

Use your mouse or computer arrow keys to toggle through presets, using the Return/Enter key to load a Preset. If you have a scroll wheel mouse, then you can move the Browser window up and down.

15.2.4 Customizing Presets

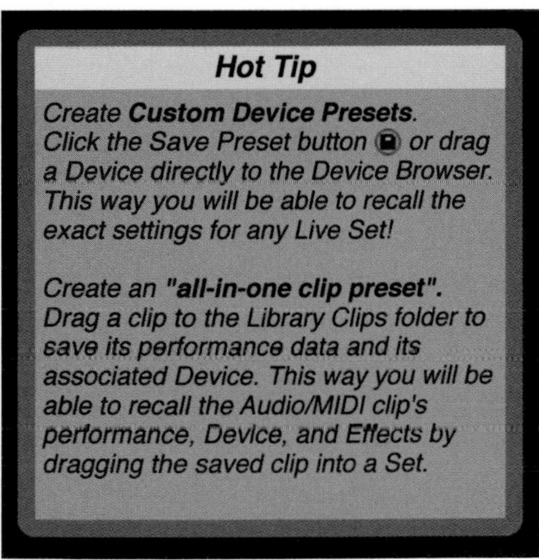

Hot Tip
Create **Custom Device Presets.** Click the Save Preset button or drag a Device directly to the Device Browser. This way you will be able to recall the exact settings for any Live Set!
Create an **"all-in-one clip preset".** Drag a clip to the Library Clips folder to save its performance data and its associated Device. This way you will be able to recall the Audio/MIDI clip's performance, Device, and Effects by dragging the saved clip into a Set.

There is a wealth of power in Live's Preset functionality, in fact, more than in most programs. One would assume that presets have limited functionality – simply load a preset and save, nothing more. With those simple features, you can recall your favorite settings. Revolutionary isn't it?...Well, not quite. Although that's a good feature that we'd all miss if it disappeared, customizing and saving presets the "Ableton way" goes above and beyond the basics. In Live, you can literally drag your device to the Browser, giving your custom device a dedicated home among the other factory presets in the Browser. We've all seen a standard save preset command, but when have you seen a drag and drop feature? The Device and File Browsers are designed to retain all your custom settings as presets and they can do it in real time.

There are a few options for saving Presets. You can click on a device's Save Preset button to save its current settings or you can drag the device from its Title Bar directly into the Device Browser. Either way, the custom Preset is stored in the Browser as a recallable preset for all your projects. You will be prompted to name your new Preset. If for some reason you don't want to save the Preset, press the Esc key on your computer keyboard and the save will be cancelled as long as you are still at the naming phase. Later on, if you decide you don't want to keep your preset, then select it and press the delete/backspace key to move it to the trash.

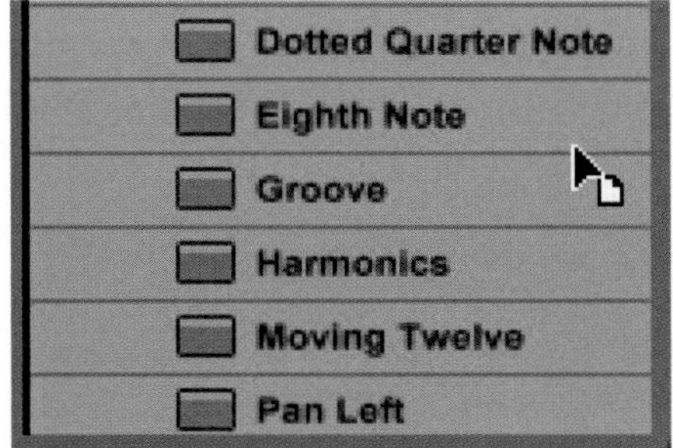

Figure 15.6 Drag your device to the Library to save as a preset.

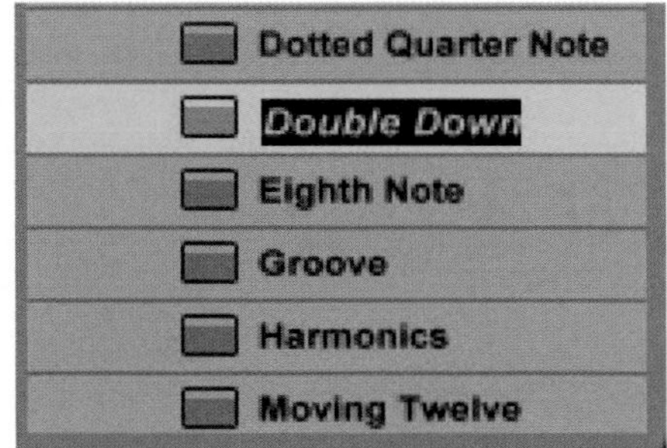

Figure 15.7 Type in a custom name.

But wait! Ableton didn't stop here with preset functionality. You can also drag your clips to the Browser to save a device and its clip data as an all-in-one preset. Clips retain device and performance data when stored in the Library or within a Set. So, quickly recall your favorite device, a performance, or both from the Library or navigate to your Set via the File Browser. Just like that, you can recall the device, effects, and your audio or MIDI clip performance data for any project. If you would like to save your current device settings as the default (Library Defaults) for that device, then right click or ctrl + click on the device's Title Bar and select "Save as Default Preset." This will store your custom settings as the default for the selected device in the Library Defaults folder. Every time you load this particular instrument, it will load with your title and custom settings. This works for all of your device types. Note, when dragging various items to or between Browser icons, you can navigate between each different Browser area by hovering over each individual Browser icons while clicking + holding the item with the mouse.

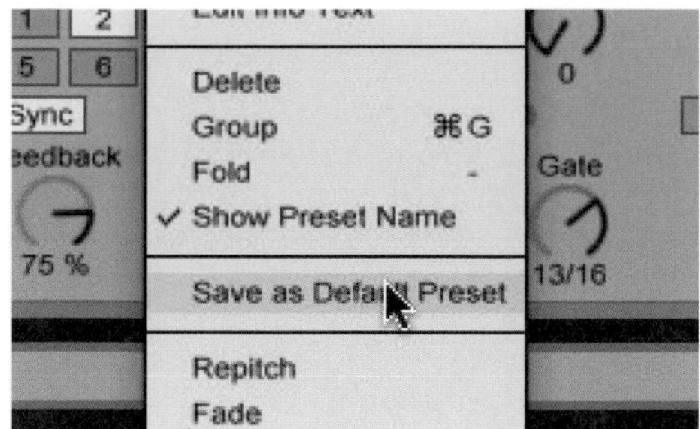

Figure 15.8 Save as Default Preset from a device contextual menu.

Figure 15.9 Navigate between File Browsers by hovering over their individual icon while click + holding a preset with the mouse.

In addition, under the Defaults, there are a number of additional Default Preset folders. Each one allows you to determine how Live should behave related to the default type. For example, "Dropping Samples" allows you to dictate how Live should behave when dropping samples onto the Track View or a Drum Rack. All you need to do is create a new empty Simpler/Sampler on a MIDI track, customize its settings to your desired default, then drag it to the "On Track View" or to an "On Drum Rack." With this set, the next time you drop a sample either onto the Track View or on a Drum Rack, the Simpler/Sampler device will load with your default settings. For more details on Simpler device, launch to ▶ 15.3.2.

15.3 Live 8 Instrument basics

Our SAE students use Impulse and Drum Racks in the Synthesis & Sampling class. First recording freestyle beatbox vocal tracks, then dragging each individually cropped audio region into Live and quickly tweaking them into incredibly legit drum sounds. Live 8's Instruments are unbelievably easy and effective to use.

Joshua Grau, Director of Education, SAE, Atlanta

As we mentioned earlier, both Live 8 and Suite 8 come with a core set of instruments and presets, the difference being that Suite 8 includes all Ableton's Software Instruments minus the Orchestral Instrument Collection. For those of you without Suite 8, be aware that Live 8 is limited to only a few instruments. For that reason, we'll point out the two main instruments located in the Library "Instruments" folder: Impulse and Simpler. These will get you up and running in no time. Impulse is primarily used for triggering and playing back sampled drums and percussion along with other one-shot audio files while Simpler is a basic sample playback instrument. These two devices make it very easy to begin making music without investing in third-party virtual instruments. Both instruments come with presets, so feel free to start exploring their capabilities as soon as you can. Although these are useful instruments in their own right, you will find yourself gravitating to Racks, which is where the real excitement is. Since Impulse and Simpler make up the core of many Racks, we'll begin here to eliminate some of the more complex variables for the sake of learning.

15.3.1 Impulse

One of the most essential elements of music is rhythm. Without it, music cannot exist. Thanks to Ableton, your music will be loaded with rhythmic possibilities because Live 8 and Suite 8 include many ways to create percussive accents and beats with MIDI-driven drum samplers and synths delivered via Drum Racks, Instrument Racks, and with the stand-alone instruments such as Impulse. We'll get to the highly anticipated Racks later ▷Scene 16. For now let's look at the basic concept of a drum sampler.

Figure 15.10 Impulse user interface.

The folks at Ableton programmed Impulse to be about as easy to use as any device can be. Simply drag a sample from the File Browser to an Impulse "sample slot" and you're ready to create a beat. No sample? No problem! There are a few preset kits that come with Live 8 and that are totally customizable with Impulse's modulation parameters. Each Preset functions within an Instrument Rack. For more information on Racks, launch to ▷Scene 16. Download "Impulse" (Impulse.alp) Live Pack from Ableton's Website www.ableton.com/livepacks for additional presets if needed. In all, Impulse provides you with the ability to program rhythmic patterns and percussive layers into your music.

Impulse is located under Live Devices>Instruments. Double-click or drag the device to a MIDI track, Track View, or Device Drop area in your set to instantiate Impulse. This automatically brings up the Impulse user interface in Track View.

15.3.1.1 User interface

Impulse is divided into two sections: sample slots and parameters. There are eight sample slots, each of which can be filtered and modulated by various parameters.

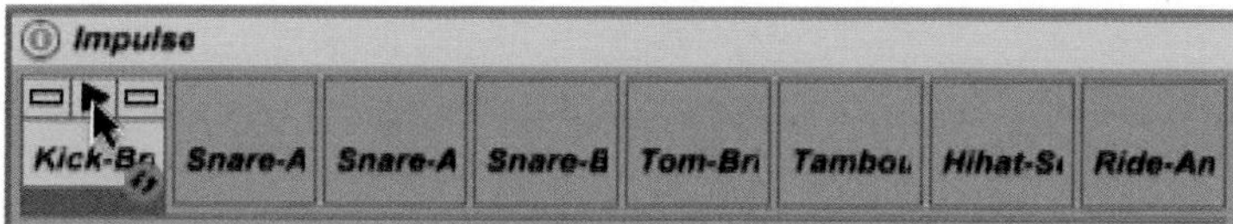

Figure 15.11 Impulse: eight customizable sample slots.

Drag samples directly to the sample slots from the File Browser or any external source. You can also Hot-Swap them on the fly. Each slot is automatically mapped to MIDI notes C3–C4 or computer keys a–k. They can also be triggered with the mouse if you click on the slot. An arrow will appear indicating where to trigger the sample. To the right and left of the arrow are the mute and solo buttons. Assign your MIDI controller's trigger pads to each slot if you've got one! Overall, these features make it very convenient to trigger Impulse via MIDI, computer keys, or the mouse. Unique to Sample Slot is a Link button that appears in the lower left corner of the Impulse interface when slot 8 is selected. This button links slot 8 with slot 7, allowing each the ability to mute the other's (7 or 8's) playback when one is triggered after the other. The main purpose of this is to emulate how closed hi-hats silence open hi-hats.

Impulse's available parameters provide some quick and useful ways to shape the sound of your sample kits. Each sample slot can be individually modified with its own set of parameters located directly below the sample slots. Click on the slot to select the sample, and then set its unique parameters as you like. You can also assign envelope automation to each parameter via the clip envelope editor. The following is a brief description of each parameter.

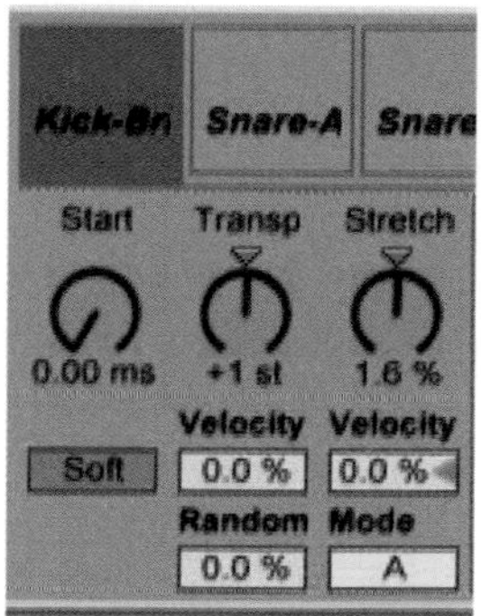

Figure 15.12 Impulse: Sample Parameters for each Sample Slot.

Below the sample slots starting on the left side of the user interface are the Sample Parameters. Each slot's parameters are adjustable. Select a slot to access its individual parameters. Start determines where the audio sample playback will begin when an incoming MIDI note triggers it. Soft creates a slight fade-in on the attack phase of the sample. Transp controls the transposition (pitch) of the sample. Random controls the percentage of randomized modulation that can be applied to the transposition to create some interesting effects. The Stretch knob controls the amount of time stretching that can be applied to the sample. The type of stretching is based on the selected Mode. Stretch Mode *A* is designed for low-frequency samples and *B* for high-frequency samples. Both Stretch and Transposition can be modulated. Stretch/Transposition can be modulated by incoming note velocities, which are controlled by the Velocity sliders. Next to Stretch is Saturation/Drive. This effect simulates the sonic characteristics of analog tubes and distortion, designed to thicken the sample's sound. In essence, making it sound warm and more "analog."

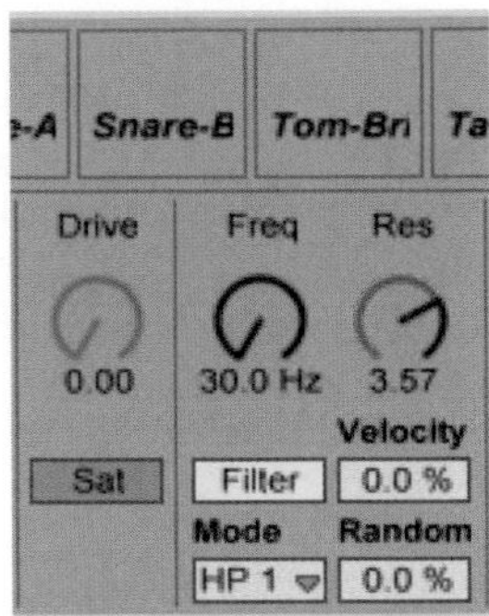

Figure 15.13 Impulse: Filter Parameters for each Sample Slot.

The center section of the interface contains the Filter Parameters. Freq/Filter activates and controls the frequency filter cutoff for the selected sample. From the Mode chooser, select between different filter types: Low pass, High pass, and Notch. Res controls the resonance amount of the selected filter type. The Velocity slider controls the amount of modulation applied to the resonance filter parameter based on a percentage of the incoming MIDI note velocities. Random controls the percentage of randomized modulation that can be applied to the frequency to create some interesting timbral effects every time the sample is played.

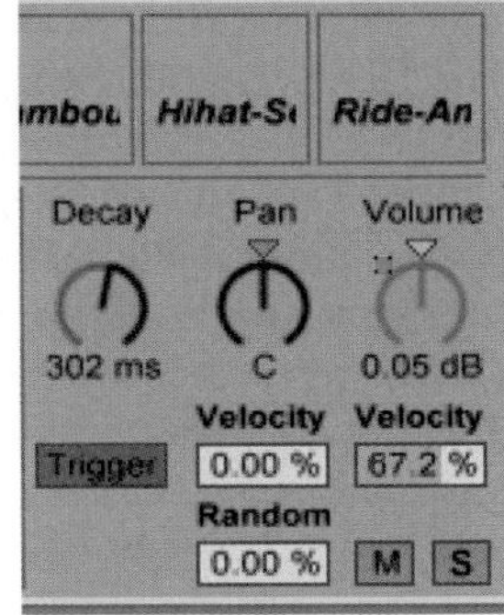

Figure 15.14 Impulse: Envelope Parameters for each Sample Slot.

On the right-hand side are the Envelope Parameters and sample controls. Decay controls the length of the sample's fade out in seconds and can be assigned one of two modes. Trigger mode applies the decay envelope immediately with the onset of the sample and applies the set decay amount. Gate mode suspends the decay envelope until the note is released (note off message). Next to decay is Pan, which adjusts the sample's location in the stereo spectrum/image. Velocity controls the amount of modulation that can be applied to the pan parameter by a percentage of incoming velocities. Random controls the percentage of randomized modulation that can be applied to the pan amount to create some interesting imaging effects. Volume controls the amplitude level of the sample and can be modulated with the Velocity slider. This controls the amount of modulation that can be applied to the sample's volume parameter by a percentage of incoming MIDI note velocities. M/S buttons mute/solo the sample.

The Global Controls on the far right of the interface affect the entire instrument. Volume sets overall amplitude level of the Impulse instrument. Time controls the overall time stretching that can be applied to the Impulse instrument and Transp controls the overall transposition factor of the Impulse instrument.

Figure 15.15 Impulse: Global Controls.

15.3.1.2 Rerouting Impulse individual outputs

Traditionally, a device's output is a summation of its entire internal sound engine or sample set. In this case, the output is directly routed from the instrument, chain of devices, or track output to Live's main outs. With Impulse, each of its eight sample slots can be individually routed to another audio track instead of being summed altogether on its own track. In this way, the sample can be broken off from the track output and essentially "auxed" to another track. This allows you to isolate the sample, let's say the snare, of a kit from the internal or track mix and add audio effects or mix it separately from the rest. Let's take a quick look at how to reroute the internal mix of Impulse. This can be done in the Session or Arrangement View.

1. Load Impulse preset "Backbeat Room" or a drum kit preset of your choice into your Set.
2. Insert a new audio track and name it "BBR-Snare-Hard" or similar.
3. Set the new track's Audio From to the instrument track "Backbeat Room."

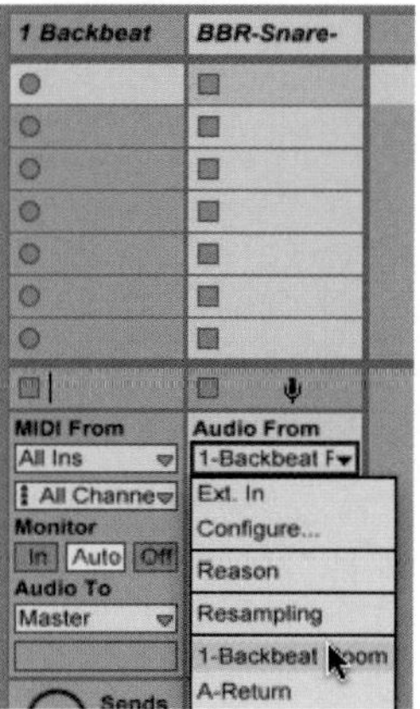

Figure 15.16 Audio From set to the Impulse instrument track.

4. Set Input Channel to ".../Snare-AmbientLoudPunch/...Hard-Impulse."

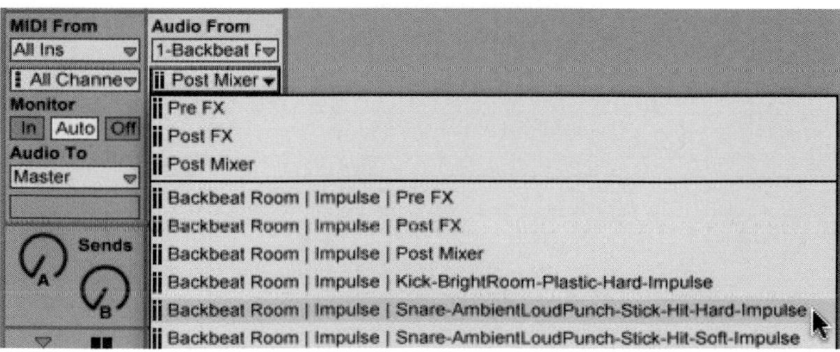

Figure 15.17 Input Channel set to Snare...Sample Slot.

5. Set the new audio track's Monitor section to "In."
6. Record-enable the Impulse track and then play it.

Now when you play your drum kit, you should still hear all the drums, except the snare you assigned will play (output) through the new audio track and not through the Impulse track. This is why we set the audio track to monitor input. If it were not set to "In," then you wouldn't hear the snare at all. The only other way to hear it would be to record-enable the audio track, thus automatically monitoring input during clip playback or record-enable both tracks to play and monitor input ([⌘] Mac/[Ctrl] PC + click on the track Record button to arm multiple tracks). Either way, having the dedicated snare track in this example allows you to insert an audio effect device directly on the "BBR-Snare Hard" track.

15.3.2 Simpler

Included with Live 8 is a basic sample player software instrument called Simpler. Before you get too excited, you should know that it is not a full-blown all inclusive software sampler, rather it covers the basic functions and synth parameters of a software sampler. That being said, its preset folder is minimal. We suggest downloading the "Solid Sounds" Live Pack by Loopmasters from Ableton's Website http://www.ableton.com/livepacks. To make complex multi-sample instruments, you need to purchase Sampler or upgrade to Suite 8, but Simpler can play multisample presets that have been made in Sampler and have been converted to a Simpler preset. This is possible because the two devices are identical under the hood. The main limitation is in parameter access. Nevertheless, Simpler is designed for quickly importing a sample (automatically mapped across a MIDI keyboard), processing it through various parameters and then playing it back as an instrument. Samples can be dropped to any part of Simpler's interface and can be either played as a one-shot or looped instrument. The good thing is that you can convert your custom Simpler presets to Sampler presets or vice versa. To do this, select the "Simpler → Sampler" command from the Simpler's Title Bar contextual menu. Launch to ▶ Website for more on Sampler.

Simpler is located under Live Devices>Instruments. Double-click or drag the device to a MIDI track or Drop Area in your Set to instantiate Impulse. This will automatically create an instance of Simpler and bring up its user interface in Track View.

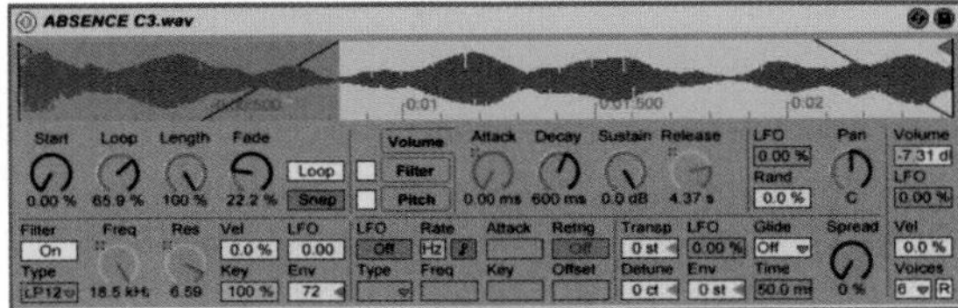

Figure 15.18 Simpler user interface.

Simpler presets can be saved or Hot-Swapped at any time just like any other Live Device. You can also swap the entire device out. Click the Hot-Swap or

Save button in the upper right corner of Simpler in the Track View. Once again, you will most likely prefer to use Simpler as part of an Instrument Rack as seen with its presets.

15.3.2.1 User Interface

Simpler is divided into two main sections: the Sample Display and Parameters. Parameters are broken down into subsections.

The Sample Display serves as a visual aid for setting the sample start, loop region, sample length, and loop crossfade. To zoom in/out in the Sample Display, place your mouse pointer over the display and click + drag the display, as you would in any other zoom-based display in Live. The rest of the interface is set aside for sample parameters and processing. Working from left to right, we'll explain each set of parameters and modulation functions.

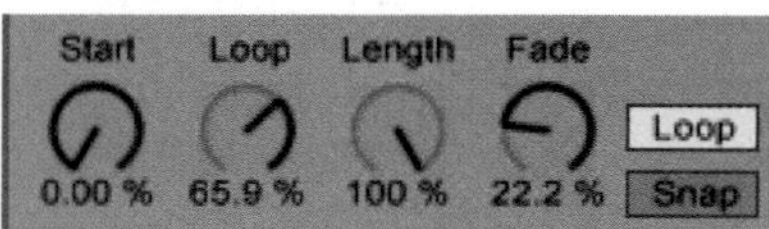

Figure 15.19 Simpler: Sample Controls.

On the upper left, below the Sample Display, are the Sample Controls. These are used to set how Simpler will play back the raw sample. Start determines where in the sample playback will begin when triggered. This is useful for many reasons. For instance, you might need a sample to start a second or two into the clip due to the silence that precedes the downbeat. Adjusting the start point to the beginning of the transient attack will do just that. Loop knob (Loop Length) controls the length of an active loop selection within the sample. If you want your sample to sustain infinitely while a note is pressed, then you want to enable the Loop Switch and then set a Loop Length. Length controls the length of the sample that Simpler will play. Fade controls the amount of crossfade applied to an active loop selection. Since a sample's loop region is often noticeable, you will likely want to disguise this effect. Fade allows for a smoothing of these audible transitions from one end of the loop to the other. Snap sets loop and length selections to the nearest zero crossings in the sample so as to avoid pops or glitches when the sample loops.

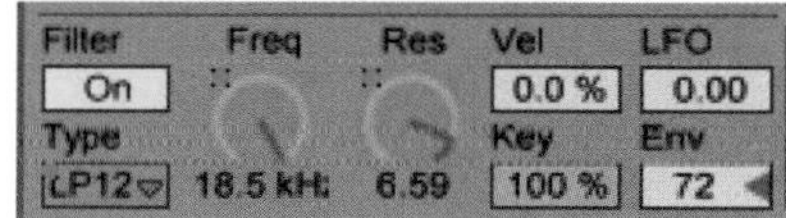

Figure 15.20 Simpler: Filter Parameters.

Filter Parameters provide the ability to shape the timbre of a sample when Filter is enabled. Use them to create unique sounds for your music. Enable Filter and choose the type of filter to be applied: LP (low pass), BP (band pass), HP (high pass), or Notch. Freq sets the filter frequency and Res sets the resonance. Vel controls how much the incoming velocities affect the filter's cutoff frequency and Key controls how much the pitch of incoming notes affects the filter's cutoff frequency. Low-frequency oscillator (LFO) slider sets the amount of modulation depth to affect the sample's cutoff frequency. Filter

Envelope Amount (Env) sets the amount of modulation depth that the filter envelope has on the filter's cutoff frequency (Freq).

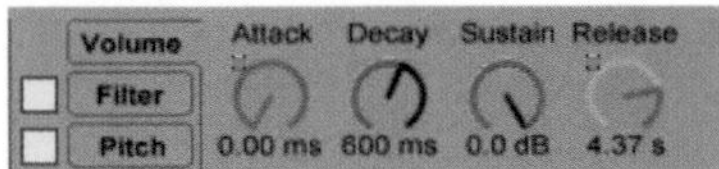

Figure 15.21 Simpler: Envelope Section.

The Envelope section contains three unique envelopes and **A**ttack **D**ecay **S**ustain **R**elease knobs for modulating the sample. Use the Envelope Chooser (Volume, Filter, Pitch) to toggle between ADSR settings for each. The volume envelope manages the amplitude of the sample over time. The filter envelope manages the Filter's modulation effect on the sample's playback over time. Filter must be enabled to use the Filter Envelope. Enable the Filter Envelope by clicking the button next to the Filter Envelope Chooser. When on, you will be able to scale the depth of modulation over the Filter's cutoff frequency. This slider is located in the Filter section. The Pitch Envelope manages the shape of modulation over the sample's pitch. Clicking on the small square button next to the Pitch tab will activate it. When activated, you can also access the Pitch Envelope's modulation amount.

Figure 15.22 Simpler: Modulation Section.

The Modulation section is built around an LFO. There are a number of parameters available for the LFO. Choose the type, frequency, attack, behavior, and how the pitch of MIDI notes influences it. The LFO can be routed to the filter, envelope, pan, and global sections via the LFO amount in each section.

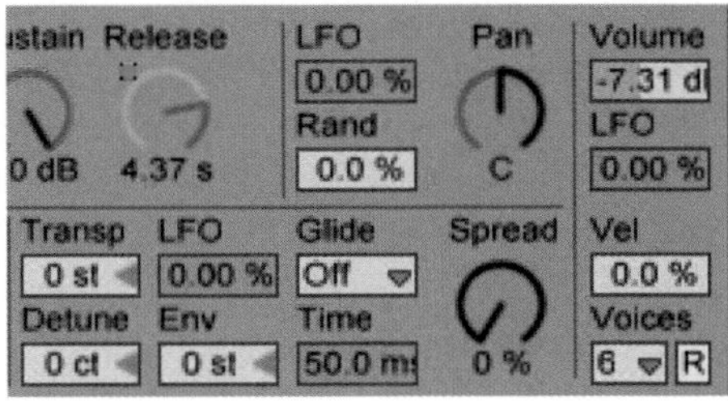

Figure 15.23 Simpler: Global Controls.

The Global Controls affect the entire instrument. From the Glide chooser, you can set the instrument to global Glide or Portamento Mode. Both modes determine how notes transition from one to the next. Time (Glide Time) sets the speed of the transitions from note to note. Glide Mode is used for adding a smooth transition between different notes/pitches, creating a monophonic (single voice) instrument. When the next note is pressed, the transition sounds like a slide (glide) between the two notes based on the Time settings. This works especially well for electronic "leads" and "basses." Portamento Mode is used in the same way, but functions polyphonically (multiple voices), meaning multiple notes can be played and overlapped while transitioning. The Pan knob adjusts the instrument's perceived location in the stereo image. Spread adds a fattening chorused effect by adding a duplicate voice that is slightly detuned – a chorused synth-like effect. The overall Volume and Pan of Simpler can be controlled from the Volume amount slider and Global

Pan knob. Both parameters can be modulated by the LFO. In addition, Volume can also be modulated by incoming note velocities (Vel) located below the LFO slider. The Voices chooser sets how many voices Simpler will use (polyphony). The "R" next to Voices is the Retrigger button. Activate this to help manage voices that sustain or overlap when trying to conserve voices and CPU. When active, currently sustained notes will be retriggered instead of overlapped with any new or additional notes that are played, which otherwise would add to the voice count.

15.4 MIDI effects

From the Live Device Browser, you can load some great MIDI Effects. Inside this folder are several built-in effects that come bundled with Live 8 and Suite 8. They are not to be confused with audio effects, which we'll talk about in the following section. First, MIDI Effects only function on MIDI tracks and only process MIDI signals (notes, data, etc). Because the MIDI data is made up purely of digital instructions to be converted to audio by an instrument, it is very easy to manipulate the data before it is converted into an audio signal. This is what happens with Live's MIDI Effects. When added to a MIDI track with an instrument, they intercept the incoming MIDI notes and redistribute them based on the particular effect. For example, if you play and hold a note with a basic instance of an instrument, you will hear one attack, then sustain, or decay depending on the instrument. Now if you insert

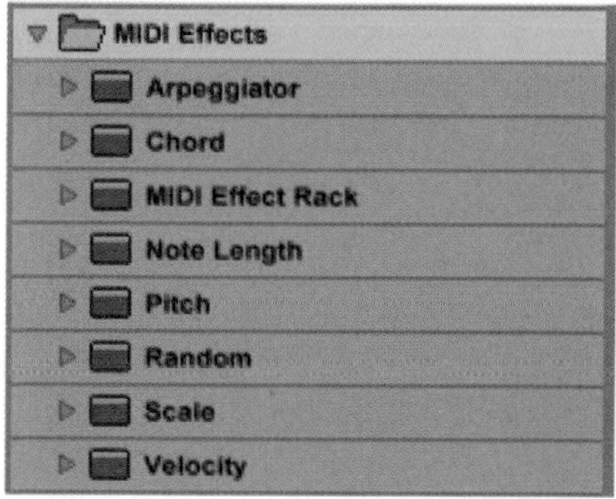
MIDI Effects
Arpeggiator
Chord
MIDI Effect Rack
Note Length
Pitch
Random
Scale
Velocity

Figure 15.24 Live MIDI Effects.

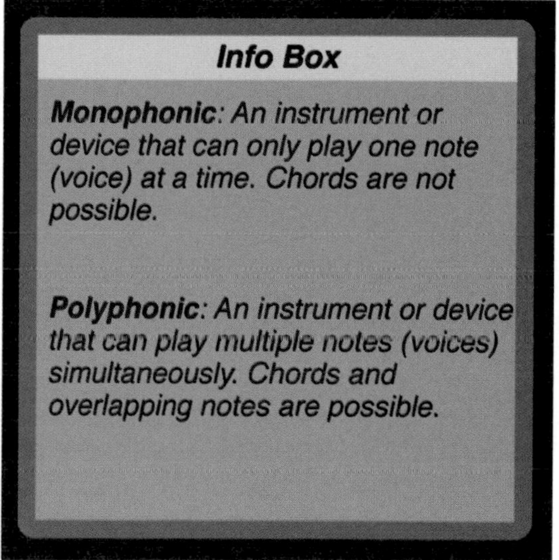
Info Box
Monophonic: An instrument or device that can only play one note (voice) at a time. Chords are not possible.
Polyphonic: An instrument or device that can play multiple notes (voices) simultaneously. Chords and overlapping notes are possible.

an arpeggiator, then play and hold two or more notes, the arpeggiator will create a repeating pattern based on the held notes. Thus, the Arpeggiator captures the incoming data and reprocesses it to repeat in a pattern based on the preset or customized settings. You can also look at this as a single chain of events. MIDI notes and data enter into the MIDI effect device and then are passed on into the instrument and then flow onward culminating as the resulting audio output.

Live's MIDI Effects include Arpeggiator, Chord, Note Length, Pitch, Random, Scale, Velocity, and MIDI Effect Rack. To learn more about Effects Racks, launch to ▷Scene 16.

A MIDI effect must be placed in front of the instrument you are applying it to since it processes incoming MIDI notes. To insert a MIDI effect or preset, select a MIDI track and then navigate through the MIDI Effects folder drop – down menu. Once you find one you like, double-click the effect in the Device Browser or drag the MIDI effect directly to the MIDI track or its Track View. This can create some very interesting results. Live will automatically insert the effect before the instrument, as you will see displayed in your device's Track View. You can add multiple MIDI Effects including multiples of the same effect type into one instrument track if you like. If there are no MIDI tracks currently in your Set, you can load a MIDI effect to your Set to autocreate a new MIDI track with the inserted effect only. The best way to become familiar with all the MIDI Effects is to load up an instrument and a MIDI clip then start dropping and swapping effects and hear what happens.

Figure 15.25 Arpeggiator MIDI Effect in Track View.

We have already described the Arpeggiator, so here is a brief description of the rest. Each effect is user-definable and comes with presets. Chord creates chords out of incoming pitches, up to six additional notes added to the original note. Note Length alters the length of incoming notes. Alternatively, it can be used to trigger a note upon the release of the note (Note Off message), rather than when the note is initially pressed (Note On message). Pitch transposes incoming notes. Random randomizes and changes which notes are actually output/played when they are inputted. Scale forces incoming notes to fall into a defined scale so that pitches outside of the scale map will be altered to the defined scale degree equivalent. For example, a major scale degree could be forced to sound its equivalent minor degree when played. Scale can also be used to transpose incoming notes. *Velocity* reassigns incoming note velocities to different velocity values based on various parameters.

15.5 Audio effects

My favorite features in Live 8 are the Multiband Dynamics (MBD) and Limiter. I don't have to use any other program to make my music sound as big or hit as hard as I want it to. Now Suite 8 and these two plug-ins allow me to get my mix to a pre-master phase without opening another DAW…that is extremely helpful.

Thavius Beck, Electronic musician and producer, Los Angeles, CA

Live 8 and Suite 8 come bundled with flexible audio effect devices for creative production and performance purposes. These devices can be found in the Live Device Browser below Instruments and MIDI Effects. As you might assume, these are used with audio tracks to process audio signals. Interestingly enough, they can also be dropped into a MIDI track as long as they are inserted after the instrument at its output. To simplify, when an instrument is inserted into a MIDI track, it is no longer strictly MIDI. Instead, it is more of a hybrid MIDI/audio track like an instrument track that you commonly find in other DAWs. In this way, the programs can handle both MIDI and audio, being that the MIDI instrument translates a MIDI input into an audio output. Therefore, this audio output can be processed in the same way an audio track output can be processed. We will see more on signal flow in the following section.

Audio effects are loaded from the Device Browser just like MIDI Effects. Select a track, then double-click the effect or drag it into an existing track or to its Track View. If you drag an audio effect to the Session View Drop Area, a new audio track will be created. To load a preset, navigate the folder drop–down menu in the Browser. You can add multiple effects including multiple instances of the same effect on a single instrument. This can create some very interesting results. Don't forget that audio effects can also be inserted on a Return track or the Master track.

The user interface of an audio effect varies from device to device depending on the type, but many of them incorporate some type of interactive controls such as the X-Y controls for shaping effects in real time. Imagine the power and creative controls when using MIDI Mapping for the X and Y when they apply to Frequency and Feedback controls! Assign X and Y to a different controller assignment to remote control each axis independently.

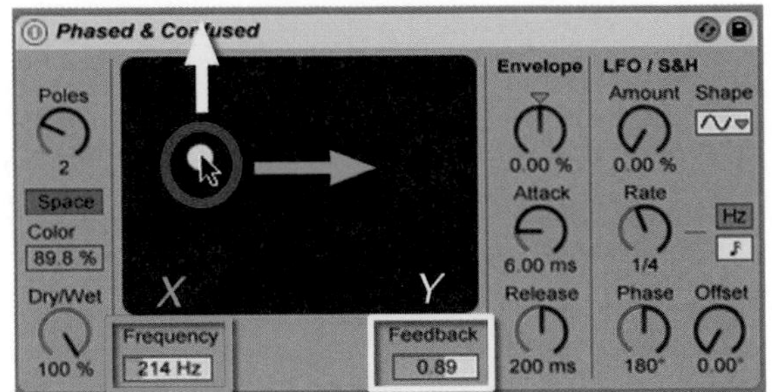

Figure 15.26 X-Y controls for manual real-time sound shaping.

Hot Tip *Assign Frequency and Feedback controls to a MIDI Map assignment to split up control of the X-Y, up/down and left/right, for real-time manipulations.*

Now that we have enticed you with a peek at some creative control, let's take a brief look at a few of the audio effect devices available for use in Live 8 and Suite 8 (depending on which one you own).

15.5.1 Corpus

Corpus is an audio effect that uses physical modelling to recreate the acoustic characteristic of resonant object. It is derived from the powerful resonator section of Collision and is integrated as a stand-alone device the Audio Effects Library. Corpus is only available with Suite 8 or by purchasing Collision. Launch to **Website** for an information on Collision, which will subsequently cover the basic concepts of Corpus as well. As an audio effect, you can apply these fantastic resonaters to the sound of any instrument or audio signal's timbre. For example, they can be applied to percussion-based sounds for a reverb type effect, or on to the Master track as a spatial effect for your whole project mix.

Some of its unique features include an LFO that directly modulates the resonator frequency, a built-in Limiter for peak level control, and an X-Y controller/display area that allows for control over a multitude of parameters in real time. Above all, Corpus has a sidechaining input for additional control by signal sources within your Live Set. See sidechaining, clip **15.6.3**.

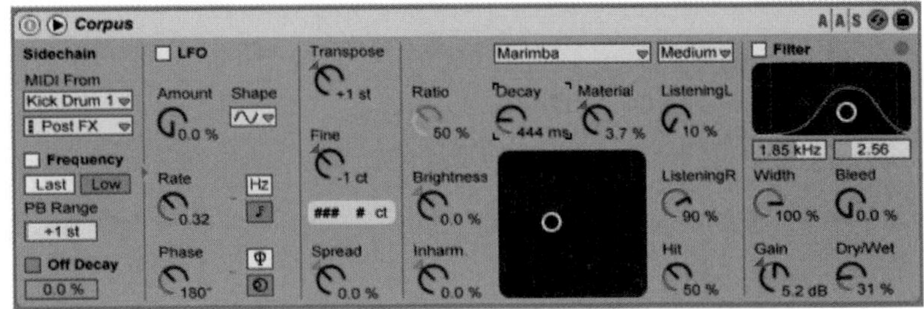

Figure 15.27 Corpus.

15.5.2 Frequency Shifter

With this audio effect, you'll be able to create and control some "out of this world" sounds from an input signal. You can control these shifts of frequencies with either a coarse or fine-tune knob. Experimentation opens up a wide array of interesting results, especially with rhythmic audio sources and their varying attack properties. There are two selectable modes of processing going on here: Frequency Shifting and Ring Modulation. Frequency Shifting adds or subtracts variable frequency ranges to your audio's input frequencies and Ring Modulation adds and subtracts variable frequency ranges to and from your audio's input frequencies. Both modes can result in some serious sci-fi style sounds. These sounds result in flange or phase type effects as well as intense metallic crystallized effects. Frequency Shifter also allows you to add LFO modulation to the entire process. Included is a Ring Modulator parameter to add even more "tweakability," resulting in surprising sonic creations.

Figure 15.28 Frequency Shifter.

Try adjusting the fine-tune frequency amounts to produce cool phasing and pitch transposition effects. For a completely different approach and result, switch the Mode button to Ring for very abstract ring modulation effects. Combine this with the LFO parameter for panning and sweeping effects. Switching your Rate button from Hertz to Tempo Sync will give you a wide variety of subdivisions to choose from in conjunction with your selected LFO amount. Also try varying the Dry/Wet mix ratio, or even automate it to hear and use the audible difference between both the dry and the processed signal.

Another useful function of Frequency Shifter is the activation of Ring Mode and Drive. This introduces a completely different effect when the Dry/Wet knob turned all the way down to 0%. The result is pure unadulterated distortion depending on how high the Drive dB level is set (located directly below the Drive button). Note that you have to balance down the overall output of your track as you add more Drive signal.

15.5.3 Limiter

This Brick Wall Limiter allows for mastering-quality signal processing that yields maximum dynamic range while reducing the overall output to a designated level. The Limiter is mainly used to control the overall level of your project and prevent clipping (overloading). Variable control over the release time provides great flexibility for fine-tuning your output signal. In addition, the stereo switch is an interesting feature that gives you the option to separate the stereo signal for the independent limiting of each channel (L/R).

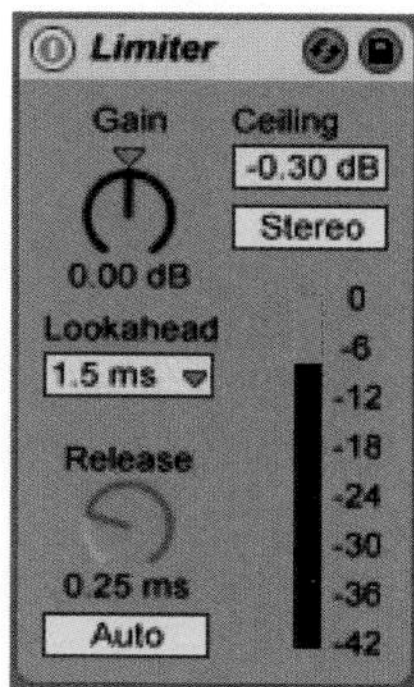

Figure 15.29 Limiter.

Other on-board features that are useful in the mastering realm with Limiter include the "Lookahead" Time chooser. This tells the Limiter how quickly to respond to and attenuate the transient spikes and peaks in the audio. This is a particularly popular feature in the engineering world of dynamic processing in that it allows the user to increase small amounts of distortion and saturation artifacts into the resulting audio. It doesn't take a whole lot of attenuation and adjustments with Limiter to hear its effects on your audio. Start out conservative, gradually adding gain and you'll quickly hear how it starts to affect output. Keep in mind that Limiter can be used at the output of any individual device to control its output as opposed to an entire track mix or the Master track output.

15.5.4 Looper

My favorite new feature in Live 8 is Looper. Being an instrumentalist, the less I have to think about technology in a live performance situation, the better. Live 8 and Looper are immensely helpful in the studio for quickly assembling polyphonic musical ideas from a monophonic source such as the Theremin.

Randy George, thereminist, Los Angeles CA

By introducing Looper, Live 8 brings back the "classic vintage style" tape-on-tape loop recording technique first made popular in the 1960s and 1970s. Ableton brings this awesome technique into the modern age of recording by designing it as a virtual Live 8 audio effect. Looper makes it very easy to not only record and loop your audio but also overdub on top of that recorded audio loop endlessly, and best of all, "hands-free" just as in the popular foot pedal "hardware" units found today.

Looper has several unique features that allow you to launch and input real-time ideas in a tight and concise manner, building endless musical passages on top of one another. You can also import and export audio files and clips to and from Looper to a track or to the Browser. It's all up to you.

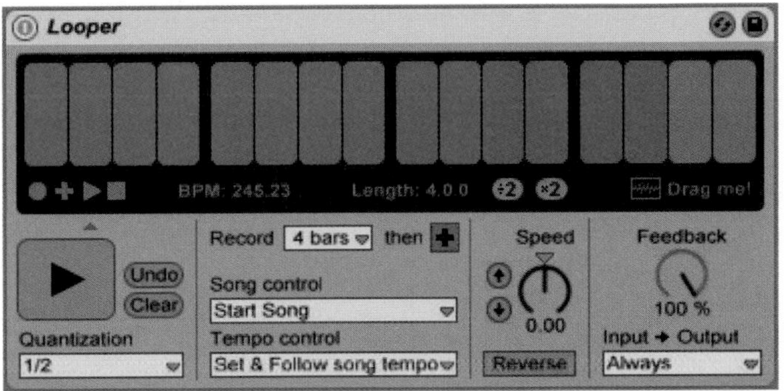

Figure 15.30 Looper.

A fantastic and fun way to use Looper is to incorporate a footswitch. This allows hands-free control, keeping the creative juices flowing take after take. By using Live's MIDI Mapping feature, you can easily assign the footswitch control to the Looper's Multi-Purpose Transport button. Simply release the footswitch when you are ready to record and you're engaged in full Looper mode. Release the footswitch again to switch to overdub mode. Now Live 8 will cycle through over and over again, playing back your recorded performance, hence the name...well, you get the idea. Once you are ready to stop "overdubbing" new parts, depress the footswitch again to activate the playback only function. Now Looper will cycle and loop your performance until you "tap twice" on the pedal to stop play or select stop with the spacebar.

Looper gives you the ability to drag your looped performance directly from Looper into a track, thus turning it into an audio clip. Don't forget that the clip's Warp Mode will be set to Re-Pitch since the audio inside the Looper is based on speed and pitch. You can change this to any other Warp Mode you wish after it's imported as a clip. Optionally, Live can calculate the tempo of your loop based on the first activated "punch out" you make on the first recorded pass. This is very useful in that Live will then globally sync all your Set's clips to this original Looper based on the tempo until you change the master tempo itself. This gives you the ability to add audio clips from your Live Browser to enhance your recorded Looper performances(s). We'll discuss this more in the Song Control Chooser portion of this section.

You can also activate and use multiple Looper devices to run across and sync several tracks with one another. Insert audio effect devices in front of Looper in the Track View as well to enhance your sound and ultimately the Looper's results. This is a great technique for guitar and other ethereal or ambient-style audio performances.

Some of the features with Looper include a large visual display of segments representing the bars and beats of your recorded and looped performance. The record button in the middle of the interface gives you the option to either overdub or simply replay your audio after every pass. There is also a Speed and Reverse function that adds these tape-style results when selected. You can also record directly in Looper with the Reverse setting engaged to create a more radical sound.

Figure 15.31 Looper: record directly in Looper with the reverse setting engaged to create a more radical sounds.

You'll quickly find that you have total control over Looper's ability to sync to Live's master tempo. This is useful when you already have other rhythmic clips loaded into your Set that may or may not be launched at a specific time. Note that Live's transport can be set so that as soon as you "punch out" with Looper, it engages and plays from that point forward. Change these parameters if needed, but it's important to know that while Live is playing globally, Looper is simply playing back just like an audio clip itself. An important feature to know about is the Song Control Chooser. By selecting "none" in the chooser pull-down window, Looper's tempo will not effect the Global Transport in any way. If you select "Start Song," the Global Transport of Live will begin playing as soon as you engage Play or Overdub mode in Looper. The latter is great for having loaded clips in standby mode ready for playing after overdubbing your first Looper performance. This is why a footswitch is so important in your interaction with Looper.

Figure 15.32 Looper: if you select "Start Song," the Global Transport of Live will begin playing as soon as you engage Play or Overdub mode in Looper.

After overdubbing a new audio loop, you can change the total length of the loop with the "divide 2" and "times 2" selector just below the bars and beats display segments and to the left of the Drag Me function. For example, this works great for situations when you have a 2-bar original loop and you wish to add several 4-bar overdub loops on top.

Figure 15.33 Looper: record length.

Several options are available for monitoring Looper's direct Input and Output of recorded audio. Just as with a track's Monitor Input and playback buttons, this lets the user decide whether or not the constant input signal can be heard at all times. This choice just depends on whether you plan to overdub more than one track, use just one take as a loop, or use multiple Loopers across several audio tracks or Return tracks. You'll find this chooser for Input-Output on the bottom right of Looper.

Figure 15.34 Looper: Input-Output Chooser.

Looper takes a bit of practice to fit your needs with both the performance and parameter settings, but after a few passes, you'll find that it's a wonderful extension to the "Live 8" experience. Especially for singer-songwriters and guitarists.

15.5.5 Multiband Dynamics

This mastering effect processor combines the power of six discrete processors into one interface. Together they offer a real-time compressor and expander with several innovative real-time control parameters not found in other dynamic processors. How does it work? Live 8's MBD, as we'll call it here, provides both upward and downward compression and expansion across three completely independent frequency bands.

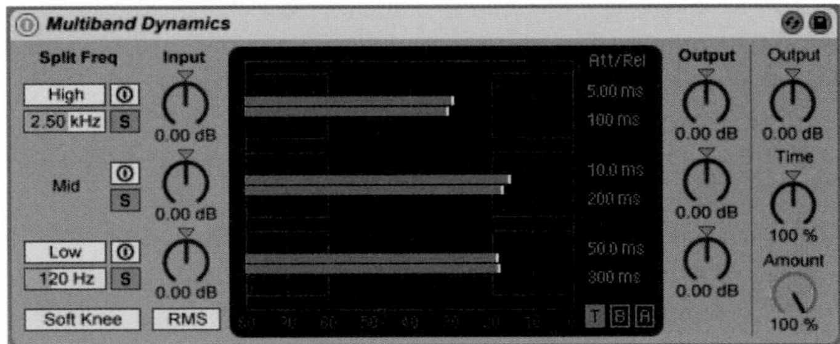

Figure 15.35 Multiband Dynamics.

It comes with individual attack and release controls plus crossover points for each band itself. You have the option for soloing and designating a bypass for each band, which promotes flexibility. MBD incorporates a user-friendly, real-time drag tool available inside the interface. By using your mouse and pointing to a specific area within the graphic display window, you can access and dramatically alter the volume and threshold of each band allowing for customized sound-shaping effects right from within the MBD interface. A good way to learn the basic variables of this "click and drag" feature is to solo a single band on the left side of the display and then start adjusting the levels and threshold, which are contained in the visual "block-style" segments in the display for that particular band. Keep in mind that when working with the bypass and solo buttons of each of the three bands, the Mid-band controls the entire audio signal when the "high" and "low" bands are bypassed. You can quickly manipulate and edit a specific band's input, output, and threshold levels within the display and get quick audible results by either dragging left or right on one of the block segments with your mouse to adjusting the threshold levels. Dragging up and down on the middle of the block will adjust the signal level for that particular compression or expansion band.

On the right side of the display, you'll find three options for viewing the MBD's Attack and Release Time parameters and Above Threshold and Below Threshold values, signified by the "T," "B," "D," buttons. These are especially useful for precision mastering and sound sculpting.

Practicing with a single band's audio will help in identifying discrete audible values and more when you first start to work with all three bands. Speaking of all three bands, you can also use drag and alter across both the compression and expansion displays all at once by holding down the [Alt/Opt] key while clicking + dragging the appropriate parameter. If you wish, you can do it across all three bands at once by selecting [shift + Alt/Opt] while clicking + dragging the particular parameter needed.

The MBD can be put to great use across all of your upcoming Live Projects as a real-world mastering tool for multiband compression, de-essing, and single-band processing techniques. It also makes for an effective sound design tool. Try working with the high and low band's frequency settings located on the extreme left side of the interface. In addition, check out the presets found in the Multiband Dynamics preset folder in the Device Browser. Some of these are tailored for mastering entire music tracks while others are dedicated to vocal and speech detail. You might consider loading up an MBD preset and experimenting with adjusting the Amount knob and Time Control knob (Time Scaling) to get started. Time Control adjusts the speed of the attack and release to all the active bands. The same holds true for the Amount knob, which adjusts the overall level and intensity of both compression and expansion across all the active bands. The audible results can be minimal or very obvious. Again, using these two functions is a great way to get acquainted with this complex device.

15.5.6 Overdrive

This in-your-face distortion effect is Ableton's answer to the notorious guitar stomp pedal, but with user control that will make guitarists jealous. Overdrive is great for adding distortion while maintaining a punchy and raw edge with great dynamics to any audio. Included within the Overdrive interface are some unique functions that allow you to engage subtle nuances all the way to hard and heavy distortion and saturation. When we say hard, we mean really saturated and dirty! From its front panel, you can control overdrive, tone, and dynamics. The Bandpass Display allows direct control over the band-pass filter's center frequency and bandwidth values by using the mouse to move the X-Y control (yellow circle). They can also be controlled via their respective slider boxes below the Bandpass Display. Be sure to experiment with using the X-Y functionality in real time. This is fantastic for exploring the possibilities of Overdrive's broad frequency spectrum filtering effects. Drive controls the amount of distortion and Tone effects the timbre of the signal post distortion for boosting the high-frequency content. The Increase Drive amount yields more distortion. The Dynamics parameter provides control over the amount of compression applied; thus the percentage of compression versus the original dynamic range of the signal. Try keeping it somewhere between 60% and 80%, which is a good starting point for introducing light compression into your mix. Dry/Wet sets the balance between the processed and unprocessed (original) signals with the exception of Tone, which always affects the sound unless Tone is set to 0%. Keep the Dry/Wet value at 50% as a starting point and gradually increase the value along with the Tone and Drive values. Be sure to set Dry/Wet to 100% when using Overdrive as a Send effect.

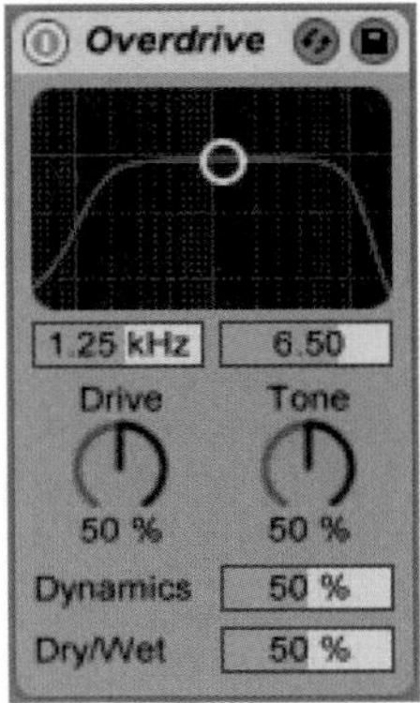

Figure 15.36 Overdrive.

Throw Overdrive on your drum tracks and get that grungy white noise snare or band-passed drum effect. It also works great in conjunction with other audio effect devices like Compressor and Dynamic Tube. Try loading up all three of these effects within a single instrument track!

15.5.7 Vocoder

First, there are different kinds of vocoding and therefore different methods of implementation. With that said, here is a general explanation of vocoding.

The general concept is that a carrier audio signal is modulated by a modulator audio signal, which effectively amplifies and attenuates the frequency content of the carrier to create different sounds. Putting this idea into practice, you can take a voice from a microphone (the modulator), pass it through a filter bank/envelope follower, and have that information trigger oscillators in a synthesizer (the carrier). Using a synth as the carrier and a drumbeat as the modulator, the result would be something like a rhythmic synth continually influenced by the drumbeat based on whatever is played on the synth.

With Live 8's Vocoder, you have the ability to generate many vocoding effects with ease, and on top of that, it can even operate with or without an external carrier signal. Vocoder can achieve effects by tapping into its own built-in internal carrier signals or even modulate its own signal.

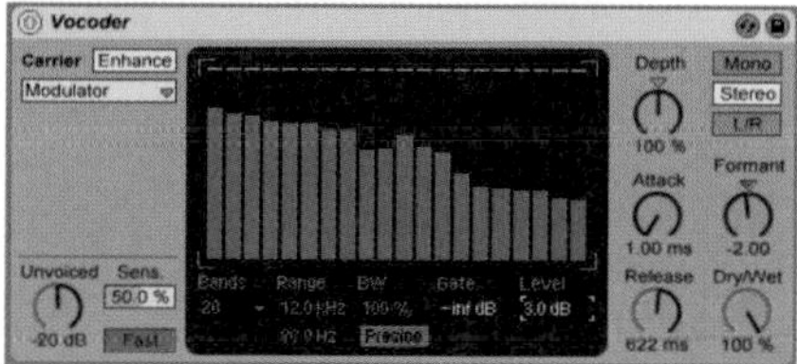

Figure 15.37 Vocoder.

Insert Vocoder on your modulator track and then assign the carrier signal with Vocoder's carrier chooser on the left side of its interface. With the intuitive routing flexibility of the sidechain input (carrier signal source), you will find Vocoder to yield a satisfying vocoding experience – which we can't say about some other third-party Vocoder plug-ins.

Vocoder's available parameters allow you to control and fine-tune variable frequency bands with format adjustments and with a multitude of band-pass filters. It also offers a carrier signal source for classic Pitch Tracking, making it possible to recreate that "Beastie Boys" sound effect. Among its parameters are Upper and Lower Pitch Detection sliders (high/low) that control ranges over the Pitch Tracking oscillator. These detection parameters limit the frequency range that will be tracked by the oscillator, thus creating a "robotic floating" vocal effect similar to a "mouth harp." Try putting Vocoder on a drum beat track and set Carrier to Pitch Tracking, then crank up the High parameter, tweak the Low parameter, and adjust the Formant to see what you can come up with. If that's not enough, you'll find four separate waveforms to choose from to help manipulate the initial characteristics of the actual Pitch Tracking oscillator.

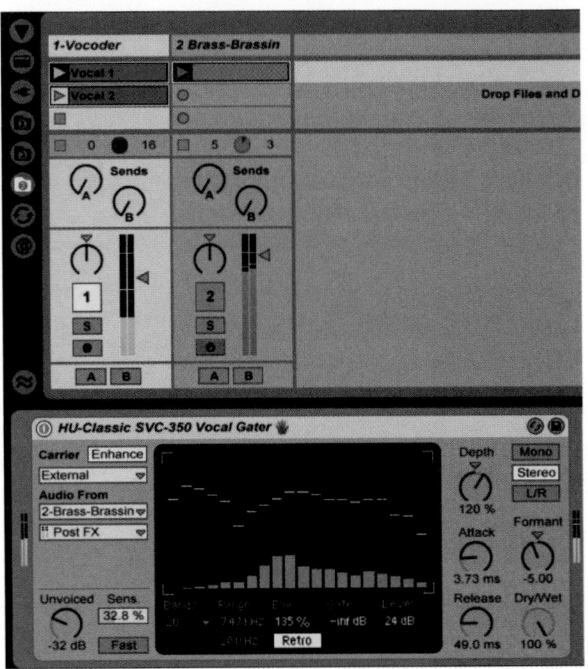

Figure 15.38 Vocoder Sidechain in action.

To really take Vocoder as far as possible, try using a microphone and your voice and follow the below given steps:

1. Using a new empty Set, insert Vocoder on an audio track; we'll call it "Track 1." This track will be the modulator.
2. Assign your voice to be input on Track 1 and set its Monitor input to "In," so you can eventually audition your vocal sound after you have completed the following steps.
3. Using a new MIDI track as "Track 2," add an instrument that has a quick attack style. Operator or Analog's "Brass style" patches work well for this.
4. Record-enable the MIDI track and play some chords. After you find a few chord progressions you like, record some MIDI clips, then when you have your clips recorded, mute the MIDI track by selecting its Activator button.
5. Launch your newly recorded MIDI clip(s) and feed the MIDI Track 2 back through Vocoder as the carrier signal by selecting MIDI Track 2 as the External input source for Vocoder's sidechain input chooser.
6. Hop back on the microphone and sing or speak to get some audio activity going in Track 1, thus into Vocoder while the chords or notes you recorded on Track 2 (your carrier instrument) play as well.

7. You should start to hear some effected vocal sounds playing through Vocoder. Now it's time to tweak and control your new sound with the Formant control (shift the frequencies of the filterbank) and Release knob parameters (release time of the filter band's sustained amplitude). You can also try selecting filter band variations from the Bands chooser menu. Try something between 20 and 32 for more obvious changes in the tone and clarity.
8. By now, you should notice that no matter the pitch of your actual voice, the instrument Track 2 actually controls the real-time pitch. Pretty sweet!
9. Additional enhancements can be achieved by selecting different filter ranges in the higher frequency bands just to the left of the Bandwidth "BW" slider and Precise/Retro Mode button just below that. By selecting a wider BW value and choosing Retro Mode, you should start to hear that familiar Vocoder sound.

The "singing synthesizer" also works great with a drum loop as the modulator. Here are some parameters to consider using for this. Try increasing the Release value on the Vocoder and you'll open up the true length of the active filter bands to hear a more sustained singing style effect. Adjusting the Attack knob will determine how fast Vocoder will respond to the dynamic changes in amplitude from the modulator signal. This is one reason drum-style clips can work well in the example above. Try adding delay and reverb along with Vocoder to create a complex device. Note that Vocoder doesn't have to be used to create a "robot voice" or "singing synth" effect, although it's a popular effect.

15.6 Device chains

Up to now, we have focused primarily on individual devices and haven't spent much time exploring how multiple devices will interact when put together on the same track, such as combining a MIDI effect with an instrument. I guess you could say we've been holding out on you. Well, that's all about to change. Before we start stringing together devices into what is called a "device chain," we should define it. So what is a device chain? A device chain is what Ableton defines as the connection and signal flow of multiple devices on the same track, whether that be on a MIDI, audio, Return, Group, or the Master track. Remember, devices are instruments, effects, or plug-ins; and just like any other DAW, you can insert multiple devices on the same track, forming a serial connection of devices. As you begin chaining devices together, you'll find there is a hierarchy to the chain. Devices always receive and pass audio or MIDI signals in succession (serially) from left to right. This is not a new concept. In fact, it is the same process that all DAWs follow except they usually pass signal from top to bottom in succession. A device chain can be any combination of instruments, effects, or plug-ins. You'll see in our example an arpeggiator, synth, compressor, and a delay creating a complex instrument. This is an example of what we'll refer to as an instrument-based device chain. For a detailed description of signal flow, launch to 16.2.3. Note in our example that the Arpeggiator and customized Compressor devices are folded within the Track View. This is a useful way to see more devices in the Track View.

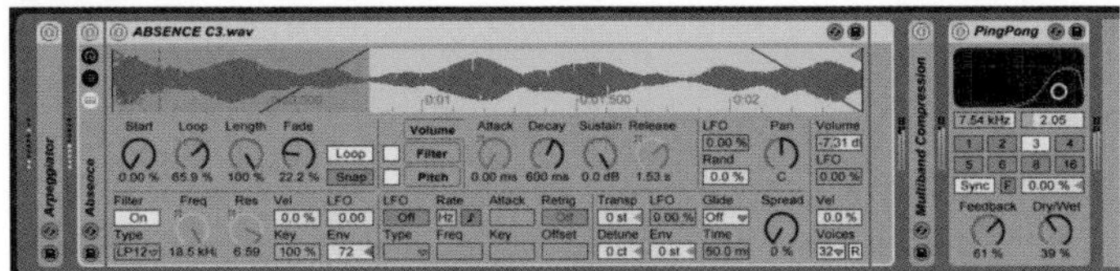

Figure 15.39 Fold devices for more viewable space in Track View.

Remember, a device chain can also be created on an audio, Return, Group, or the Master track. For example, a Phaser, Chorus, and Flanger effect make up an effect-based device chain (effects processor) in our example. Note, once again that our Chorus device is folded in between the Phaser and Flanger.

Figure 15.40 Chorus "Slow Pan" folded between devices.

15.6.1 Chaining effects

There are two ways to go about chaining effects together. The concept is fairly simple. The key is to understand the signal flow and the difference between an "insert effect" and a "send effect." First, when you are building custom instruments in Live, you will generally make use of audio effects as inserts. This means that they are inserted directly into the signal path of instrument track or audio track containing some audio file source, creating a chain. We saw this when we added a compressor and delay to an instrument. In that way, inserted effects devices make up a part of the instrument and its sound. The alternative to this is to create a chain of effects on a Return track, creating an effects processor to be used as a Send effect such as a reverb. In this scenario, the Return track receives a copy of an audio signal from an audio or MIDI track via a track Send(s), hence "send effect." This signal is received and processed through the Return track's device chain – left to right in the track display – then output to the Master track. At the Master track, the effected ("wet") signal received from the Return track is blended with the original track source ("dry") signal.

Device chains may also be created on the Master track where the audio signal received is the sum of all the tracks in your Set. The Master track then processes the "Mix" through its effects chain. This is the common scenario for self-mastering your mixes or productions. Live's MBD and Limiter are great choices to experiment with for this purpose!

We have just described effects chain examples consisting of basic devices that can be created and chained one by one, without the assistance of an Effect Rack or presets. Effect Racks on the other hand add a whole new level of complexity and available performance parameters. For more on Effects Racks, launch to 16.4.2.

15.6.2 Interfacing with device chains

As you begin to build device chains, you'll notice that the Track View gets filled up quickly. Since devices visibly extend left to right, it's quite possible that some will already be off the screen and out of view after only a few devices have been added. For this reason, Live offers a couple of ways to view and navigate the track display to maximize efficiency. Ableton is notorious for keeping screen real estate in mind and seems to always find a way to add features without interrupting the workflow.

When you first add a device to a track, it will appear in full view, meaning that you will see its entire user interface. As we have mentioned before, use the Track View Selector to scroll back and forth through the device chain.

This way you can view and work with the device in full view. You can also navigate left and right with the keyboard arrow keys or with a horizontal scroll wheel on your mouse if you have one. If you prefer to have all your devices in view at once or have no need to view the details of a specific device, you can collapse or "fold" them. Double-click the Title Bar of any device in the chain to fold it and it will be reduced to a thin vertical bar, like a book on a shelf. On the spine of the device, you will see the Title, Save Preset, and Hot-Swap button. Collapsing devices creates a ton of space in the Track View for you to focus on the effects that you are working with. Double-click again to unhide (expand) the device back into full view. Collapse all of them if you really run a tight ship.

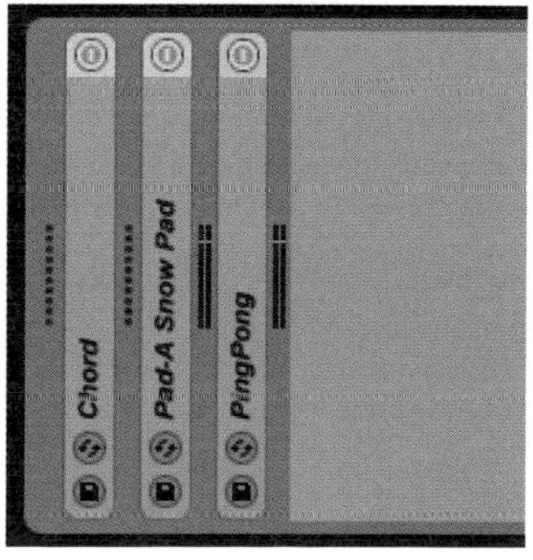

Figure 15.41 All devices folded.

Moving effect devices around in a chain is very easy. Remember that signal flows left to right through the chain. With that in mind, you can reorder your effects, however you prefer. Just click + drag a device to the desired location in the chain and that's it.

Hot Tip *When working with active clips and effect devices at the same time on the same track, you can easily toggle between detail views (Track/Clip View). If you are working in Clip View, you can switch to Track View by double-clicking a track's Title Bar or simply pressing shift + Tab on your computer keyboard. To return back to the Clip View, double-click directly on the clip you wish to view or use shift + Tab again. This applies to both the Session and the Arrangement Views.*

15.6.3 Sidechaining

Sidechaining is a popular process in audio production, yet has remained elusive in the general user's quest to master the concept in the DAW. Ableton has now made this easy to comprehend and execute. The practice of sidechaining audio consists of taking the audio signal from one track and using it to affect (modulate) another that contains an audio effect device that supports sidechain input. In essence, using one sound to change another sound. One very common use is to effect the dynamics of another sound. For example, you may hear an on-air radio commercial using "ducking," where, when a voice speaks the music dims, or the classic "French house" trance chord breathing-suction effect (sidechain compression effect). Let's not forget synth-pad gating effects, where sidechaining results in the carving out of rhythms within a sustained pad or chord!

Four of Live's audio effects contain the sidechain input feature: Compressor, Gate, Auto Filter, and Vocoder. In all these audio effects with the exception of Vocoder, you will see a small arrow button located next to the Device Activator as shown in our Compressor audio effect example. This is the Sidechain Toggle.

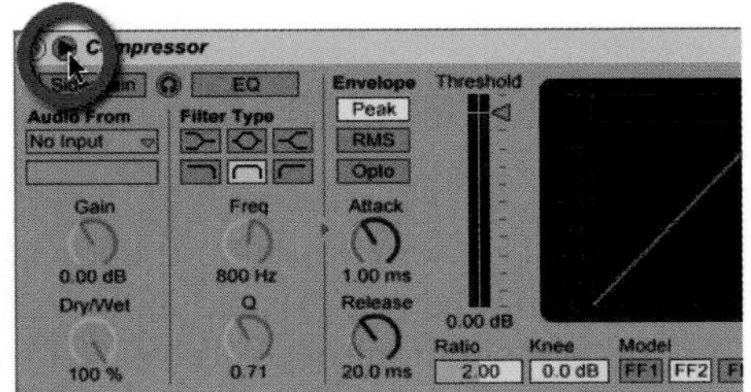

Figure 15.42 Sidechain Toggle.

By clicking on the Sidechain Toggle, you'll see all the components of the sidechain feature contained in Compressor, Auto Filter, and Gate. Vocoder's

Sidechain Input parameters, as mentioned above, are already open and ready to work with. For this demonstration, we will reference three examples depicting the Gate audio effect.

First, you will see the Gate as a standalone without sidechain open or activated. The second example shows the sidechain parameters open and ready to be activated.

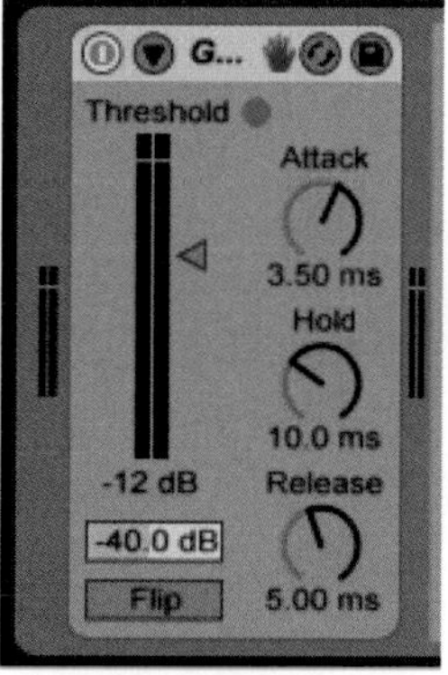

Figure 15.43 No side-chain active.

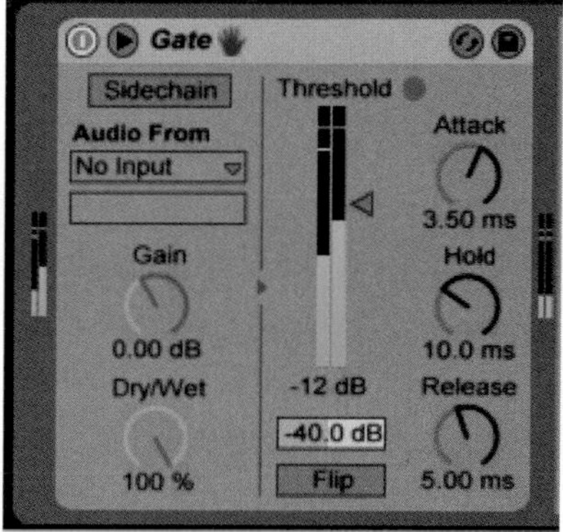

Figure 15.44 Sidechain parameters.

Finally, the third example shows the sidechain feature fully activated with the Audio From input chooser set to a source track of audio/MIDI that is activating the Gate sidechain feature.

Figure 15.45 Sidechain activated with the Audio From input chooser set to a source track of audio/MIDI.

Once a signal is represented in the sidechain process, it's time to set the appropriate parameters to achieve the desired sound you are looking for. Let's dive into a quick run-through on working with the sidechain feature within the Gate audio effect in Live 8.

With Gate, it's possible to achieve the popular "stutter" effect (rapid ducking effect). The audible effect is very fast silencing in the audio that makes it sound glitchy or gapped but at a speed that doesn't result in real silence. To achieve this effect, you need two MIDI tracks: "Track 1" with a synth pad and Gate audio effect; "Track 2" with a drum kit. Use presets from the Live 8 Library. Track 1 should be a legato pad-like synth. A synth pad or drone-style instrument patch will work perfectly. Track 2 "source track" can be any drum kit you like.

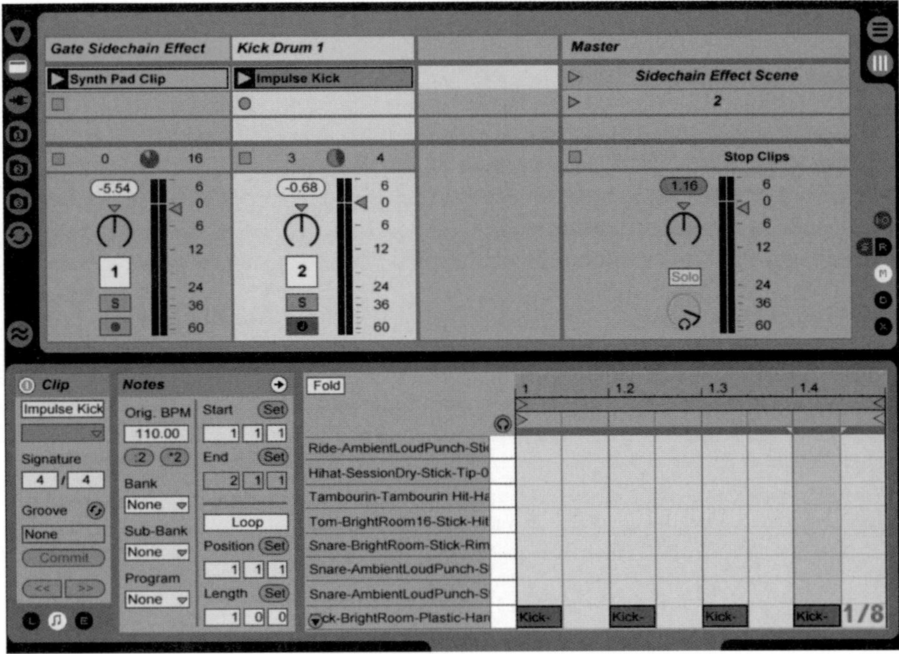

Figure 15.46 Gate Sidechain effect with Drums for getting that "stutter" effect. Kick or Snare sample selected for drum source.

1. Select a solid kick drum or snare drum sample for your drum "source" Track 2 to feed into the sidechain input chooser window.

Figure 15.47 Source fed into sidechain input chooser of Gate.

2. Program MIDI notes for the kick drum to play back a frequent note pattern for two bars at 120 BPM.

3. Now using your legato synth pad you selected for your Gate sidechain track, create a two-bar sustained chord loop or simple melodic phrase into a clip in the same row as the kick drum clip if not already.
4. Launch your scene row with both clips playing the synth pad "sidechain" track and the "source" track (your kick drum pattern).

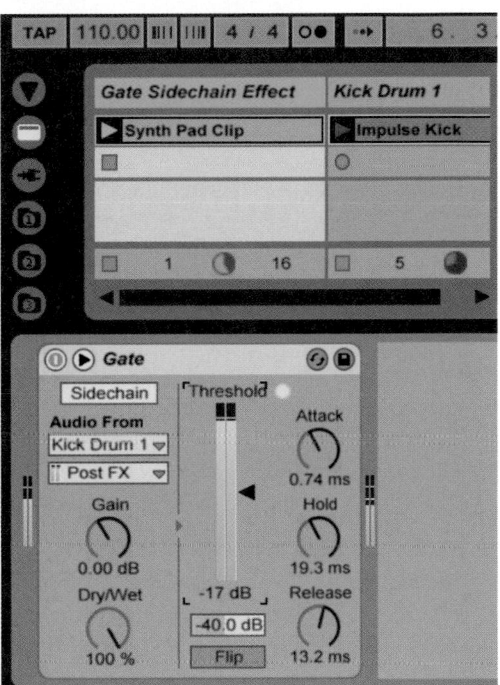

Figure 15.48 Both clips launched. Synth pad track gated by kick drum.

You'll notice that your signal level on your Gate effect is activated. That's what you are looking for. Now control Gate's threshold parameter to begin dialing in the perfect balance between a ducking gate style effect and the original synth pad melody or note chord. If you don't hear an immediate effect after adjusting the threshold level, make sure that the Sidechain button is selected (yellow) in the Gate sidechain window. Generally, the higher you raise your threshold level, the more obvious the gating/sidechain effect will be. Here's a tip: adjust your Release and Hold parameters to give a natural decay to your effected sound. Although the clips are playing back and gating, fade down the drum "source" track volume to hear just the effect of the Gate. Pretty cool! If you want to get creative, you can also add a Ping-Pong Delay effect after the gate sidechain effect to animate and bring the new sound to life. Now try dropping in a full drum loop clip on a third audio track and hear the creative possibilities start to unfold.

15.7 Plug-in devices

Plug-ins, by definition, are applications that function within another software application, enhancing the host software's functionality. In our case, these are instruments and effects that Ableton does not develop. Plug-in devices are designed by third-party developers to operate within any DAW. As you know, Live 8 and Suite 8 include many well-integrated instruments and effects. Once you have exhausted all of them (and that may take awhile), you should consider branching out and investing in some additional third-party instruments and effects. Although Live comes bundled with several useful instruments and effects, it also supports third-party Audio Units (AU – Mac only) and VST Plug-ins' virtual instruments and effects. This will allow you to work with thousands of additional unique instruments and effects accessible within the Live Browser, giving you seamless integration of your third-party plug-ins just like your Live 8 and Suite 8 instruments and effects. You can even mix and match Live's effects and instruments alongside your newly acquired third-party plug-ins within your tracks and Device Racks. Now, before you spend all your hard earned cash on the hottest new plug-ins, understand that no matter what plug-ins you own, your creative ideas, arranging skills, and perception of sound are far more important. Just like any other hobby, profession, or sport, it's easy to get caught up in the technological glamour of it all. With that in check, the more resources you have, the greater the potential for realizing your music. Once you upgrade or buy Suite 8 and max out your Ableton resources, then stock up on some cutting-edge third-party plug-ins!

15.7.1 Plug-in Device Browser

There are two types of plug-ins: Instruments and effects, both of which are accessed from the Plug-In Device Browser. Before you can use third-party Audio Units or VST Plug-ins with Live, you'll need to turn them "On" in your Live Preferences menu (Preferences>File Folder>Plug-in Sources), so Live can locate them. Once activated, Live will scan for all your AU and VST Plug-ins, so they are readily available in the Plug-in Device Browser for use.

The Browser organizes virtual instrument and audio effects together into one single folder structure, so you may have to do some digging to find what you want – but it's well worth it!

15.7.2 Third-party instruments

Third-party instruments come in all shapes and sizes and are often called "virtual instruments," "soft synths," or "plugs," etc. To clear up any confusion, these titles are generic, "on the street" labels, but there is a difference. First, "virtual instrument" is a fair name but beyond that, the rest are slightly inaccurate. Software synths are instruments that generate sound through synthesis or a synthesizer if you wish. By definition, that leaves out samplers and sample players, which are designed to store and play audio samples as opposed to generating digital tones through synthesis techniques. So, we are back to "virtual instruments." Add the word "plug-in" to the end of virtual instrument and it's settled. As you know, Live uses the term Plug-in Device. This is because they don't delineate between instruments and effects. Unfortunately, you will have to memorize what plug-ins are of which type. This is not that big of a deal, but it can slow you down when you accidentally add what you think is an instrument and have it turn out to be an effect, or vice versa. Thankfully there is the undo command ([⌘+z] Mac/[Ctrl+z] PC or Edit Menu>Undo) that will save your life, time, and time, and time again.

15.7.3 Third-party effects

As for effects, they are what they are: digital signal processors. They too come in all different shapes and sizes covering an array of dynamic and time-based processors. Effects serve one of two purposes: "sound shaping," e.g., dynamics, timbre, pitch, etc., and 'performance enhancing,' e.g., spatial perception, timing/tempo. Common sound-shaping plug-in effects include EQs, Compressors, Limiters, De-essers, and Saturators, etc. Some of the more complex and bold third-party effects come packaged as an "all in one" type effect featuring multiple effects working together within one dedicated user interface. Common performance-enhancing plug-ins include Delays, Chorus, Flanger, Modulators, and Reverbs to name a few. Of course, many effects border the line between both purposes. In the end, the classifications don't matter, just as long as you know what each effect does and that you like the result.

15.7.4 User interfacing and layout

Before you can start using AU and VST Plug-ins in Live, you must make sure that they have been activated in Live's preferences. If you haven't already done so, go to Live>Preferences>File Folder Tab. From the Plug-in Sources section, turn on the plug-in format you wish to use. You may consider both if you are using a Mac and have both types installed on your computer.

When it comes to using third-party plug-ins, as opposed to Ableton instruments, there is a difference in the way the user interface is handled. Load the plug-in the same way you do your Live Devices. When loaded, a plug-in device's user interface or graphical user interface (GUI) will appear in a new floating window instead of the Track View unless it has no custom GUI. In this situation, its parameters will only show up in Track View. You can always close it by clicking on the small red button along the Title Bar in the upper left of the interface window of the visible effect. Once closed, it will then appear in Track View. This functionality is available in both the Session and the Arrangement Views.

Figure 15.49 Plug-in user interface appearing in a Floating Window.

Figure 15.50 Assign any two specific plug-in parameters to the X-Y Control Window for real-time manipulation.

When the plug-In device floating window is closed, access the plug-in via Track View, where you'll see its assignable X-Y Control Window. This lets you assign any two specific plug-in parameters at a time to the Control Window for free-flowing control over the plug-in's parameters in real time. If your plug-in has 32 or fewer automatable parameters, they will be displayed as sliders to the right side of the Control Window and will be ready for assignment in the X-Y choosers at the bottom of the interface in Track View.

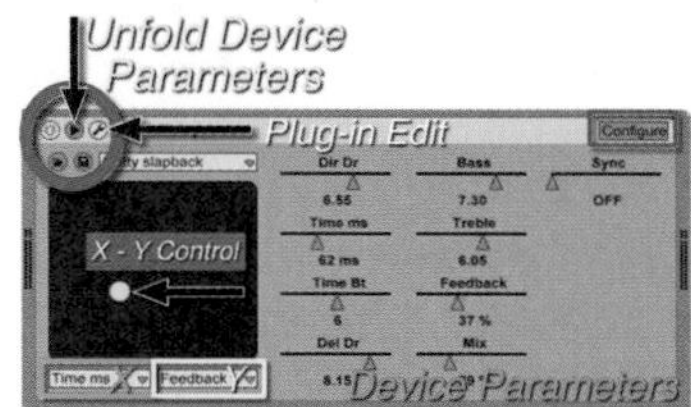

Figure 15.51 Plug-in device X-Y Control/Parameter layout.

If it has more than 32, this area will show nothing. Instead, you must manually add parameters to be displayed in the device. On the Title bar, there is an Unfold Device Parameters button, and just below that is the Preset Chooser and Load Preset button when presets are available. Unfold device parameters to view the available internal parameters for the plug-in. If none are available, then add parameters by selecting the Configure button on the right side of the Title Bar. You can also rearrange and delete parameters. Click "Configure" (green = On), then reopen the plug-in's main interface floating window by clicking on the Plug-In Edit "wrench" button (if closed) and click on a parameter on the plug-in's interface to add it to the listing of available device parameters.

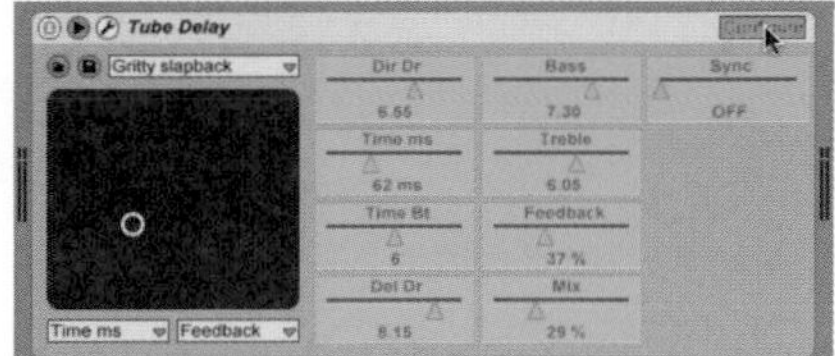

Figure 15.52 Configure Mode.

To delete, click on a green-lit parameter in the device and press the delete/backspace computer key. Not all plug-ins are compatible with the Configure feature, so don't be disappointed if you can't assign any controls or access presets from the Track View. For example, Native Instruments Kontakt Sample Software is not compatible with this feature. Instead, you would assign automation through its own internal automation window.

Once you have selected parameters of a device, they will become available for MIDI/Key mapping in Live as well as assignable in the X-Y choosers. Choose parameters from the chooser that you wish to control in the Control Window or map them as you see fit. Next to the Unfold button is Plug-in Edit button. Click this to open and close the plug-in's floating window. As for the rest of the Track View, controls remain the same.

Figure 15.53 Kontakt floating window with Sample Logic sample libraries.

Figure 15.54 Must use the third-party automation assignments for some devices to make assignments to the X-Y control parameters in Live.

15.8 External (MIDI) Instruments

Yes it's true, people still use external MIDI instruments to make music. The most common instruments are generally synth keyboards, virtual analog synths, drum machine/sequencers, and digital keyboard workstations. We won't go into the specific manufactures or devices here, but we'll look out how they function in Live.

15.8.1 Routing

Live's MIDI tracks can not only be routed internally between tracks and instruments but they can also be sent to the external world for use with your external MIDI hardware devices. A configuration such as this requires a MIDI track to send out MIDI data and an audio track to monitor the external device's audio output so that you can hear the sound of your gear and record it to track. Another way to set this up is to use Live's External Instrument device located in the Live Devices>Instruments folder. This useful instrument is a gateway to your external MIDI devices. There is no actual instrument associated with it, but what it does is it acts as a bridge to your device by converting the audio track into a hybrid MIDI/audio track. This means that you can trigger an external MIDI device and monitor its output in Live, all on one track. This is similar to instrument tracks in other DAWs. External MIDI Instruments are not only usable for an external MIDI controller or sound module that generates sound, they can be used for handling the output of multitimbral plug-ins and other outside of Live audio sources. For now, we'll focus on external MIDI. Therefore, these devices must be connected to your computer via a MIDI interface or USB connection, and the audio output should be routed back into the computer or connected to some type of studio monitors, speakers, or headphones to here their output. When routed back into Live, you can record and process this output.

The following exercise shows how to set up the external MIDI device:

1. From the Devices>Instruments folder, load the External Instrument device onto an empty MIDI track or drag it into the Device Drop area. This will open the Track View and you should now see a device interface with "MIDI To," "Audio From," and "Gain knob."

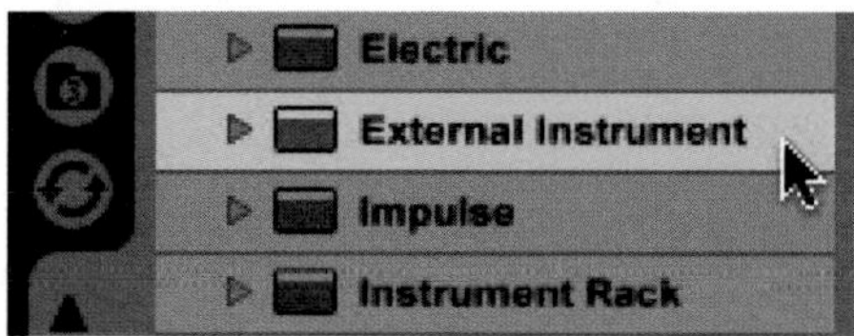

Figure 15.55 Live Devices: External Instrument.

2. Set "MIDI To" to the MIDI port your device is connected to and set Audio From to the audio input channel the external device is connected to. This will vary depending on the signal flow of your studio. The only way to monitor through Live is to route the audio back into the computer.
3. Set your MIDI track to receive MIDI from all inputs and all channels.
4. Arm your MIDI track and begin playing.

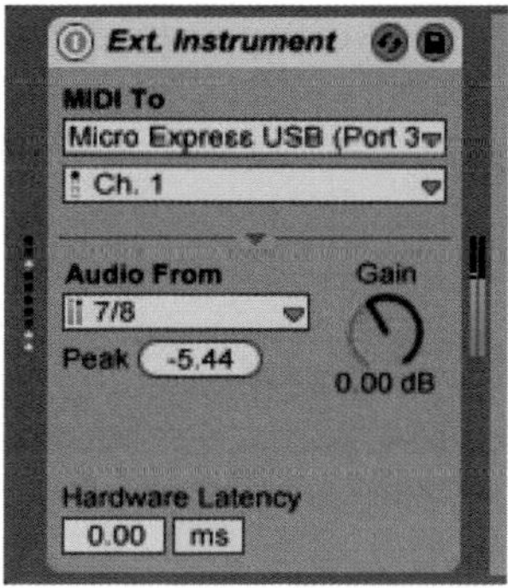

Figure 15.56 External Instrument routed to the outside world and back.

You should now hear your external MIDI instrument through Live. If so, then you are good to go. If you don't, make sure the Monitor section is set to "Auto" and that the MIDI input channel indicator is lighting up when you play. If you still aren't hearing anything, check all your connections, inputs and outputs for MIDI, and audio. Finally, consult your MIDI instrument's manual. Other common issues can be related to the instruments Local Control, channel settings, volume fader, etc.

15.8.2 Rendering external MIDI

Live is so intuitive that you can spend more time thinking about music than computer technologies. This is especially evident with rendering MIDI to audio, whether an external MIDI instruments or Live-based MIDI devices. Most of the time you will render audio "offline," meaning that the process happens faster than the actual time it takes to play through the song and you won't hear playback during the rendering process. With external audio, the process cannot be done offline because the audio must be converted from analog to digital since it lives inside the External Instrument. This means the external device must physically play the notes for it to be rendered. The good thing is that Live handles this for you. When you render, Live will automatically determine whether the content contains internal or external audio. If it does contain an external audio source, it will render in real time.

Whenever you want to render your external MIDI source (or native), you have a few choices. In 9.3, we talked about Freezing and converting MIDI to audio, so that is one option. One of our favorites! Your second option is to do as we said and render your MIDI (Export Audio/Video). The third choice is to "print" your MIDI as audio to a track by recording the external device's audio output onto a new audio track in Live. You will have to decide what you prefer. Launch to 9.5 for more about exporting and printing audio and MIDI.

15.9 Working with devices

There are many exciting ways to create, produce, and perform with third-party and native Ableton Devices. With the popularity and convenience of software-based samplers and instruments, it has become ever so important to understand how to work with "multi-instruments," also called "multitimbral," or "multis" for short, in your productions. What we are talking about is harnessing the power of a virtual instrument plug-in that can handle multiple sample-based virtual instruments encapsulated into one interface. This is commonly referred to as a "sample player." Instrument devices such as these allow for various instruments to be addressed over one or multiple unique MIDI channels so that each instrument can be triggered separately or all together in the case of one channel. Let's take a moment to look at a plug-In device as a multi-instrument in detail.

15.9.1 Produce: multi-instrument plug-in

For this example, we'll use EzDrummer (plug-in instrument) by Toontrack. EzDrummer allows you to selectively assign its built-in mixer outputs to dedicated channels on a host sequencer or DAW; in this case, Live's MIDI tracks.

Here's how to do it:

1. Select EzDrummer (or your favorite multi-instrument plug-in) from the Plug-in Browser and insert it on a MIDI track in Live.
2. Turn monitor from "Auto" to "In."
3. Create eight new MIDI tracks. You'll have nine MIDI tracks in all.
4. Using the multiparameter, select function (shift + click), highlight tracks 2–9 and then select (in our case here) "1-EzDrummer" in the MIDI To input chooser on Track 2.

Figure 15.57 Select "1-EzDrummer" in the MIDI To input chooser on Track 2.

5. Being that you have multiselected your MIDI tracks, tracks 2–9 will be automatically set to 1-EzDrummer input across all the selected tracks at once. A great time saver!
6. Set the MIDI To output chooser for tracks 2–9 to each discrete EzDrummer MIDI channel. You'll find that with your specific instrument plug-in that this chooser window's menu choices will match up accordingly. If not, check with the manufacturer and make sure that the instrument has the actual multi-instrument output capabilities.

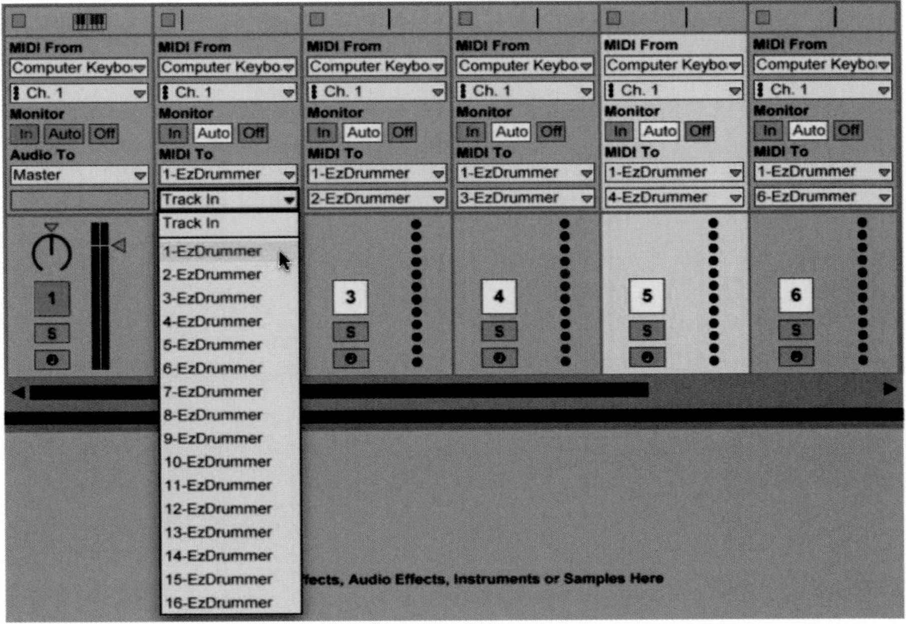

Figure 15.58 Set the MIDI To output chooser for tracks 2–9 to each discrete EzDrummer MIDI channel.

7. Now open your instrument plug-in device on Track 1 and set the outputs within the user interface to match your Live MIDI track assignments on track's 2–9 starting with 1-EzDrummer as in this example.
8. Arm Track 2 and record some MIDI notes to create a MIDI clip. Repeat this across the remaining MIDI tracks until you have a scene row of MIDI clips. You'll soon see and hear your performance as a multi-instrument performance.

Figure 15.59 Make the appropriate MIDI routing assignments to connect your plug-in and MIDI tracks.

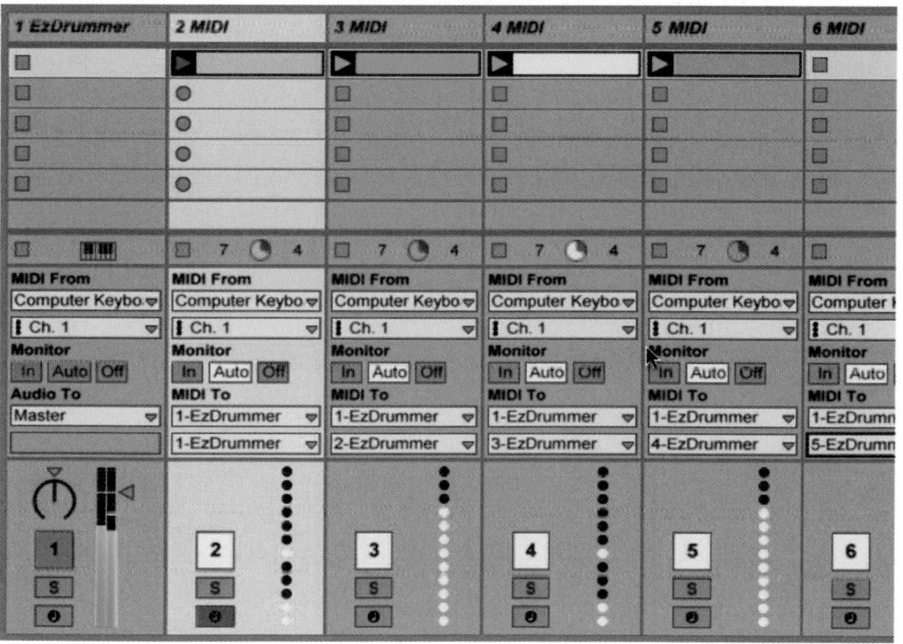

Figure 15.60 Multi-instrument in action.

9. Remember that Track 1 is serving two purposes. It is acting as the host track for your dedicated plug-in instrument device and also as a dedicated monitoring MIDI track. There is no need to record on this track for any reason. With the proper assignments as shown here, you have multiple tracks to record to discretely and a very flexible way to take your production to the next level in Live 8.

In this chapter

16.1 Introduction to Racks 345
16.2 Rack interface and layout 346
16.3 Drum Racks 350
16.4 Creating Device Racks 353
16.5 Racks in Session View 365
16.6 Working with Racks 367

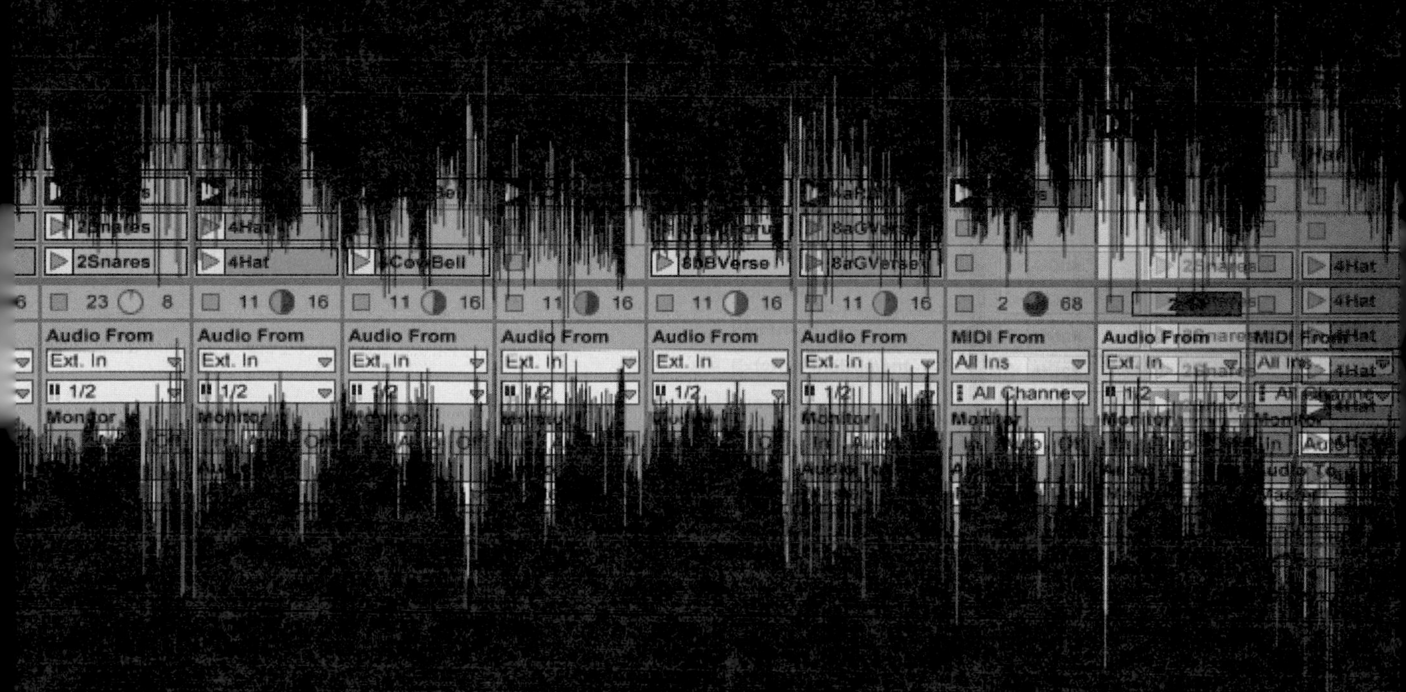

SCENE 16

Device Racks

For one, the Beat Repeat plug-in and Drum Racks have made some of my songwriting really stand out. I'm no longer trapped for hours moving little dots to create a flowing groove. Live 8 continues to be my go to writing tool as nothing else allows for such spontaneous creation. From hooky drum patterns to arrangement ideas I am always surprised and excited by what I'm able to create with this program. If I can think it, Live 8 can do it!

Zac Baird, keyboardist with the band KoRn

16.1 Introduction to Racks

Up until now we have only looked at basic instrument devices, e.g., Impulse, Simpler, and various MIDI/Audio effects. These have all been very exciting and no doubt have sparked your creative juices, but they are all examples of fairly traditional Live devices (often enhanced with a few effect devices in a simple chain), that can be created without the assistance of a Rack. If you thought our early examples were fun, then you are in for a treat. Instrument Racks, Drum Racks, and Effects Racks take Live devices to the next level – a good reason to graduate to Suite 8 if you haven't already!

A Device Rack provides you with the ability to build complex audio or instrument devices by combining effects and instruments with flexible hands-on parametric controls into one device chain. This pools the power of all of Live 8's and Suite 8's devices to create unique custom effects processors, multilayered synths, and intricate, effectual, and interactive performance-ready instruments. This all sounds great, but you're probably wondering what this all really means, and how you can take advantage of its theoretical power.

16.1.1 What is a Device Rack?

Think of a Device Rack as a group of individual device chains that function in parallel (individual rows). Each independent device chain is self-contained and processes its own signal as it flows from left to right, not interacting with the signal of any other device chains. At the end of the rack, the chain's output signal is summed with the other chains in the rack to the track's output. Simply put, a rack is a device made up of a combination of multiple device chains, although a rack can consist of only one device chain as commonly found in Live 8/Suite 8 presets. Now, before we move on let's point out that Drum Racks function slightly different than the rest of the Device Racks. Each chain has an input/output section that allows it to be routed to a designated MIDI note that will trigger it, whereas with all other Racks, all chains receive the same input signal except those that are affected by a specific set of rules established within the Rack. Launch to 16.4.3 for more on Drum Racks.

16.2 Rack interface and layout

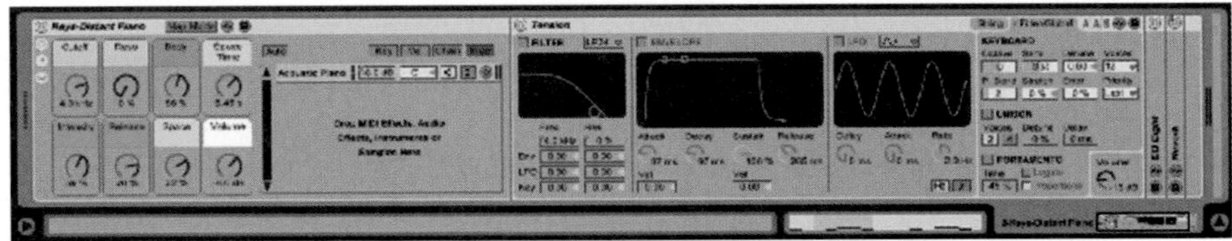

Figure 16.1 Device Rack: Macros, Chain List, Device.

On the surface, all racks are made up of at least three main areas: Macro Controls, Chain List, and Device interfaces. The buttons show up on the left of the device in black and are highlighted in yellow. Drum Racks include more areas, most notably Pad View, which we'll get to a little later. For now, let us take you on a guided tour of the Rack's interface and layout. At the top of the device Rack display are the usual buttons: Device Activator, Hot-Swap, and Save Preset. Next to Hot-Swap you will see a button called Map Mode. This is used to map macro controls to parameters in the device Rack to control various assignable parameters, as described in the coming section. On the left-hand side is the View Column containing the three Rack specific selectors mentioned earlier: Show/Hide Macro Controls, Show/Hide Chain List, and Show/Hide Devices. Use these buttons to unfold and manage your device's view in the track display.

16.2.1 Macros

To familiarize you with Racks, we are using an Instrument Rack preset that comes with Live 8 and Suite 8. The first thing that sets a Device Rack apart from a nonrack Device is the set of Macro Controls, when shown in view. Their visibility depends on whether they were in view when a preset was saved. Macros are designed for controlling the internal parameters within a Device Rack.

To show Macros, click the Show/Hide Macro Controls button just below the Device Activator.

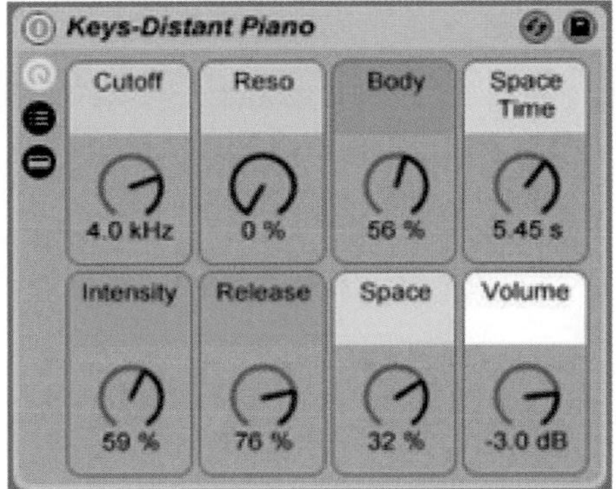

Figure 16.2 Macros for controlling internal parameters within a Device Rack.

These eight controller knobs introduce great power and control over the contents of a Device Rack. Once your instrument or effect has been dialed in, you have the option to hide all of your devices and only use the Macro Controls if you wish. The key is customization. Set your own colors for each knob and assign their control however you like. In the same way that Macros control the device Rack's internal parameters, they too can be assigned to MIDI/Key maps. Now think of the possibilities on and off stage with unlimited controls over your instruments and effects! Refer to ▷Scene 12 for in-depth mapping. We'll come back to Macros in a moment.

16.2.2 Chain list

Racks handle signal flow in a similar manner as a standard instrument or effect device, except that they employ a more complex internal routing system that handles which chains can receive signals. With that in mind, Racks can hold multiple chains that operate in parallel with each other, something that standard devices do not do.

As mentioned earlier, each chain in a Rack (except for Drum Racks) generally receives the same incoming MIDI or audio input signal, although this is determined by the setup of the Zone and Chain Select Editors. More on these features in a minute. This becomes obvious when you look at the interface and see that the Chain List houses each device chain in a horizontal layout. Not only does this represent each device chain in the Rack, but it also serves as an internal mixer and Rack view organizer for navigating each device's parameters and settings. Each device chain in the list has a title, overall volume, pan slider, Chain Activator button, Solo button, and Hot-Swap button – very powerful tools in their own right. When used in conjunction with its Macro controls you'll open up a whole new world of options and real-time control.

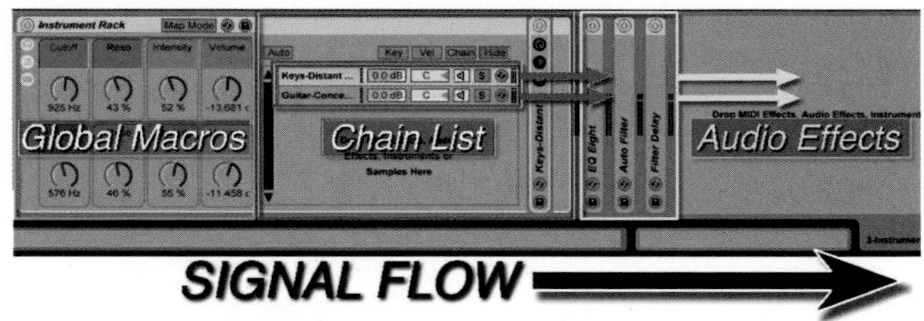

Figure 16.3 Instrument Rack: signal flows left to right. Chains operate in parallel to one another, receiving and passing signal simultaneously through the rack to the outputs.

Below the Chain List is the Drop Area where you can add effects, instruments, samples, and more Racks. Don't forget that Audio Units and VSTs can be dropped here as well. At the top of the Chain List there are several buttons. Auto sets the Chain List to automatically select (highlight) a chain when it is receiving input. This way you can see when chains are triggered, allowing you to watch what is happening with your current device chain while your Set is playing.

The Key button brings the Key Zone Editor into view. This is where you can specify the Key Zones for each chain. Key zones are the areas or ranges of the MIDI keyboard from where a chain is assigned to be triggered. If a note is played outside of the assigned Key Zone the chain will not be triggered and therefore will not play back. This is often used to split the keyboard note range so that two different instruments can be triggered via their own note range.

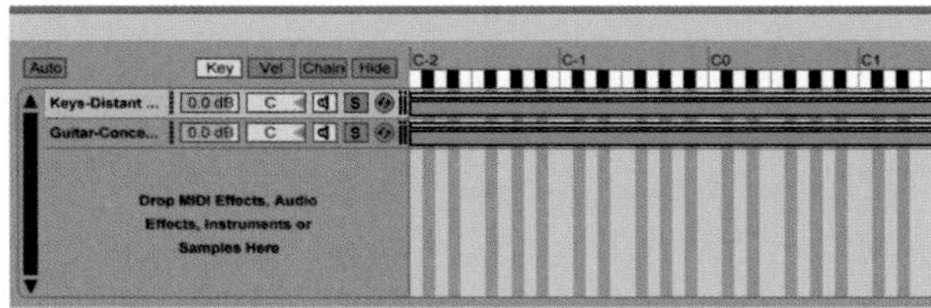

Figure 16.4 Key Zone Editor.

"Vel" button shows the Velocity Zone Editor, which represents velocity values spanning from left to right. From this editor you can assign a chain to particular velocity values that when played will trigger that chain if the incoming MIDI note velocity values fall within the assigned range. For example, a chain assigned to velocity values 1–40 will be triggered when notes that fall in that value range are played. Notes that fall outside that range will not trigger the chain, resulting in the chain not playing back for that note. Velocity zones are often used for triggering various timbral layers of an instrument so that hard velocities trigger an aggressive or excited sound and softer velocities a milder, somber tone. You'll find this is not the only use for these zones.

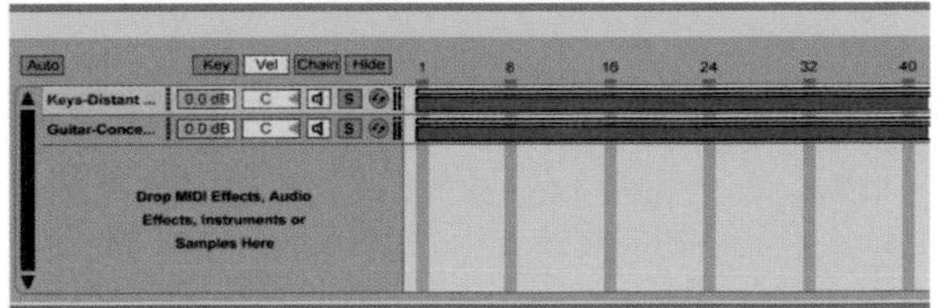

Figure 16.5 Velocity Zone Editor.

Audio Effects Racks do not have "Key" or "Vel" buttons, instead they have only a Chain button to display the Chain Select Editor. The Chain Select Zone determines what chains can be triggered. As long as a chain falls within the orange Chain Selector it will be triggered. This functionality is ideal for establishing the ability to choose between different chains or layers of chains to be played back when they fall within the assigned zone ranges. Because each chain can be assigned to a specific chain select zone or to overlapping select zones, you can use them with Live's MIDI Mapping Functionality to vary your Rack Device in real time as an all-in-one multi preset instrument or effects Device. To change the position of the selected chain(s) drag the Chain Selector located in the Chain Select Ruler at the top of the Chain Select Editor.

Note that when any zones overlap (here or with the other Zone Editors), consider setting a Fade Range so transitions between zones can be smooth and less obvious to the ear. The Fade Range for each zone is located directly above the zone and can be dragged from either end of the zone.

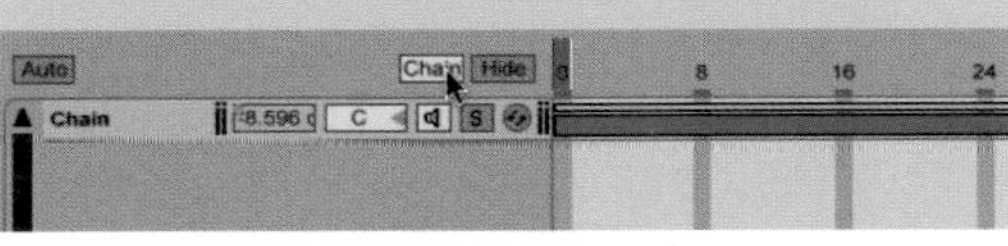

Figure 16.6 Chain Select Editor.

Click the Hide button to close out Key, Velocity, or Chain Select Zone Editors from view.

16.2.3 Macro mapping

To set the colors of each knob, simply right+click or ctrl+click on a knob and select a color from the palette. There are two ways to assign a Macro to a control parameter and vice versa. You will first need to enter Macro Map Mode by clicking the Map Mode button at the top of the device. This is only visible when the Chain List is in view (button below the Macro Controls button). Click

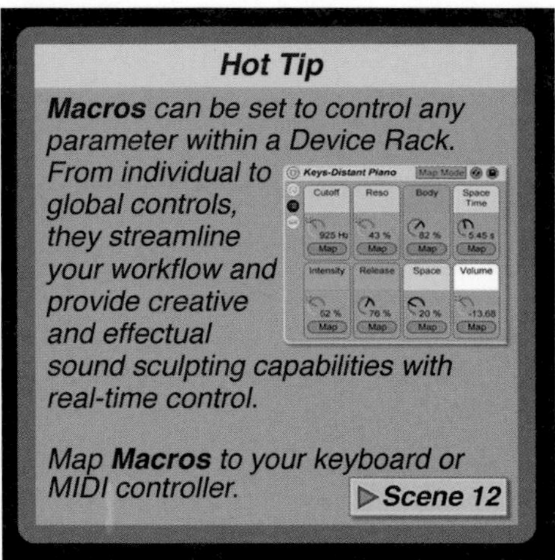

the Show/Hide Chain List, which will open up the Rack in Track View, then click Map Mode. This opens up the Macro Mappings Browser in the left side of the main window of your Set. There you will see a list all of the current mappings for the selected device Rack as well as additional control parameters. In the Track View all of the mappable parameters and controls will now be highlighted green. To assign control, select any parameter then click Map on the Macro of choice.

Hot Tip
Macros can be set to control any parameter within a Device Rack. From individual to global controls, they streamline your workflow and provide creative and effectual sound sculpting capabilities with real-time control.
Map Macros to your keyboard or MIDI controller.
Scene 12

16.3 Drum Racks

Although a Drum Rack is a Device Rack, it gets special attention because it differs in its user interface, controls, and fundamental operation. The most obvious difference, from an interface standpoint, is the presence of Pad View, located next to the Macros.

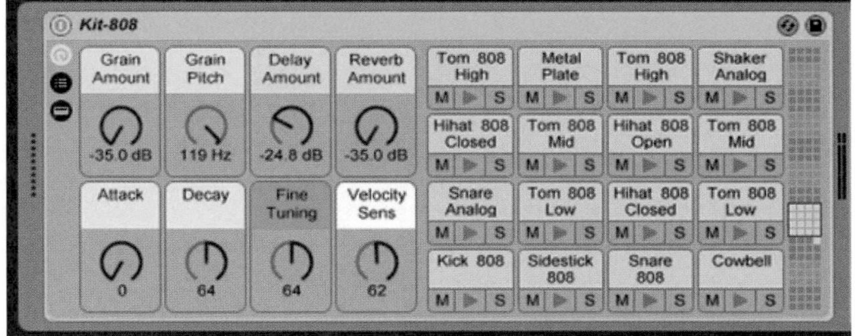

Figure 16.7 Drum Rack: Macros and Pads.

16.3.1 Pad View

Pad View is used to interface with samples and various devices. Pads are triggered like a drum pad. Each pad is assigned a unique MIDI note (1–128) ranging from C-2–G8. This is exactly how Live 8 handles MIDI sliced audio clips created from the Slice to New MIDI command. Refer to Clip ▶ 14.3.2. Drag your favorite samples, effects, instrument, Ableton Preset, and even audio loops from the Live Browser onto pads. Ableton Preset Folders are unique to Drum Racks as opposed to traditional presets. Such folders contains multiple presets intended to be loaded simultaneously onto multiple Drum Rack Pads, and they can also be selected and loaded individually to a single pad as one component of the preset. If you like, you can unfold them directly in the Browser first to determine which contained preset component you want to use. On a more basic level, Pad View can also be used to trigger, mute/solo, and Hot-Swap a pad's associated content. Understand that a device in this situation can be a synth or sample-based instrument so long as it creates sound, i.e., Analog, Operator, Collision, Simpler, Sampler, etc.

Just to the right of the pads section is Pad Overview. It shows the pads that are currently in view, and allows you to navigate to additional sets of pads, 16 at a time, 128 in all. Pad Overview is a vertical display beginning with the lowest Pads/MIDI notes at the bottom ascending to the highest on top. Drag the View Selector (outline box on the right over the Pad View) or use keyboard arrow keys to navigate the view in increments of an octave.

Looking inside the Drum Rack you will once again find a Chain List along with its mixer section that has volume, pan, activator, solo, and Hot-Swap.

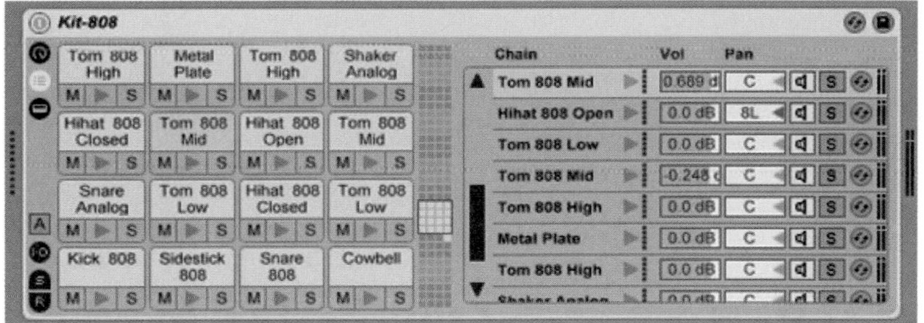

Figure 16.8 Drum Rack Chain List.

Everything is here within a Rack as expected, except that it is missing the three editors for Key Zone, Velocity Zone, and Chain Select. Instead, the Auto Selector has been moved to the View Column along with three additional View Selectors on the bottom left of the pads:

- I-O – Show/Hide Input/Output Section
- S – Show/Hide Sends
- R – Show/Hide Returns
- A – Activate Auto Select

Auto Select, as mentioned, sets the Chain List to automatically select the chain currently processing MIDI input. You will see the selections change during playback. The I-O selector displays the MIDI Input and Output section for each chain.

The Send Selector displays the chain's Send Level Controls and the *Return Selector* displays the Racks' return chain(s). Sends and Returns handle internal audio effects routing within the Rack, something that is unique to Drum Racks. The Returns Chain List appears directly below the Device Chain List. This is where audio effects will be added (dropped) to the Drum Rack. There can be up to a total of six return chains in all. Sends will appear in the Chain List mixer section next to volume when an audio effect has been added to the Return Chain List (creating a return chain). The Chain Send Level slider determines how much signal is sent to another return chain in the list as indicated above the slider. Also included in the Return Chain List is a mixer section and an Audio To chooser for output routing.

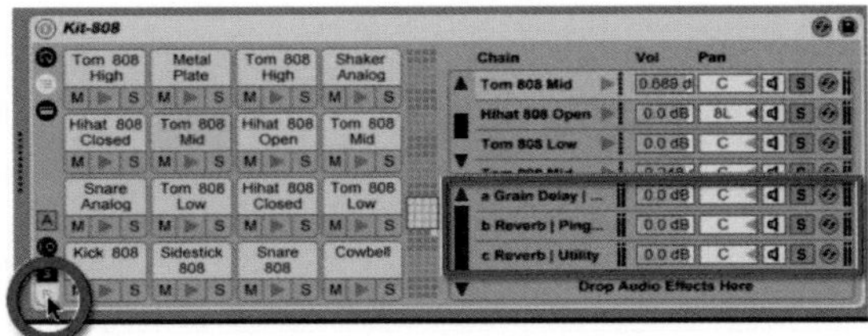

Figure 16.9 Return Chains.

As you can see, Drum Racks offer more advanced signal routing than the other device chains, whether it be MIDI input/output or internal effects.

16.3.2 Routing

Moving forward with this concept, each Drum Rack Pad responds to a unique MIDI note. In turn, each "parallel" chain is generally triggered by an individual MIDI input signal, although it can also be set to receive input from multiple notes if desired. Therefore, to keep things fairly simple, we'll assume each chain within a Drum Rack's Chain List has been assigned to respond to a unique MIDI note. Such routing is determined by Receive and Play assignments established in the Chain List's Input/Output section. As you can see, Drum Racks are fundamentally different in how their chains process MIDI input.

Time for a rundown!

To fully wrap your head around the Drum Rack routing concept, remember that signal flows left to right in a Rack. That being said, each drum Pad is assigned to its own MIDI note, e.g., C1, C#1, D..., etc. Each Chain is triggered by a Pad, meaning when you play a MIDI note, its related Pad will trigger a chain in the Chain List that is assigned to respond to it. This assignment is made via the *Receive* chooser. Whichever MIDI note is assigned here is the note that will trigger the chain, unless set to receive "All." For example, if a chain's *Receive* is set to MIDI note C1, then C1 will trigger the chain when it is played. If a chain's

Receive is set to "All" than all incoming (played) MIDI notes trigger it. The Play chooser located next to Receive determines what note is sent into the device(s) within a chain when the receive note is played. To recap, Receive assigns the MIDI note that will trigger a chain and Play determines the MIDI note that is output to the device(s) in a chain.

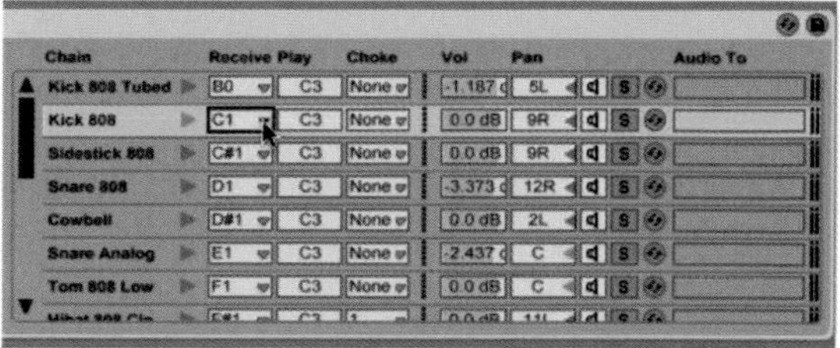

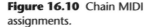

Figure 16.10 Chain MIDI assignments.

Next to the Play is Choke Group chooser. Choke Groups are designed for the purpose of having one chain silence another chain when both are played in succession. Chains assigned to the same Choke Group will function in this way. There are 16 choke groups. This feature is designed to emulate real-word acoustic drum kit hi hats, where closed hats silence open hats.

16.4 Creating Device Racks

Fortunately, you don't have to start creating your own Drum Racks right away. There are tons of presets already designed for you to use. Just browse through the Instrument, Drum, or Effects Racks in the Device Browser. To identify Rack presets look for their double-pane icon in the Device Browser next to preset names. All other device presets will have a single pane icon. Most of the time this will be obvious, but some device presets are in fact Racks even though they are not in a folder labeled as such. Once you have exhausted yourself with the presets, it is time to create your own. Not only will this bring great flexibility to your music, it will also help hammer home the concepts we have been discussing. Let's take a look at each of the Rack types, one by one. There are a few ways to create device Racks. You can start off with a new empty Device Rack (individual audio effect or instrument), or you can build a device chain, then group it into a Rack after the fact. We'll start with the latter. For this demonstration you will want to open up a new Live Set.

16.4.1 Instrument Racks

We are going to create an instrument from scratch, sculpt the timbre, add effects, group it into a Rack, add a second parallel device chain to the Rack, and finish by mapping the Macros to the chains. Sounds fun! You will need to download the Impulse and Solid Sounds by *Loopmasters* Live Packs if needed from Ableton's Website for the following exercises (http://www.ableton.com/downloads). If you wish to use different instrument presets, then try your best to use ones that are similar to those we describe in the exercise.

1. Go to Live Device Browser > Instruments > Simpler > Components > *Spectral* and double-click or drag "Digital Spike" into your Set. You should now see a new track and Simpler/Digital Spike in Track View.

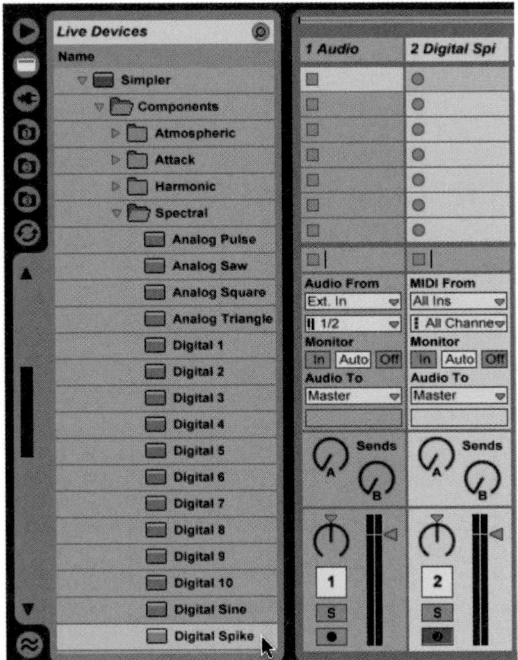

Figure 16.11 Build an Instrument Rack from Scratch. Start with a standard Device Preset.

2. Go to Live Device Browser > Audio Effects > *EQ Eight* and double-click or drag "E Bass Vintage" to your Simpler Track View.
3. Go to Live Device Browser > Audio Effects > *Saturator* and double-click or drag "A Bit Warmer" to your Simpler Track View after the EQ Eight in the chain.
4. Go to Live Device Browser > Audio Effects > *Compressor* and double-click or drag "1976" to your Simpler Track View after the Saturator in the chain.
5. Go to Live Device Browser > Audio Effects > *Ping Pong Delay* and double-click or drag the default *Ping Pong Delay* to your Simpler in Track View after the compressor. Be sure not to drag a preset.

6. Go to Live Device Browser>MIDI Effects>*Arpeggiator* and drag "Arpeggiator" to your Simpler Track View and insert before Digital Spike in the chain.
7. Set the Arpeggiator "Style" to Up, "Rate" to 1/16, and "Steps" to 1.

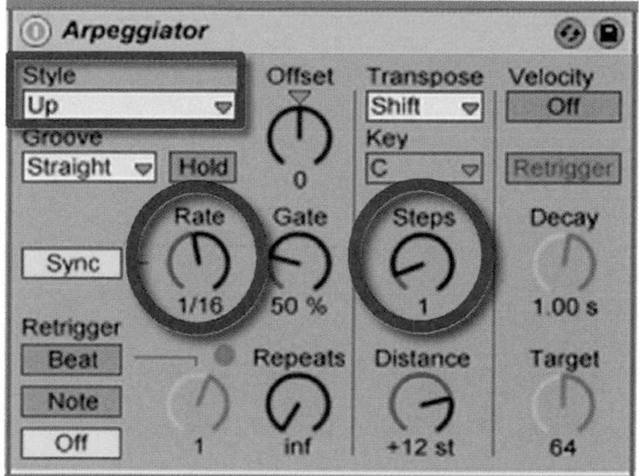

Figure 16.12 Arpeggiator "Style" to Up, "Rate" to 1/16, and "Steps" to 1.

Test drive your instrument. Use the computer MIDI keyboard if you don't have a controller and press a few keys. Now that we have an instrument we can play with, let's group it into a Rack.

1. Select all of the devices in the chain we just created. Click on any device title bar in the Track View to select it and then select all with [⌘+A] Mac/[Ctrl+A] PC.
2. Right click or [⌘+click] Mac/[Ctrl+click] PC to bring up the device contextual menu. Select Group to create a Rack. You can skip the contextual menu if you wish and press [⌘+G] Mac/[Ctrl+G] PC instead.

Figure 16.13 Select Group to create a Rack.

There you have it, your first Instrument Rack. Now lets add a parallel device to the chain, just for fun.

3. Click the Show/Hide Chain List selector to bring the chain list into view.

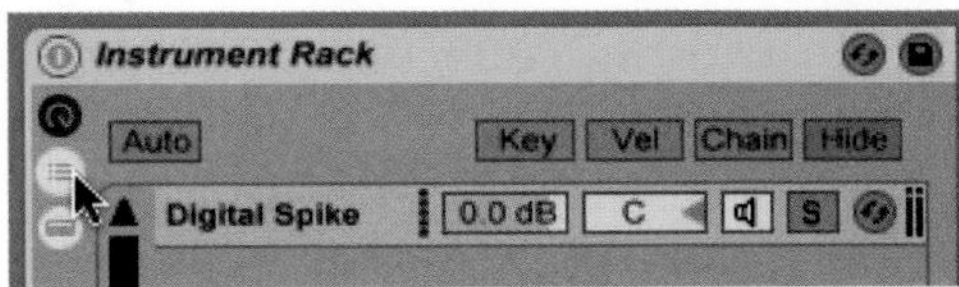

Figure 16.14 Show/Hide Chain List.

4. Go to Live Device Browser > Instruments > Simpler > Loopmasters > *Bass* and drag "Abandon" into the Rack's Drop Area. You should now see a new chain called Abandon below Digital Spike. You have just loaded another Rack inside your Rack. This is a powerful way of building complex instruments.

Figure 16.15 Nesting Racks inside Racks as Chains.

5. In the chain list mixer section, mix the two chains so they are nice and balanced. You will most likely need to lower the volume of the abandon bass.

Hot Tip *Group Rack Chains together to consolidate them into their own "sub" Device Rack for more discrete control. Click and highlight all the chains in your Rack Device then select Group from the contextual menu.*

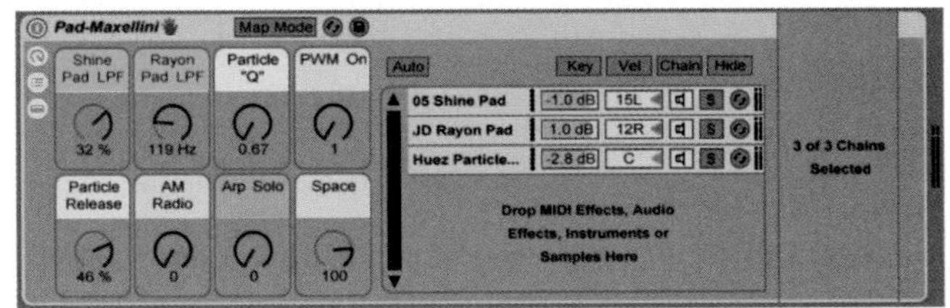

Figure 16.16 Group Rack Chains.

You'll quickly see that you have a nested Rack Device within your main Rack Device. This has several advantages depending on your needs, one being that you can create very complex multilayer instruments that are contained in one Rack.

Now let's map our Macros to a few parameters within the chains.

1. Click on the Show/Hide Macros selector to bring the Macros into view.
2. Click on Map Mode to make assignments.

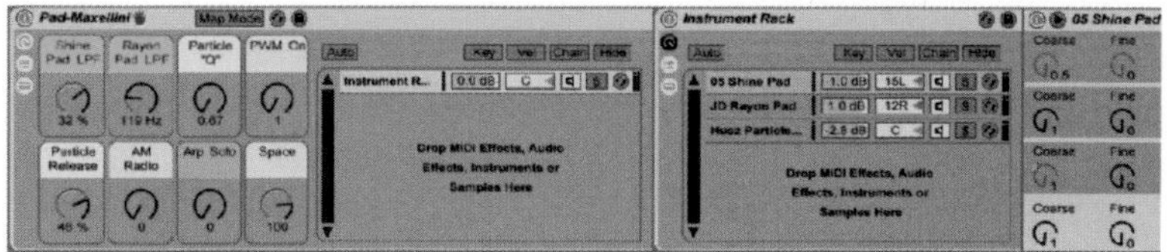

Figure 16.17 Nested Rack Devices become a single chain within the main Rack Device.

3. Click on Digital Spike Chain Volume, then click Map on Macro 1 or right click or ctrl+click and select Map to option. Do the same for Abandon and click on Macro 2.

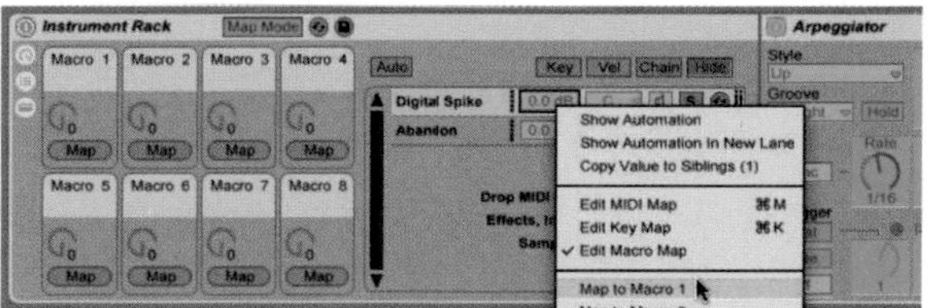

Figure 16.18 Map Macros 1 and 2.

4. Map Digital Spike's Envelope Release to Macro 3.

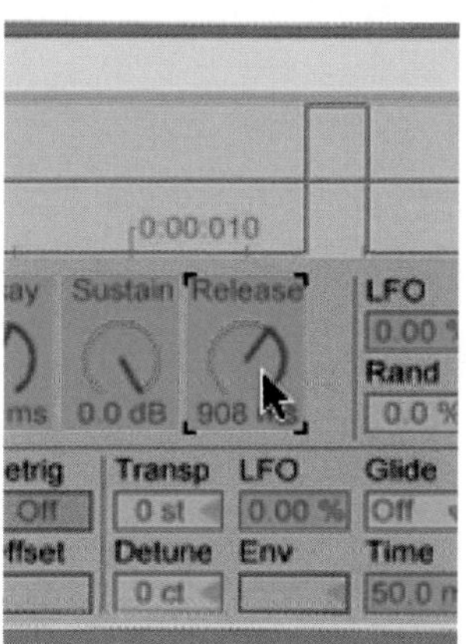

Figure 16.19 Envelope Release to Macro.

5. Map Digital Spike's Ping Pong Delay Dry/Wet knob to the Instrument Rack Macro 4, Spike's Filter Freq to Macro 5, and Res to Macro 6.
6. Map "Abandon's" Filter Freq to the Instrument Rack, Macro 7, and Res to Macro 8. Because they are already mapped to Abandon's own Macro Controls, you will need to assign them from the Instrument Rack Macro to

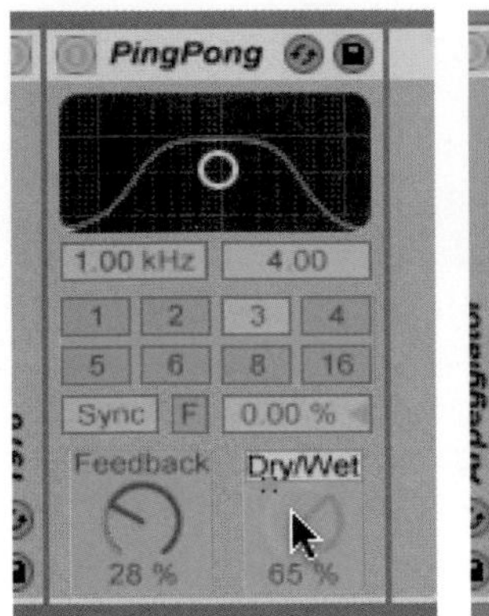

Figure 16.20 Ping Pong Dry/Wet to Macro 4.

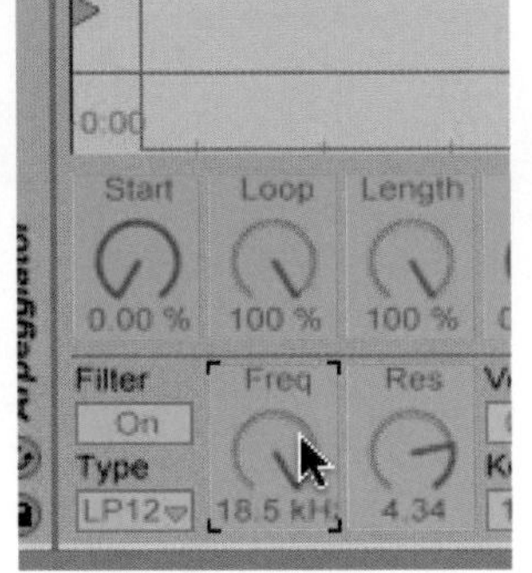

Figure 16.21 Filter Freq to Macro 5.

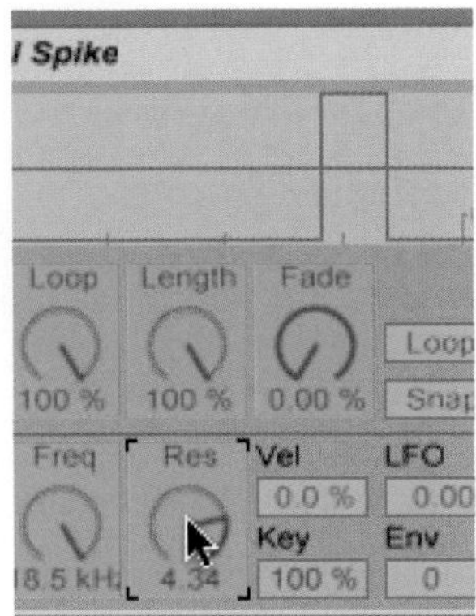

Figure 16.22 Res to Macro 6.

Abandon's Macro as opposed to assigning them via the actual Filter and Res knobs on the device interface.

Figure 16.23 Filter Freq to Macro 7 and Res to Macro 8 via Macro to Macro mapping.

Figure 16.24 Custom Macro assignments.

7. You will usually have to reset each parameter setting using the Macro controls after you have mapped them, because Live automatically sets new Macro assignments to the value of "0" when assigned to a Macro. Feel free to mix up the parameter settings until you get something you like, or just match ours below. Give your instrument a whirl and see how it sounds. Don't forget to balance your track output if you start overloading the Master track. Once you are done, take a look at some pre-assembled Instrument Racks on

our Website. Download "CPP_InstRack-Pad-Lead-Arp Project" and check out some more examples.

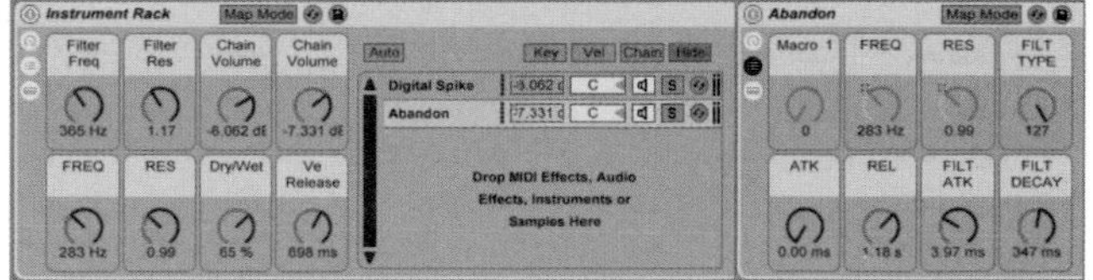

Figure 16.25 Instrument Rack, two chains, with custom Macros.

16.4.2 Audio Effects Rack

There are many ways to use Audio and MIDI effects Racks, so let's continue with the Instrument Rack we just created and see how we can use them. Keep in mind that MIDI effects handle MIDI information and audio effects handle audio signals. The following examples represent only a few possibilities. If you are unable locate the device presets demonstrated below, find your own suitable supplements.

Adding an Audio Effect Rack to a track device chain:

1. In Track View, hide the Chain List so you are just looking at the Macros.
2. Go to Live Device Browser > Audio Effects > Audio Effect Rack > *Bass* and drag or double-click Bass Chorus to the audio effect at the end of the chain (Drop Area at the back of the Instrument Rack separated by the Level Meters). This applies the Audio Effect Rack to the output of the entire Instrument Rack and forms a new part of the track's device chain.
3. Try it out, and then delete what you just created.

Figure 16.26 Instrument Rack and Audio Effects Rack making up a device chain.

Let's try another.

Adding an Audio Effect Rack to a chain within a Device Rack:

1. Select the Abandon from the chain list inside your Rack.

2. Go to Live Device Browser > Audio Effects > Audio Effect Rack > *Bass* and drag "Bass Chorus" to the back of the Abandon chain just outside of the Level Meters. This applies the Effect Rack only to this chain within the Rack. You will see a yellow highlighted vertical line when you are dragging over a drop area. If you were to select the other chain you would see that the Audio Effect Rack is not a part of any other chain.
3. Try it out then delete what you just created.

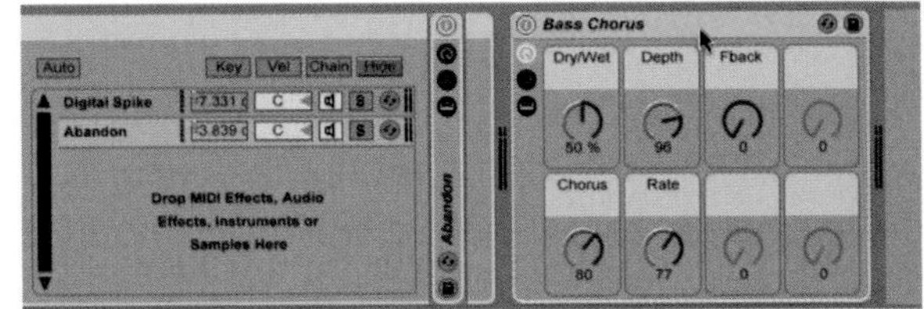

Figure 16.27 Audio Effects Rack applied to only the "Abandon Rack" chain.

Let's try another.

Adding an MIDI Effect Rack as a track device chain:

1. In Track View, hide the Chain List so you are just looking at the Macros.
2. Go to Live Device Browser > MIDI Effects > MIDI Effect Rack > *Arps* and drag or double-click "Three Octave Converge" in front of the Instrument Rack. This applies the Effect Rack to the entire Instrument Rack forming a new device in the chain.
3. Open the Effect Rack Chain List and change the Arpeggiator "Rate" to 1/16.
4. Try it out. Now delete what you just created.

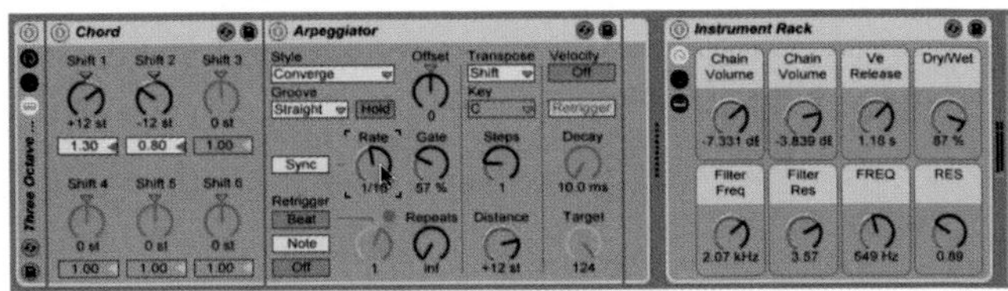

Figure 16.28 MIDI Effects Rack as a part of the track's device chain.

Let's try another.

Adding a MIDI Effect Rack to a chain within a device Rack:

1. Select the Abandon from the Chain List inside your Rack.

2. Go to Live Device Browser > MIDI Effects > MIDI Effect Rack > *Arps* and drag Three Octave Converge and drop on the Abandon Chain List Title. This applies the Effect Rack only to this chain within the Rack.
3. Show Arpeggiator Effect Rack chain list, click on the chain, and then change the arpeggiator rate to 1/16.
4. Try it out then delete what you just created and let's try another.

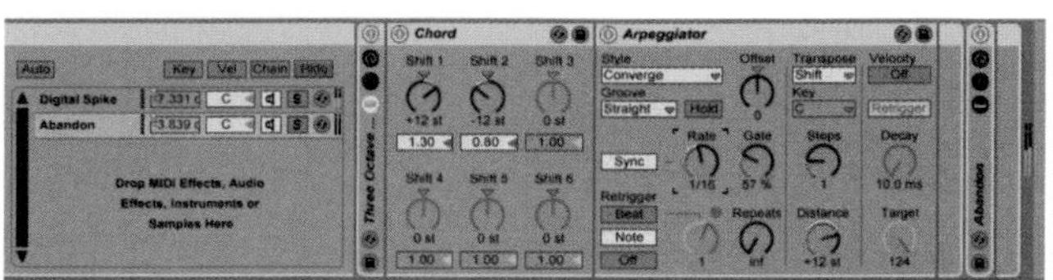

Figure 16.29 MIDI Effects Rack as a part of Device Rack chain – Abandon chain.

16.4.3 Drum Racks

Creating Drum Racks can get as complicated as you want, but we'll try to keep it simple. For this example we'll use individual drum kit samples to create our own Drum Rack. Feel free to use your own kit samples or the samples from the Live 8 Library. On our Website, we have also provided some additional samples and clips to accompany this Exercise. Download the samples **CPP_Drums**.

1. Create a new Set then go to Live Device Browser > *Instruments* > and double-click or drag Drum Rack into your Set. Be sure not to select a preset. You should now see a new track and an empty Drum Rack in Track View.
2. Bring the Drum Rack's Pads into view, and then go to File Browser> Library > Samples > Waveform > *Drums* and select a kick drum and drop it on Pad C3.

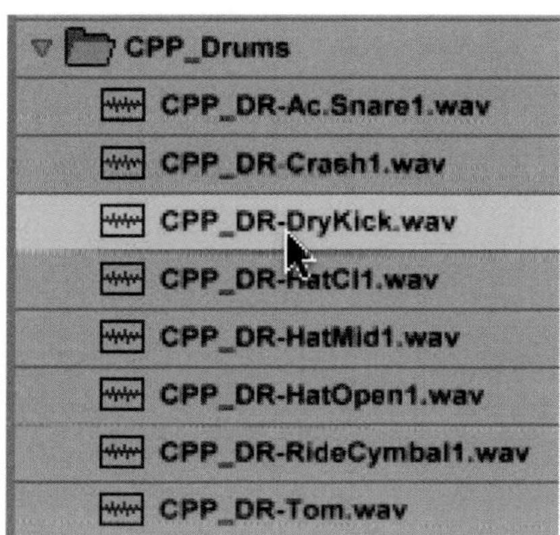

Figure 16.30 Select individual samples to load in a Drum Rack Pad.

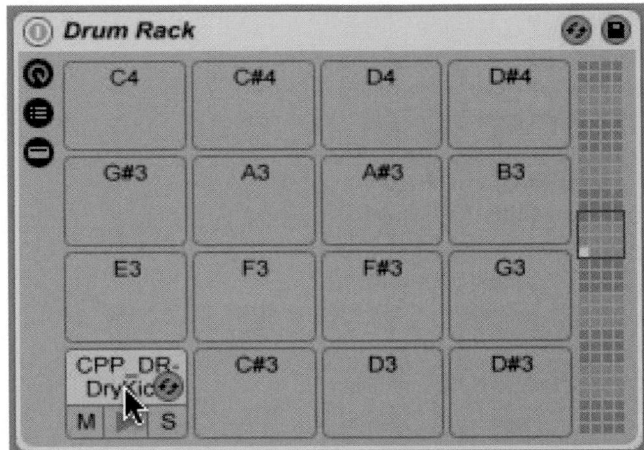

Figure 16.31 Drag + drop a sample on a Drum Rack Pad.

3. Choose a snare sample and drop it on D3. Live will automatically treat each sample as a chain and map it to the pad where you've dropped the sample.
4. Choose a closed hat sample and drop it on B3.
5. Choose a tom sample and drop it on F3.
6. Now you need a MIDI clip. Go to File Browser > Library > Clips > Acoustic > Backbeat-120 and select and drag "Backbeat-120-Sidestick" to the first Clip Slot of your Drum Rack. You can also download generic MIDI files for your Drum Rack from our Website called DR-…Groove.mid. Please note that for some of these MIDI files, you may need to reassign your samples to different Pads to match the MIDI note assignments in the file.

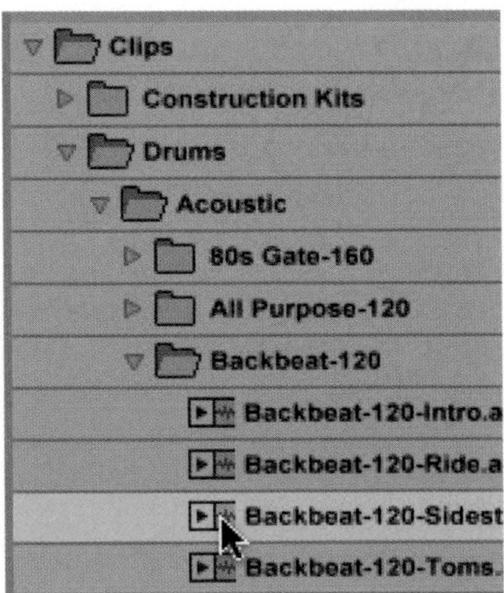

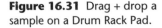

Figure 16.32 Drag in a MIDI clip to trigger your Drum Rack.

7. Activate the Clip Launch button and listen to the kit. Notice how the Chain List and Pad behave during playback.

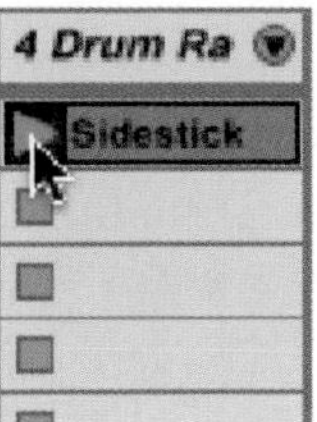

Figure 16.33 Launch your Drum Rack Clip.

8. While the clip is running, open up the Chain List and select the snare chain. From the Simpler instrument associated with the snare sample, adjust the volume "Release" knob to extend the release so the snare doesn't sound choppy.

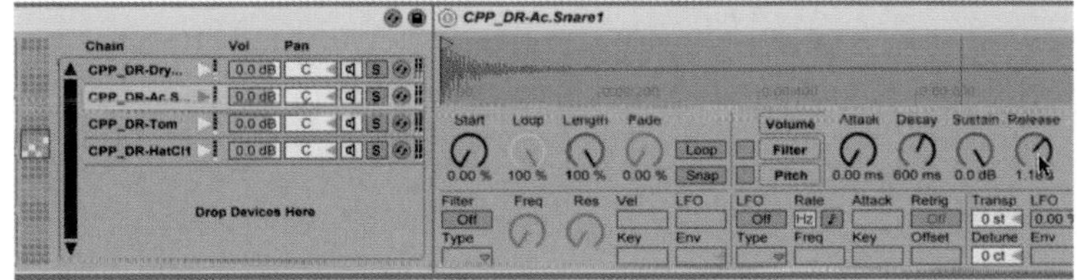

Figure 16.34 Adjust individual chain parameters to your liking.

Continue to tweak as you like. Your final product should look similar to our example.

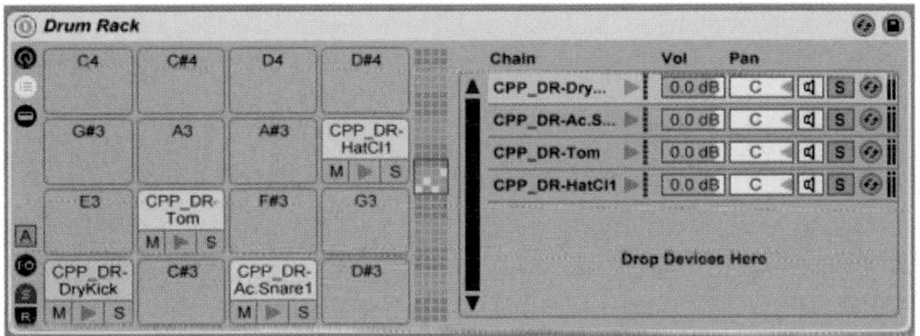

Figure 16.35 Custom Drum Rack.

That's all you need to get started, but if you want to take this further, you could easily add audio effects or an Effect Rack to any chain or to the entire Rack just like we did with the instrument and effects. You can even drag effects onto Pads in Pad view. For some preassembled Drum Racks, download **CPP_DrumRack-Ac-Elec-Kits Project** from our Website.

16.4.4 Drum Racks and Audio Effects

By integrating exclusive sends and returns for audio effects, you can further develop the capabilities of a Drum Rack Device. Let's look at how we can achieve this.

1. Load up a Drum Rack preset or create your new Drum Rack Device on a MIDI track and add your own samples (sounds). For this example we are going to use the "Kit-606 Classic" preset from Suite 8. Remember you can simply double-click on the Drum Rack Instrument Device in the Device Browser to load it into a new MIDI track without dragging and dropping.
2. Click on the Rack's Show/Hide Chain List Selector to open the Chain List. You'll notice the vertical row of four buttons on the extreme left-hand side of the Rack and are now visible displaying: **A**, Auto; **I-O**, In/Out; **S**, Send; and **R**, Return.
3. Click on the "R" (Show/Hide Return Selector). The Return Chain List labeled "Drop Audio Effects Here" will open immediately beneath the standard Chain List.
4. Drag a Reverb audio effect (or any audio effect) directly into the Drop Area of the newly opened Return Chain List. It will now be displayed in the Return Chain list with the letter "a" in front of the effect's title.
5. Click on the "S" selector (Show/Hide Sends) on the far left of the device to open the Chain Send Level window for each chain. They reside directly beside the volume and pan controls in each device chain.
6. Double-click on your audio effect in the Return Chain window to bring it into view then change the Dry/Wet mix knob to 100%. You may need to scroll through the Track View Selector to bring the effect into View.
7. Play some Drum Rack pads or launch a MIDI clip pattern to create some sound from the Drum Rack.
8. On one of the active device chains increase the Send level to around "−5.0 db." You should begin to hear the reverb effect level increasing along with your Drum Rack mix. Try this with another chain's Send value. You can also adjust the Return Chain's level as much as 6.0 db with the volume slider located in the middle area of the return chain.
9. Rename or Hot-Swap the Reverb or selected audio effect device at any time if needed. You can also assign the Send value to a Macro Selector in the Macro section to the left of the Drum Rack Device.
10. Load another audio effect into the Return Chain Drop Area and instantly create another Return Chain just below the first one.

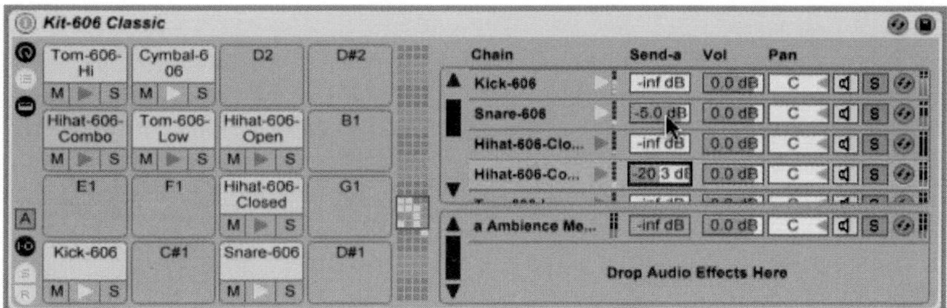

Figure 16.36 Return Chains.

You can see how your Drum Rack Device comes alive with this great feature of Sends and Returns with audio effects, not to mention the many creative possibilities it facilitates. Don't forget to save your Drum Rack as a new preset. You'll be able to recall it anytime with all your newly included audio effects plus all the sends and returns you created!

If you still desire more creative power, then Group individual chains within the Drum Rack to new Drum Racks, nesting them inside the current Drum Rack. This provides even more Macro control over each chain in the Rack. To do this, open up a contextual menu from any device Title Bar and select "Group To Drum Rack." As a matter of fact, this can also be done to any device in your set at any time if you want to convert it into a Drum Rack.

16.5 Racks in Session View

A Rack functions as a self-contained device, made up of a combination of instruments, effects, and even more Racks. It's a nesting process, where instruments, effects, and Racks are encapsulated in a container functioning as a single device, i.e., instrument or effect. As we already discussed, each parallel device chain within a Rack is independent of the others, except that each chain output is summed together. The internal mix of a Rack is generally managed from its chain list, but it can also be conveniently mixed in a more traditional sense via Session View Tracks. What we mean is that each individual chain in an Instrument Rack or Drum Rack is viewable on its own "sub" track allowing for each chain to be viewed and submixed just like any other track type. Unfold your device chain on a track in Session View by clicking on the small black triangle button (Chain Mixer Fold button) in the track's Title Bar. When unfolded, you will see that each chain is on its own subtrack within the Device Rack track.

On each chain in Session View you'll find: Chain Volume fader, Activator (mute), Solo, and Preview button when the Mixer section is in view. Remember, this is purely a view for submixing/signal routing of Instrument/Drum Racks in the

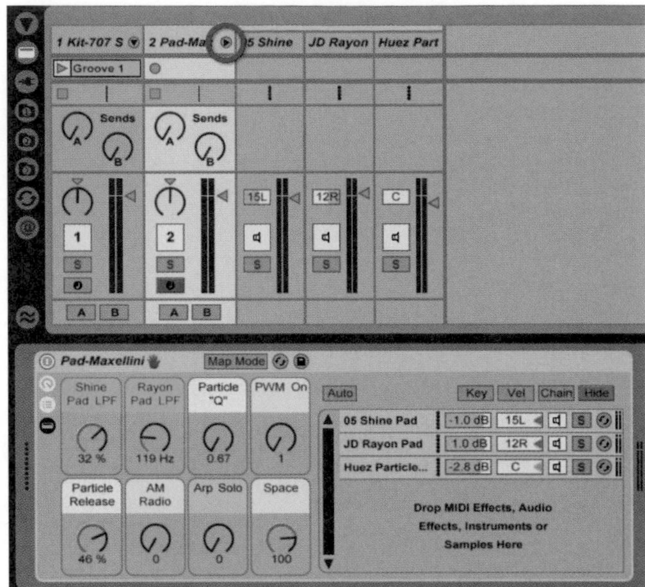

Figure 16.37 Chain in Session View.

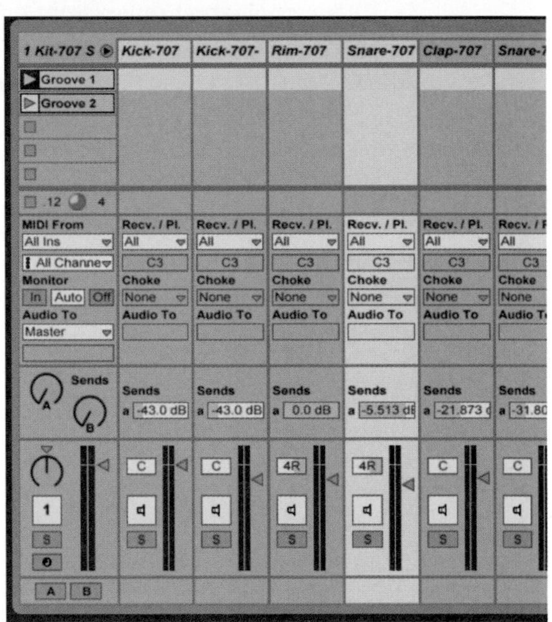

Figure 16.38 MIDI I/O choosers and Send Level sliders are exclusive to Drum Racks.

Session View. No clips can be added directly to chains nor can you record-enable them. They are simply a vertical display of the Rack's chain list. The MIDI I/O choosers and Send Level sliders are exclusive to Drum Racks.

One very interesting feature is that you can add audio effects or additional instruments to a chain just by dragging and dropping the device anywhere in the Session View over a designated chain. As soon as a new device is added in this way, you'll see the chain list and new device – for that specific chain – displayed in Track View. This helps to avoid clutter or confusion.

Figure 16.39 Drag and drop a device to a designated Session View chain.

This precise way of viewing and working with your Instrument Rack and Drum Rack Devices in Session View provides a more accessible and traditional way of mixing, especially when you group chains within a Rack. There are so many possibilities with Racks that we'd have to launch you to our next book to describe.

16.6 Working with Racks

As with almost every feature in Live 8 there is a clever way to convert or isolate one aspect so as to use it independently or to build-up something new. Ableton has done a great job of coming up with answers to the cliché comment "That's great, now wouldn't it be cool if you could…?" Odds are they have gone that extra step to imagine what that next desirable step or attribute might be. Now, they can't always add everything we want, but there is a whole lot more there than you think. With Device Racks this becomes ever so prevalent.

16.6.1 Produce: *Convert Chains into MIDI Tracks*

A unique way to work with chains in Session View is to turn them into discrete MIDI tracks, isolated and separated from the rest of the nested chains. See our example using the Rack Chain titled "Tamb-707."

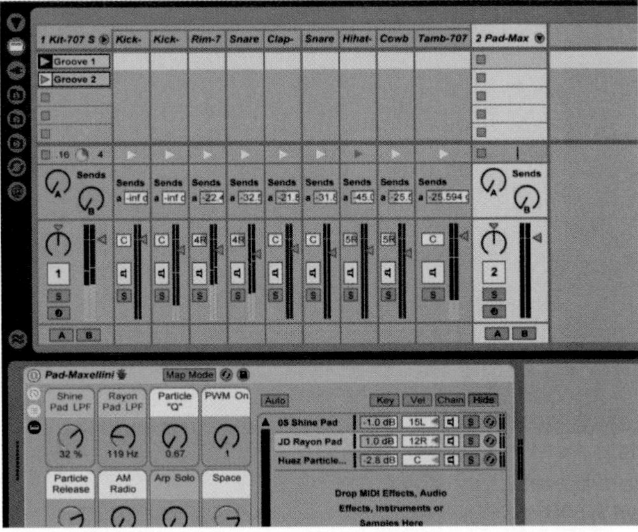

Figure 16.40 Drag a chain to the Session Drop Area or a new empty MIDI track in Session View.

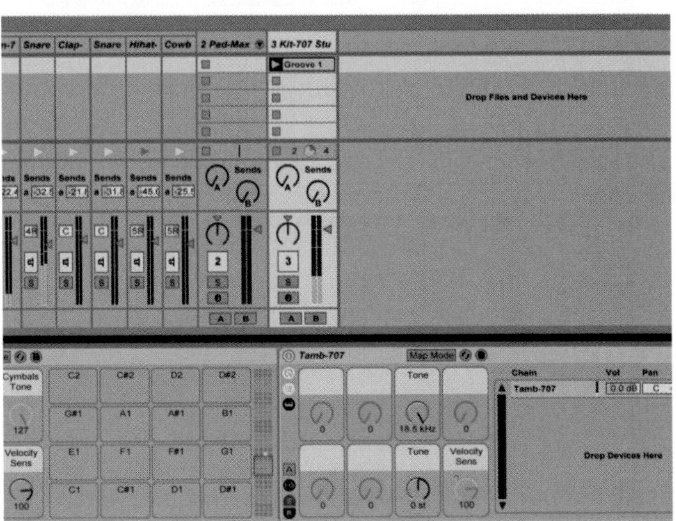

Figure 16.41 By separating a single chain from the nested chains in the Session View, you now have an isolated and editable MIDI clip.

Unfold your Rack in the Session View to view the chain then click + hold on one individual active chain and drag it to a new empty MIDI track or into the Session

View Drop Area to the right of all your tracks. Live will convert your selected chain into a single editable MIDI track along with its included performance as if it has MIDI assigned to trigger it. This conversion includes any audio effects from that chain, and best of all, the individual instrument sound itself from the original chain. All of the important parameters and devices themselves will be visible below in the Track View. Don't be confused by the new MIDI Track Title. Live creates the new MIDI track by copying the name of the main Rack Device you pulled it from. You can easily rename it "tamb-1" or any other name you wish.

By separating a single chain from the nested track chains in the Session View, you now have an isolated and editable MIDI clip that can be edited and saved for future use. Not only does this feature work with both Instrument Racks and Drum Racks, but it can also happen in real time while Live is playing. This is a great way to experiment with changing out the current "snare," for example, with a different sound, or controlling its volume more selectively. It's important to note that when you extract a chain it will be removed completely from its instrument. The basic design of this feature is to isolate MIDI data. One other use for this functionality is to freeze and convert the separated chain's MIDI clip to audio for warping or additional editing. Launch to Clip 9.3 for more on Freezing and converting MIDI into Audio.

Look at this as a way to split out your MIDI notes and corresponding instruments from a complete Instrument Rack into separate MIDI tracks. Try this production tactic with a couple of your track chains and see what happens.

In this chapter

17.1 Overview of Suite 8 373

SCENE 17

Suite 8

"Suite 8 is a monster! Creatively and practically, it's my go to piece of software."

Steve Ferlazzo, Producer and Programmer, Keyboardist for Avril Lavigne

17.1 Overview of Suite 8

Suite 8 is Ableton's flagship "XL" software package that includes Live 8, the large compliment of Ableton Software Instruments, and sample-based instrument collections. This includes items such as a full-fledged sampler instrument, five virtual software synths, three sample-based drum and percussion instruments covering the essential acoustic and electronic instruments created in cooperation with SONiVOX, Chocolate Audio, Puremagnetik, and a selection of custom clips and sounds by Zero-G, Cycling '74, and SoneArte. Suite 8 bolsters an enormously expanded instrument preset sound library of over 1600 customized presets created by top sound designers, and additional Construction Kits, Effects Racks, and user templates. Above all, Suite 8's large black box holds the 480-page hardbound reference manual.

After that, we must still emphasize that Suite 8 is strictly a bundle that includes Live 8 as a part of its all in one package. This means that when you buy Suite 8, it also comes with Live 8. Live 8 is also sold separately and comes with a basic set of instruments and samples. For an additional cost, you can buy or upgrade to Suite 8. You can also purchase individual Ableton instruments to enhance your experience with Live 8. These are available for individual purchase from Ableton's Website. Suite 8, on the other hand, conveniently combines all these Ableton Instruments (Analog, Collision, Electric, Latin Percussion, Operator, Sampler, Tension) and sample collections (Drum Machines, Session Drums, Essential Instrument Collection 2 [EIC 2]) along with Live 8 in one box as mentioned above.

Be aware that Suite 8 is also available for download from the official Ableton Website but does not include Session Drums or the EIC2 as of this writing. These sample libraries are contained on a separate DVD package and are currently only included in the boxed version of Suite 8. Of course, you can buy them later if you want. Boxed up or downloaded, this bundle is truly ready to be used for creating, producing, and performing your music!

For a general overview of the instruments included in Suite 8 (which when installed will appear in the Device Browser), visit our Website: http://booksite.focalpress.com/AbletonLive8. Now that you are salivating at all the excitement packed in Suite 8, download the 14-day trial while it's still available and have a taste of some of these fantastic instruments! www.ableton.com.

In this chapter

18.1 The possibilities of video with Live 8 377

18.2 Working with video 377

18.3 Video clips 379

18.4 Synchronizing music with video 382

18.5 Aligning video clips on the track display 385

SCENE 18

Video with Live 8

18.1 The possibilities of video with Live 8

Creating and synchronizing music to the moving image is a prominent medium in the music, entertainment, and digital media arts community. Video is everywhere, especially with the abundance of Web-based movies, online video tutorials, and advertisements. This reinforces the importance that a legitimate digital audio workstation (DAW) must be able to support and import QuickTime movie formats. Live 8 is no exception. Whether you're a veteran multimedia composer or new to film/TV scoring, Live makes working with video a fun and gratifying experience. Of course, when it comes to Ableton, concepts always seem to be taken one step further with the addition of an unlikely and unique performance feature; in this case, it's the ability to use Warp Markers to manipulate and help synchronize your audio and video in the Arrangement View.

18.2 Working with video

Live 8 imports movie/video files in the QuickTime format. This includes movie files with or without embedded audio tracks. To import a video into your Set, drag a QuickTime movie file directly from the File Browser or any Finder/Explore Window into the Arrangement View or into a track display, and be sure to slide it to start at the beginning of the timeline (Beat Time Ruler) if not already. Note that video clips can only be used in the Arrangement View. In the Session View, they are imported as audio clips. Once you've imported a video/movie, a floating Video window will pop up, where you can view your video as you work. Place it anywhere on your computer screen(s) that you like. As you move your cursor to different locations along the timeline, Live will automatically follow along with the video as seen in this window. For the more advanced users, you'll be happy to know that Live also displays a movie clip's QuickTime markers, which are built into the QuickTime itself when applicable.

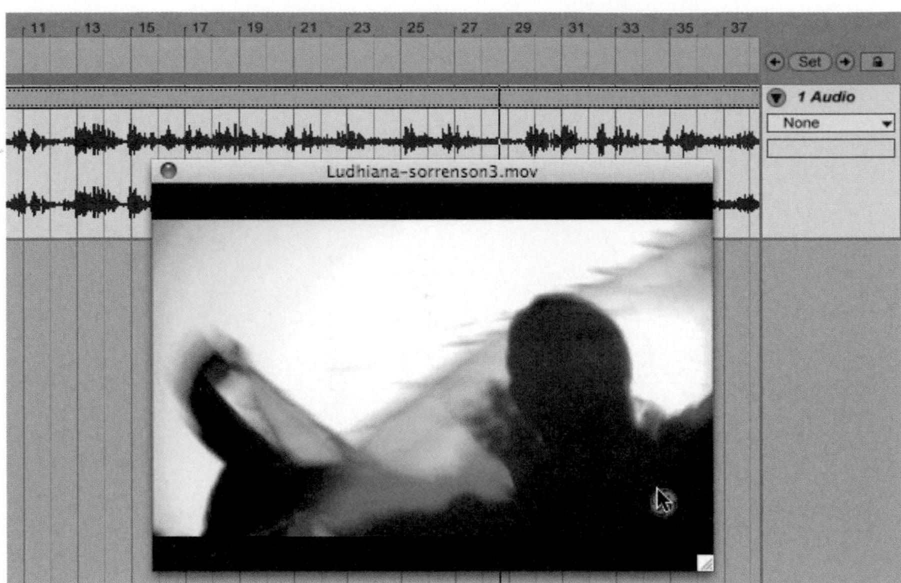

Figure 18.1 Video in the Arrangement View.

Feel free to resize the Video Window by dragging from the lower right corner or double-clicking in the Video Window to expand it to full screen. If you close it out, you can reopen it from the View Menu.

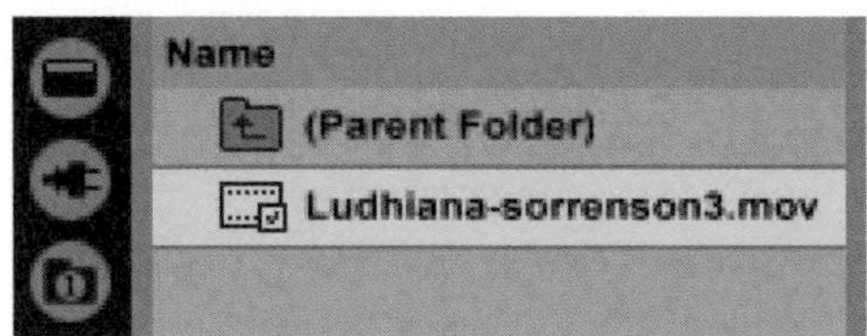

Figure 18.2 Drag a QuickTime movie file from the File Browser into the Arrangement View.

There are two standard settings that you should consider when working with video in Live. First, it's very important to understand that you will normally want to work at a 48 k sample rate. This is the standard sample rate for working to picture and should be established before you start a new project. The last thing you want is for your audio to be out of sync when you send it out for postproduction, etc. You can change the sample rate from Live Preferences. The other consideration to make is what timecode to use. When scoring to Film or TV, you should set Live to display SMPTE timecode in the time ruler as opposed to standard time. This is the industry standard for syncing your music to the start and endpoints of a film and for identifying hit points. SMPTE is displayed as hours:minutes:seconds:frames. Right click or ctrl + click on the Time Ruler at the bottom of the track display to choose a frame rate to display.

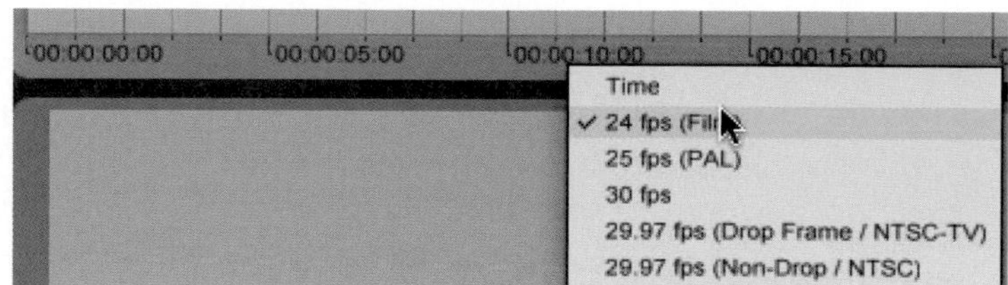

Figure 18.3 Choose custom frame rates.

A frame rate of 24 FPS (frames per second) is standard for Films, but this is dependent upon a few production factors such as the dub stage, postproduction, and music editor. You need to double-check the exact specs with one of these departments or with the director if all else fails before you start scoring to your QuickTime video.

18.3 Video clips

Figure 18.4 Notice the film-like sprocket design of a movie clip.

Video clips look and behave very much like audio clips, but you'll notice that their Title Bar has little holes ("sprockets") reminiscent of an actual film reel. As far as editing these clips is concerned, feel free to move them and trim their lengths as you see fit.

18.3.1 Editing video clips

Editing a video clip is just as easy as editing an audio or MIDI clip in Live. You can shorten its length and cut it at specific points if needed. In other words, Live's video capabilities make it a video editing tool as well. There are, of course, a couple of traditional clip editing functions that are exclusive only to audio clips. Video clips cannot be cropped, consolidated, or reversed. Live treats these commands as audio commands and will inadvertently turn your video clip into an audio clip if done so.

In regards to a video clip's audio, it's important to know that Live uses the existing audio track of a video clip to align Warp Markers for stretching and adjusting your new video clip. If the audio in your video track is not needed, you can always turn down the volume of the video by either accessing the track's fader or silencing it directly inside its Clip View by turning the volume slider all the way down. This is a useful way to work, especially when it comes to adding your own sound design and music. With that said, Live's Browser really comes into play for placing your own audio to video. It's a simple matter of selecting your clips in a File Browser and dragging them into the Arrangement View. By dragging and dropping them precisely where you want along the timeline, you can quickly get an idea going.

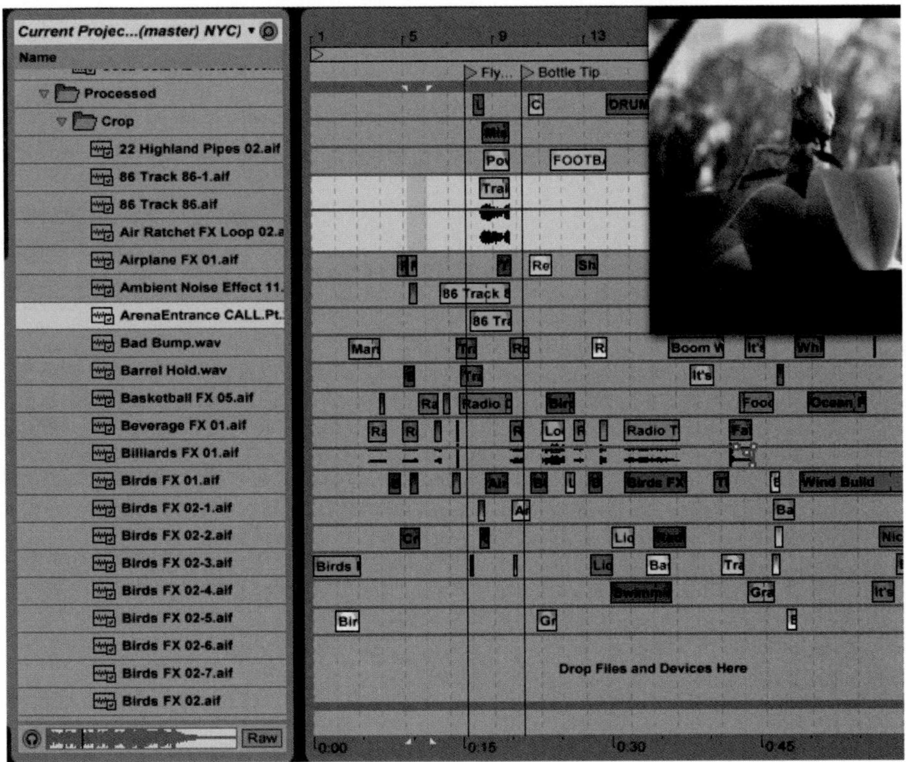

Figure 18.5 Live's Browser is extremely useful for placing audio to video.

You'll be happy to know that Live offers two ways to create and align your music in relation to your video track and vice versa. Once you have a video clip loaded on a track in the Arrangement View, bring an audio clip into a new track and position it directly below and in line with the video clip. Make sure that both clips' sample waveforms are viewable along the Arrangement View's timeline – unfold them if not. Now, double-click on the video clip to bring up its Sample Editor/Display in Clip View. Activate the Warp button in the video clip's Sample Box, which should now display a Warp Marker at the start of the clip in the Sample Display/Editor. Since you have your audio clip track unfolded in the track display and positioned directly beneath the video clip, you can easily manipulate the video clip's warp markers in the sample editor so as to adjust and position your video in relation to the timeline to match specific audio reference points.

Figure 18.6 Place Warp Markers anywhere you wish along a video clip in the Sample Editor to align your video and audio clips together.

Place Warp Markers anywhere you wish along the video clip length in the sample editor. You'll quickly see that this process opens up some interesting ways to work with both audio and video together. So, what's actually happening? Well, we'll get to that exciting information in a minute. First, let's look at this process without using warp markers.

When working with audio and video without the assistance of warp mode, simply drag and drop your audio/MIDI clips from the Browser along the timeline and then move them to the desired point with your mouse as seen in our example. You'll see that when you are clicking and dragging a clip into place, the video display updates the location of the video in real time. Think of this as a "video scrubber." By zooming in, you can edit your clip placement to the finest detail in effort to sync it up to your video.

Use the Arrangement Loop Brace and activate the Loop Switch to maintain focus on a specified area without stopping.

Figure 18.7 Move clips along the timeline to line up with spotting or hit points in the video.

18.4 Synchronizing music with video

Working to picture is all about timing and synchronization and in Live 8 that boils down to using video clips in one of three ways, the first way being the traditional method and the other two involving Live's warping feature.

The traditional approach: Import a video into an audio track and then starts scoring and laying in music as you see fit. This is a simple and traditional method to using video in any DAW. Just choose a tempo that is suitable for the music you want to create and from there you can work with audio and MIDI just like you always have. Understand that the Song tempo has no bearing on the video clip.

The warping approach: Live's warping feature is a totally unique method for working with video. Simply put, video clips can be the Tempo Master or a Tempo Slave. As the tempo master, Warp Markers are used to establish the tempo of your entire Set or for generating automated tempo variations throughout the duration of the video. This is useful for conforming your audio and MIDI clips to the tempo of the video's preexisting audio or visual cuts as determined by you with the placement of your own Warp Markers. When set to tempo slave, you can manipulate the speed of the video to creatively alter and affect its playback by using tempo of your Set and the automation of the tempo in the Master track. If you decide to use the warping feature, then this will all take place in Clip View.

18.4.1 Clip View and video

Whether your video clips have audio embedded or not, you will use Clip View and the Sample Display/Editor to handle their properties and manipulations. The Sample Editor is especially important when using the warping feature with video clips. In general, your music or sound design is intended to lock to specific visual cues or signifiers called "hit points." As we mentioned earlier, this can be handled in a traditional way, composing around and to these specific hit points as necessary. As an alternative, this can also be done using Live's warping feature. If you, in fact, decide to work with warping, then you want your video clip to be the tempo master. Activate the Warp Switch and the Tempo Master feature in the Sample Box. Once activated, Warp Markers can be added to a video clip in the Sample Editor.

18.4.2 Video clip Warp Markers

Warp Markers and video present some very interesting and unique possibilities as far as tempo and timing are concerned.

Master: When working with video clips that are set to "Master," you can use Warp Markers to adjust your Set's tempo. This serves two purposes: (1) to align specific points – start and hit points – with bars and beats and subsequently the metronome click to the video; and (2) to alter the playback of all your other clips (MIDI or audio) in your Set to match their speed to the defined tempo map that is derived from the Warp Markers. This could also be done manually by adjusting the tempo and tempo automation map when needed to fit hit points, etc.

To take advantage of warping, simply add and move Warp Markers in your video clip's sample editor. This will permit you to establish multiple alternative tempos to sync up to various hit points. You will notice when you add and move Warp Markers that the BPM in Control Bar Tempo adjusts to reflect the changes in the video clip's Warp Markers' positions. This is also seen in the Song Tempo automation control chooser on the Master track in Arrangement View. Right click or ctrl+click on the Song Tempo and select show automation.

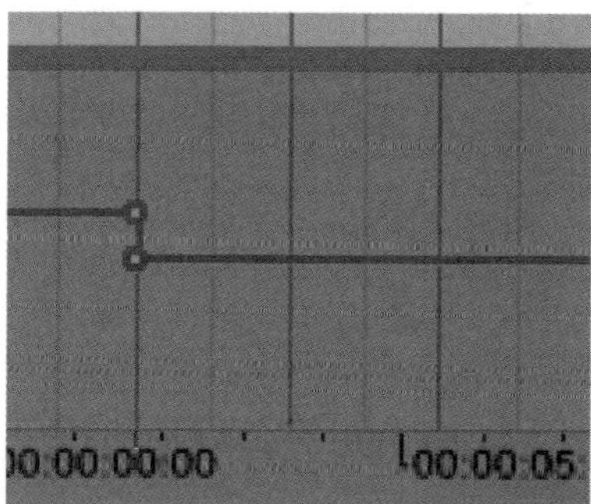

Figure 18.8 Song Tempo is automatically created to reflect changes in a video clip's Warp Markers.

For example, a video clip may have a music track already embedded in it such as a temp score, music, dialog, and sound effects. Warp Markers will visually align the video clip to the bars and beats of the timeline so that you can record or edit against a logical tempo map. It's rather refreshing when you hit the Metronome button and the click lines up with the music track or exactly with a hit point in the film. In addition, Warp allows you to transpose the pitch of the video clip's audio without changing the playback speed of the image. If the video clip is set to Slave, then any alterations such as transpositions or tempo changes affect the playback speed of the video clip and its audio.

It doesn't take many Warp Markers to lock in a tempo when you're dealing with a video that already has music such as a commercial or a music video. In our example, it took two Warp Markers: one to anchor the first downbeat with the grid (timeline) and the other to anchor the second beat. Since our music track was at a consistent tempo, only two Warp Markers had to be placed. On occasion, you may have to periodically add more Warp Markers if the audio drifts out of sync over time. This usually means that your transients are slightly off the beat and that you need to adjust them. Zoom in tight to see minor misalignments. When adding Warp Markers, try not to have multiple breakpoints in the Song Tempo automation unless you are simply trying to warp audio and you can afford subtle tempo variations, otherwise you want it perfect so when you record to picture or lay in new audio, there are no tempo fluctuations.

To begin, place the first Warp Marker at the first Transient that you know to be a downbeat; then drag it to the closest bar. Place the second Warp Marker at the second Transient and warp it over to the closest beat. Finally, delete Live's

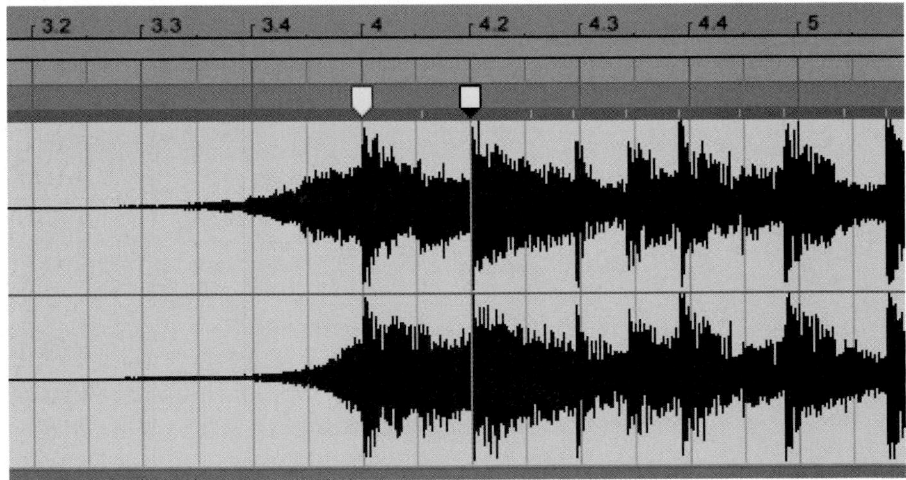

Figure 18.9 Here, we will anchor the first two beats or transients with Warp Markers to the downbeats in the grid.

auto-created Warp Marker at the beginning of the clip since you don't need it anymore now that you have created your own. To review warping concepts, launch to ▶Scene 13.

Hot Tip *When scoring to film, you may find that an otherwise perfect tempo does not sync up with an important hit point in the cue. When faced with this dilemma, use Warp Markers to make a slight and indistinguishable tempo change in a selection of the music, as to push or pull back the audio to hit the visual marker.*

18.5 Aligning video clips on the track display

Video clips can be moved and trimmed just like any other clip. When working to picture, it's desirable to have the bars of the Arrangement View's Beat Time Ruler match with what would logically be the beginning or bar 1 of your music. It's very unlikely that any video you work with will start on the first image, rather there is usually a lot of blank video or a "two-pop" that you must wait for before the first image appears. In the professional world of Film and TV, this is used to align audio and video in the post production, or simply to keep everything aligned – dialogue, sound effects, music, etc.

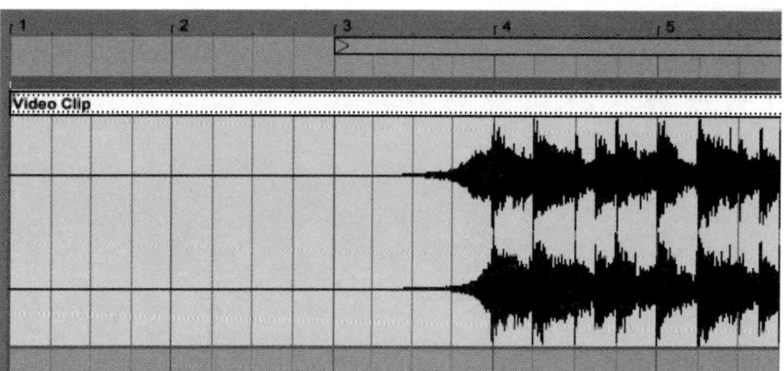

Figure 18.10 Video clip with silence at the start.

To set the video clip so that there is the least amount of silence at the beginning of the clip, it will have to be trimmed. We'll trim it down to the first downbeat temporarily (bar 4) since it's a good visual reference for where the video starts.

Let's move it to bar 2 because we need some space for the lead-in.

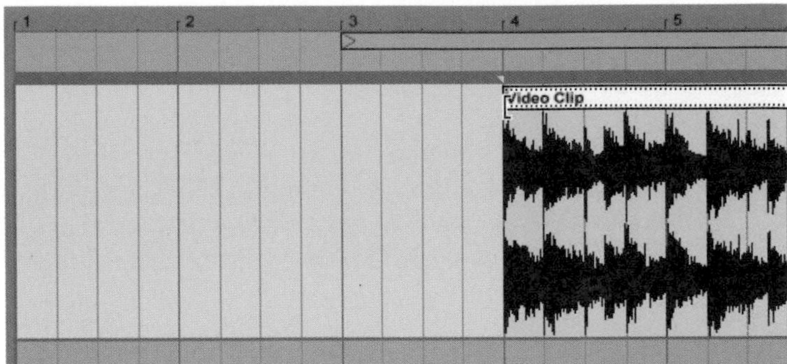

Figure 18.11 Trim the video clip back to the first downbeat for a visual reference.

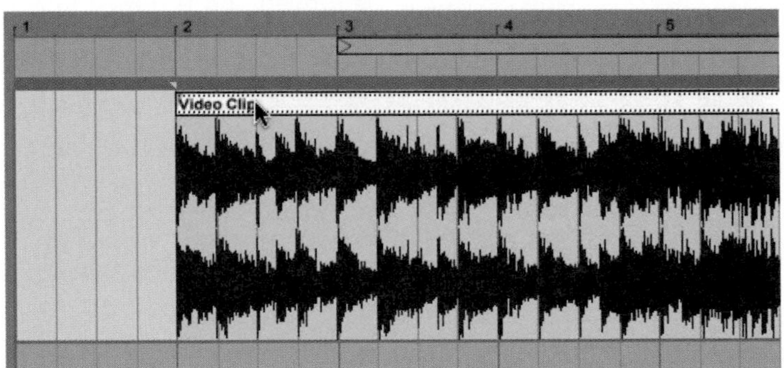

Figure 18.12 Move video clip to bar 2.

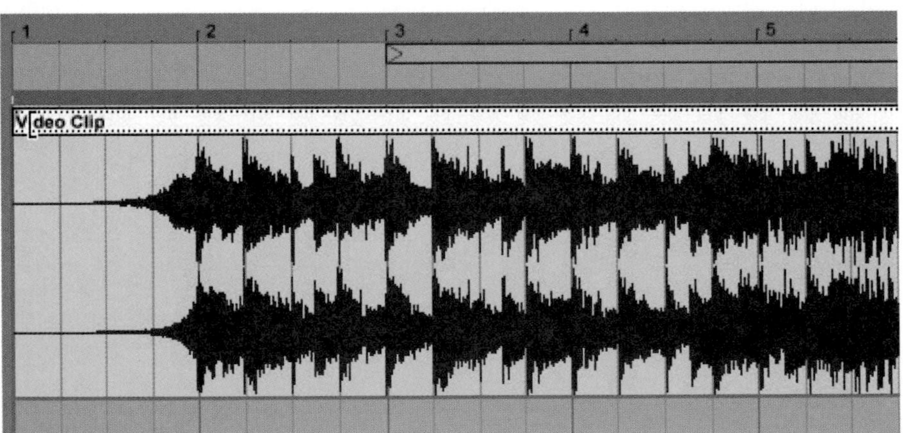

Figure 18.13 Expand (trim) clip edge to start at bar 1.

Once beat one is sitting on bar 2, we'll expand the clip back out to start at bar 1. Now, it is in sync and on the right bar grid.

Now, it's up to you, but it's a good idea to do the same thing for the video clip itself in Clip View since the Arrangement adjustments don't affect the actual contained clip's time ruler; just the start point is changed. To do this in Clip View, select all Warp Markers, then drag from the first marker and place it on bar 2 just like we did in the Arrangement View.

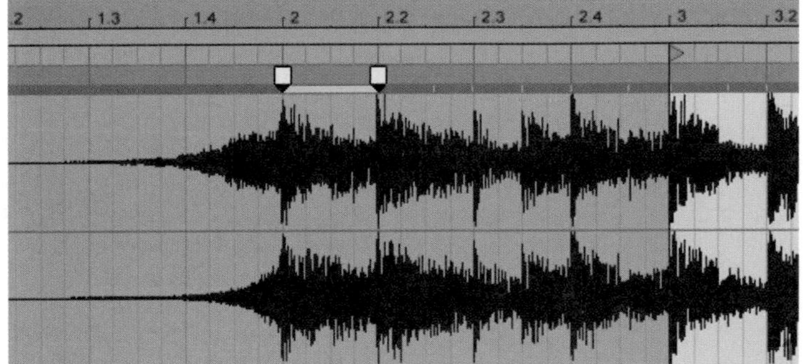

Figure 18.14 Select all Warp Markers, then drag from the first marker and place it on bar 2.

Once in place, all we need to do is adjust the position of the start markers to bar 1. Simply type "1" into its start field input box.

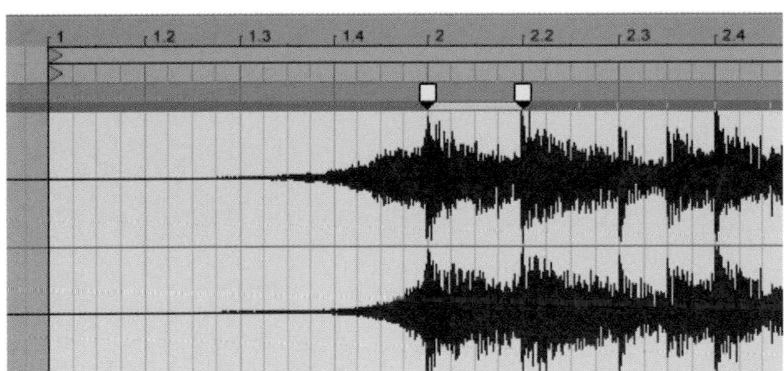

Figure 18.15 Adjust the position of the start makers to bar 1.

Index

1 Bar quantization, 125, 126

A

Ableton Performance Controller (APC40), 243, 244
Adaptive Grid, 92
All-in-one clip preset, 301, 302
Arpeggiator, 54, 312, 355
Arrangement clips
  exporting of, 56
  playback of, 92
  Tempo Master/Slave switch for, 150, 257
Arrangement looping, 50–51
Arrangement Position display field, 90
Arrangement View, 68–70
  audio clips recording, 201
  automation, 101–105
  Beat Time Ruler and Time Ruler, 86–87
  creating in, 105
  editing clips by
    consolidating, 97
    cut, copy, and paste, 97
    moving, 95
    resizing/trimming, 96
  fades and crossfades in, 99–101
  function of, 68
  Group Tracks in, 231–232
  importing audio to, 35–37
  importing MIDI file to, 42–46
  layout, 85–86
  linear, 76
  linear timelines in, 86
  looping in, 93–94, 192–193, 289–290
  as musical timeline, 14, 85
  navigation of, 86–92
    by locators, 90–92
    by zooming and scrolling, 86–89
  overview display of, 87
  performing in, 106
  producing in, 106
  record session performance to, 54–55
  recording audio to, 37–41
  recording MIDI to, 46–48
  rendering/exporting audio from, 208–209
  and Session View, 74–76, 131
  video in, 378
Audio, 34–35
  exporting of, 55–56
  importing, 35–37
  loop slicing, 286–287
  preferences, 17
  printing, 209–211
  recording to Session View/Arrangement View, 37–41
  rendering/exporting, 56, 208–209
  routing in Group Track, 231
  tracks, 111
    rendering of, 56
Audio clips, 70, 125–128, 148, 163
  MIDI clips into, freezing and converting, 196–198
  recording, 198–206
    Arrangement View, 201
    Session View, 199
  vs. MIDI clips, 141
Audio effects, 21–22, 51–53, 298, 313–314
  and Drum Racks, 364–365
  Overdrive, 322
  user interface of, 313
  Vocoder, 323–325
Audio Effects Rack
  as Device Rack chain, 359–360
  as track device chain, 359
Auto Filter, 52
Automation, 101–105
  of control parameter, 102
  drawing envelopes of, 103
  recording, 102
Automation Lane, 102–103
Auto-Warp Long Samples, 257

B

Back to Arrangement button, 132
Beading effect, 149
Beat Time Ruler, 86–87
Beats, producing, 79
Beats Mode, 262–264
Breakpoint envelopes, editing, 104, 164–165

C

Chain Select Zone Editor, 349
Choke Groups, 353
Chord, 312
Chorus device, 326
Chorused effect, 149
Clip Box, 142
  groove settings, 142
  nudge, 142–143
  properties of, 142
Clip envelopes, 163
Clip Gain fader, 149
Clip Launch button, 71, 72
Clip Quantization, 82, 144–145
Clip RAM Mode, 148
Clip Slots, 115–116
  and clips, 185–189
  recording, 188–189
Clip View, 138–141, 383
Clips, 70–72, 115, 137
  arrangement. *See* Arrangement clips
  audio. *See* Audio clips
Clip Slots and, 115–116, 185–189
  dragging, 211–212
  editing, 95–97, 119–120
  launching, 116–119
  looping, 93
  MIDI, 41, 42, 70, 128–129
    freezing and converting, 196–198
    recording, 185–189
    vs. audio clips, 141
  selecting, 95
  Session Grid, 115–116
  track and, 125–127
Commit groove, 179–181
Comping, 204–205
Complex Mode, 265
Complex Pro Mode, 265–266
Contextual Warp Commands, 267–269
Corpus, 314
CPU load, 15
Crossfade Switch, 130
Crossfader Section, 129–131
Crossfades, 99–101
Custom device presets, 301
Cut/Copy/Paste/ command, 97–99, 104

D

DAWs. *See* Digital audio workstations
Detune, 149
Device chains, 325
  chaining effects, 326–327
  interfacing with, 327–328
  sidechaining, 328–331
Device presets, customizing, 301–303
Device Racks, 345–346
  creating, 353
Digital audio workstations (DAWs), 14, 55, 109, 185, 255, 256, 279
Direct-from-disk mode (DFD), 16
Disk load, 15–16
Drop Area, 114
Drum loops, 93
Drum Racks, 350
  and audio effects, 364–365
  Chain List, 351, 352
  creating, 361–363
  Pad View, 350
  routing, 352–353
Dummy clips, 166–170
Duplicate function, 97
Duplicate time command, 97–99

E

Editing
  arrangements, 95–97
  breakpoint envelopes, 104, 164–165
  clips, 95–97, 119–120
  MIDI notes, 159–161
  video clips, 379–382
Envelope Box, 162–163
Envelope commands, 104
Envelope Editor, 164
  editing breakpoint envelopes, 164–165
  link/unlink envelopes, 164–165
  loop position/length, 166
External MIDI instruments
  rendering, 338
  routing, 336–338
Extract groove, 181

F

Fades, 99–101, 148
  in track's waveform display, 100
  volume, 100
File Browser, 23–28
File Manager, 58–61

Files
  management of, 58–59
  REX, 280, 281
  searching for missing, 59–60
Filter Envelope, 310
Filter Envelope Amount (Env), 310
Filter parameters, 306
Fixed Grid, 92
Flatten command, 16
Follow action, 145–146
  rhythmic loop creations with, 170–172
Freeze Tail Clip, 198
Freezing tracks, 16
Frequency shifter, 315

G

Gate Sidechain effect, 330–331
Glide Mode, 310
Global Quantization, 82, 118
Global Record button, 131
  features of, 66
Global Record concept, 65–67
  by linear approach, 74–80
  by nonlinear approach, 80–82
Groove, 175, 176
  commit, 179–181
  on drum track clips, 182
  extract, 181
  presets in library, 176
  settings of, 142
  Swing, 180, 181
Groove Pool, 176–177
  parameters, 178–179
Group clips, launching, 232–235
Group Rack Chains, 356
Group Slot, 235
  colored, 230–231
Group Tracks, 111
  advantage of, 232
  in Arrangement View, 231–232
  audio routing in, 213
  Group Slot in, 230
  kit, 233
  melodic, 234
  mixing with, 235–236
  printing, 237
  rhythmic, 234
  vs. scenes, 233–235
  in Session View, 229–231
Guitar-wide acoustic instrument, 53, 54

H

Hard disk overload indicator, 15
High-Quality Mode (Hi-Q), 148
Hot Tip box, 6, 8–9
Hot-Swap Groove button, 176, 177

I

Impulse
  envelope parameters, 306
  filter parameters, 306
  rerouting individual outputs of, 306–308
  sample parameters, 305–306
  sample slots, 305
  user interface, 304–306
Index cards of musical phrases, 67
Info Box, 5
Insertion marker, 156
Instant Mappings, 242
Instrument devices, 20
Instrument Racks, creating, 353–359
Invert Range feature, 250

K

Key Map Mode, 247, 248
Key mapping, 247–248
Key Zone Editor, 348
Kit Group Track, 233

L

Launch Box, 5, 6, 143
Launch Modes, 143
  types of, 143–144
Launching
  clips, 116–119
  group clips, 232–235
  preferences, 18
  scenes. *See* Scene launching
  session clips, 93
Legato Mode, 144
Level meters, 300
Library clip presets, 41
Limiter, 316
Linear approach, Global Record concept by, 74–80
Linear Arrangement View, 76
Linear recording, 66
Live Browser, 19
  device, 20
    Audio Effects, 21–22
    instrument, 21
    MIDI effects, 22
  File Browser, 23–28
  Plug-in device Browser, 22–23
Live devices, 298–299
  presets, 301
    customizing, 301–303

Live pack, creating, 60–61
Locators
  mapping, 251–252
  for navigating Arrangement View, 90–92
Loop Brace, 49, 51, 140
Loop Switch, 153
Looper, 316
  features of, 317–320
Looping, 48
  approaches for, 289–290
  in Arrangement View, 50–51, 289–290
  clips, 93
  of playback, 93
  in Session View, 48–49
Loops, 279–280, 288–289
  custom library of, creating, 293
  fixing, 270–271
  recording
    Arrangement View, 192–193
    Session View, 190–192
  resequencing, 287–288
  slicing, 286–287
  with unlinked clip envelopes, 290–292
  warping, 267
Low-frequency oscillator (LFO) slider, 309, 310

M

Macros
  for controlling internal parameters, 346–347
  mapping, 349–350, 356–359
Mapping Browser, 249–250
Marker Snap setting, 91
Master track, 110, 111, 113
MBD. *See* Multiband dynamics
Melodic Group Tracks, 234
Metronome, 40
MIDI, 41–42
  bank, 155
  clips, 41, 42, 70, 128–129
    freezing and converting, 196–198
    recording, 185–189
    vs. audio clips, 141
  controller, 243, 244, 246–247
  data, 311
  drum kit, 223, 224
  effects, 22, 51, 298, 311–313
  files, importing
    multi-tack, 43–46
    single-tack, 42–43
  mapping, 225, 243–246
  notes, writing and editing, 159–161
  preferences, 17–18
  recording, 46–48
  routing, 45
    multitrack, 46
  slices, 282–283
  step recording, 161–162
  tracks, 111, 298
    converting chains into, 368–369
    in Session View, 368
  velocity values of, 145
MIDI Effect Rack
  as Device Rack chain, 361
  as track device chain, 360
MIDI Map Mode, 241, 244, 245
MIDI Note Editor, 138–140, 156
  commands in, 156
  Note Ruler, 157–158
  writing and editing MIDI notes in, 159–161
  zooming in, 157
MIDI Velocity Editor, 158–159
Mixer, 112–114
  Drop Area, 114
  section, 86
  Session View as, 73
Mixing, 235–236
  with Group Tracks, 235–236
Multiband dynamics (MBD), 320–321
Multitrack instruments, 222–225
Multitrack recording, 222–225
Musical foundation, 215–217
Musical instrument digital interface. *See* MIDI
Musical phrases, index cards of, 67
Musical structure, 215–217
Musical timeline, 85

N

Navigation
  by Beat Time Ruler and Time Ruler, 87
  scrolling and zooming, 86–89
Nonlinear approach, Global Record concept by, 80–82
Nonlinear recording, 66
Note Ruler, 157–158
Notes Box, 154
  loop position/length, 155–156
  MIDI file start/end, 155–156
  original tempo, 154–155
Nudge, 142–143

O

One-shot audio file, 149
Overdrive, 322
Overdub (OVR), 47–48, 188
  recording, 190–196

P

Phasing effect, 149
Ping-Pong Delay effect, 331
Pitch Envelope, 310
Pitch shifting algorithms, 255
Plug-in Device Browser, 22–23, 332
Plug-in devices, 332
  multi-instrument, 339–342
  third-party effects, 333
  third-party instruments, 332–333
  user interfacing and layout, 333–335
  X-Y Control/Parameter layout, 334
Plug-ins
  AU and VST, 332, 333
  parameters, 334
  performance-enhancing, 333
  sound-shaping, 333
  types of, 332
Portamento Mode, 310
Project folders, 57
Pseudo Warp Markers, 150, 260, 261
Punch recording, 205–206

Q

Quantization Resolution, 217–218
Quantizing audio, 274–275
  in real time, 276
QuickTime format, 377

R

Racks, 345, 367–369
  Chain List, 347–349
  interface and layout, 346
  macros, 346–347
  in Session View, 365–367
RAM Mode, 16
Real-time launching base, 109
Record File Type, 18
Record on Launch, 218–219
Record preferences, 18
Recording
  Arrangement View, 192–193
  audio, 37–41
  audio clips, 198–206
  MIDI, 46–48
  MIDI clips, 185–189
  overdub, 190–196
  punch, 205–206
  Session View, 190–192
Relative Session Mapping Strip, 248–249
Remixing, 79, 81
Remote control, 241
  setting up, 242–243
Re-Pitch Mode, 265
Reset fades, 101
Return tracks, 111
  submixing with, 133
Returns Chain List, 352
REX
  files, 280, 281
  loop, 280–281
    slicing, 283–286
  mode, 281–282
Rex/ReCycle technology, 255
Rhythmic Group Track, 234
Rhythmic loop, creating using Follow Actions, 170–172
Ring modulation mode, 315

S

Sample Box, 146
  detune, 149
  edit, 147
  fade feature, 148
  loop position/length, 153
  original tempo, 151
  properties of, 147–151
  reverse, 148
  sample file start/end, 152–153
  save, 147–148
  transpose, 148–149
Sample Display Editor, 138–140
Save Preset, 300
Scene launching, 122, 216
  preferences, 217–219
    Record on Launch, 218–219
    Select Next Scene on Launch, 217–218
Scenes, 71–73, 127–128, 216
  capturing and inserting, 220–222
  vs. Group Tracks, 233–235
  mapping, 225
  tempo and time in, 219–220
  vs. tracks, 120–123
Scrub Areas, 90, 140
Select Next Scene on Launch, 217–218
Session clips, 288
  launching, 93
  rendering of, 56
Session Grid, 115–116
Session looping, 48–49
Session View, 14, 20, 34, 70, 74, 76, 109, 110
  and Arrangement View, 74–75, 131
  audio clips recording, 199
  dragging and dropping arrangement into, 77–78
  editing clips in, 119–120

Session View (*continued*)
  Group Tracks in, 229–231
  importing audio to, 35–36
  importing MIDI file to, 42–46
  loop recording, 190–192
  as mixer, 73, 112–114
  recording audio to, 37–41
  recording MIDI to, 46–48
  working in, 124–131
Set *vs.* project, 57–58
Show fades, 101
Sidechain Toggle, 328
Sidechaining, 328–331
Simpler
  envelope section, 310
  filter parameters, 309
  modulation section, 310
  sample display, 309
  user interface, 308–311
Singing synthesizer, 325
Slice to MIDI track, 282
Slicing
  audio loop, 286–287
  REX loop, 283–286
Software synths, 332
Start Marker, 140
Submixing, 111
  of Group Tracks, 236–237
  with Return tracks, 133
Swing groove, 180, 181

T

Takeover Mode, 243
Tempo, signature, 219–220
Tempo Master/Slave switch, 150, 257, 258
Texture Mode, 265
Time Ruler, 86, 87, 139–140
Time signature, 219–220
Time-stretching algorithms, 255, 256
Timing of audio files, adjusting, 274–275
Tones Mode, 264
Track Activator, 86
Track Freeze command, 16
Track Status Display, 124
Track View, 299–301
Track View Selector, 140–141
Tracks, 72, 76
  and clips, 125–127
  freezing, 16
  group, 229–232
  MIDI, 298
  vs. scenes, 120–123
  types, 110–111
Transient Loop Mode, 263, 264
Transient Mark, 259–261, 270
Transients, 258–259
Transport control, 89–90

U

Unlinked clip envelopes, loops with, 290–292
User interface, 67–73
  Arrangement View of, 68–70
  impulse, 304–306
  Session View of, 70
  simpler, 308–311

V

Velocity Zone Editor, 348–349
Video, 377–379
  in Arrangement View, 378
  and Clip View, 383
  synchronizing music with, 382–385
  warping feature, 382
Video clips
  aligning on track display, 385–387
  editing, 379–382
  Warp Markers, 383–385
Virtual instrument, 332, 333
Vocoder, 323–325
Volume fades, 100

W

Warp From Here command, 267–269, 271, 274
Warp Markers, 179, 258, 260–261, 270, 382
  Pseudo, 150, 260, 261
  video clips, 383–385
Warp Modes, 150, 151–152, 262
  Beats, 262–264
  Complex, 265
  Complex Pro, 265–266
  parameters, 152
  Re-Pitch, 265
  Texture, 265
  Tones, 264
Warp preferences, 18
Warping, 150–151, 382
  audio clips, 256–257
  audio loops, 267
  master/slave, 149–150, 257–258
  samples, 266
  songs/tracks, 271–273